RENEWALS 458-4574

Powerline Communications

Powerline Communications

Klaus Dostert

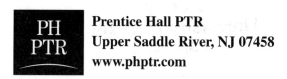

Prentice Hall PTR
Upper Saddle River, NJ 07458
www.phptr.com

Library of Congress Cataloging-in-Publication Data
Dostert, Klaus.
 [Powerline Kommunikation. English]
 Power line communications / Klaus Dostert.
   p. cm.
 Translation of: Powerline Kommunikation.
 Includes bibliographical references.
 ISBN 0-13-029342-3
  1. Telecommunication lines. 2. Electric lines--Carrier transmission. 3. Communication systems. I. Title.

TK5103.15 .D6813 2001
621.382'3--dc21   2001021229

Editorial/Production Supervision: *Laura E. Burgess*
Acquisitions Editor: *Bernard Goodwin*
Marketing Manager: *Dan DePasquale*
Manufacturing Manager: *Alexis R. Heydt*
Art Director: *Gail Cocker-Bogusz*
Cover Design Director: *Jerry Votta*
Cover Design: *Design Source*

© 2001 Prentice Hall PTR
Prentice-Hall, Inc.
Upper Saddle River, NJ 07458

Prentice Hall books are widely used by corporations and government agencies for training, marketing, and resale.

The publisher offers discounts on this book when ordered in bulk quantities. For more information, contact Corporate Sales Department, Phone: 800-382-3419; fax: 201-236-714; email: corpsales@prenhall.com or write Corporate Sales Department, Prentice Hall PTR, One Lake Street, Upper Saddle River, NJ 07458.

All rights reserved. No part of this book may be reproduced, in any form or by any means, without permission in writing from the publisher.

Powernet-EIB and TIMAC-X10 are trademarks of Busch-Jaeger Elektro GmbH. iPLATO is a trademark of IAD GmbH. All other products and company names mentioned herein are the trademarks or registered trademarks of their respective owners.

Printed in the United States of America

10   9   8   7   6   5   4   3   2   1

ISBN 0-13-029342-3

Pearson Education LTD.
Pearson Education Australia PTY, Limited
Pearson Education Singapore, Pte. Ltd.
Pearson Education North Asia Ltd.
Pearson Education Canada, Ltd.
Pearson Educatión de Mexico, S.A. de C.V.
Pearson Education—Japan
Pearson Education Malaysia, Pte. Ltd.
Pearson Education, Upper Saddle River, New Jersey

Copyright Information from the German Edition

The German Library - CIP Unified Recording

A title data record of this publication is available from the German Library

© 2000 Franzis publishing house GmbH, D-85586 Poing, Germany

All rights reserved, including photo-mechanical reproduction and storage on electronic media.

Most of the hardware and software product designations, as well as company names and brands that are used within this work, may also be registered trademarks and should be regarded as such. With the product designations the publisher essentially follows the notation of the manufacturers.

Phototypesetting: Fotosatz Pfeifer, D-82166 Gräfelfing, Germany
Printing: Freiburger Graphische Betriebe, 79108 Freiburg
Printed in Germany - Imprimé en Allemagne.

ISBN 3-7723-4423-2

To Katharina

Contents

Foreword — xi

Chapter 1 Introduction — 1

Chapter 2 The Electric Power Supply System and Its Properties — 5

  2.1 Topological and Electrical Structures — 5
    2.1.1 The High-Voltage Level — 6
    2.1.2 The Medium- and Low-Voltage Levels — 18
    2.1.3 The Interference Scenario — 32

Chapter 3 Historical Development of Data Communication over Powerlines — 43

  3.1 Possibilities and Limits of Classical Usage Types — 43
    3.1.1 Carrier Transmission over Powerlines (CTP) — 44
    3.1.2 Ripple Carrier Signaling (RCS) — 54

Chapter 4 New Usage Possibilities of the Low-Voltage Level Based on European Standards — 73

  4.1 The European CENELEC Standard EN 50065 for the Frequency Range Below the Long-Wave Broadcast Band — 73
    4.1.1 The Impact of Limiting the Signal Level — 76
    4.1.2 The Impact of Limiting the Signal Spectrum — 77

4.2	Signal Coupling, Signal Attenuation, Access Impedance	81
4.2.1	Signal Coupling	81
4.2.2	Transmission Attenuation	91
4.2.3	Access Impedance	92
4.3	Modulation and Access Methods for Use under EN 50065	92
4.3.1	Narrowband Modulation Schemes and Their Properties	94
4.3.2	Spread-Spectrum Techniques	99
4.3.3	Pseudonoise Direct Sequencing (PN-DS)	100
4.3.4	Frequency Hopping (FH)	101
4.4	Examples of System Implementations	107
4.4.1	Matched Filtering and Synchronization—Introductory Remarks	107
4.4.2	Digital Signal Synthesis by Wave Table	137
4.4.3	Digital Optimum Receiver Technology for Powerline Communication	144
4.4.4	Powernet-EIB in Building Automation	176
4.4.5	Error-Correction Coding Against Impulsive Noise	188
4.4.6	Behavior of Coded Transmission under White Noise (AWGN Environment)	198
4.4.7	Communication System for Energy-Related Value-Added Services of PSUs	200
4.4.8	Integrated Microsystem Concept as a Mixed-Signal ASIC	206
4.4.9	The Symbol-Processing Multicarrier Scheme—Summary	219
4.4.10	Modem Implementations for Energy-Related Value-Added Services	220

Chapter 5 Innovation Potential from Deregulation—Possibilities and Limits of Signal Transmission — **229**

5.1	Deregulation of the Telecommunication Markets	229
5.2	Fast Data Transmission over Building Installations (Last Meter Solutions)—the HomePlug Powerline Alliance	230
5.2.1	Mission and Vision of the HomePlug Powerline Alliance	230
5.3	Telecommunication Access over the Low-Voltage Grid (Last-Mile Solutions)	232
5.4	Energy Market Deregulation (Free Electricity Trade)	234
5.5	Bandwidth Requirements and Frequency Allocation	236

5.6	Channel Characteristics, Coupling and Measuring Techniques at High Frequencies for PLC	237
	5.6.1 Coupling of High-Frequency Signals	239
	5.6.2 HF Measuring Systems and Measurement Results	241
	5.6.3 Transmission Characteristics at High Frequencies	251
5.7	The Powerline Channel Model	257
5.8	The High-Frequency Interference Scenario	263
5.9	Access Impedance	268
5.10	Estimating the Powerline Channel Capacity	269
	5.10.1 Shannon's Theory for the Powerline Channel	269
	5.10.2 Access-Channel Capacity Estimation	272
	5.10.3 Indoor Channel-Capacity Estimation	273
5.11	Electromagnetic Compatibility: Problems and Solutions	276
	5.11.1 Compatibility with Wireless Services	276
	5.11.2 Compatibility of Different PLC Systems	284
	5.11.3 What to Expect from Regulations in the Near Future	285
5.12	Measures of Network Conditioning—Analysis of Feasibility and Efficiency	286

Chapter 6 Appropriate Modulation Schemes for PLC and Communication System Concepts — **293**

6.1	Introduction	293
	6.1.1 Basic Consideration of Broadband Techniques	294
6.2	Single-Carrier Modulation and CDMA	294
6.3	PLC Signal Characteristics and Level-Limit Measurements	297
6.4	OFDM—A Multicarrier Scheme for High-Speed PLC	299
6.5	OFDM Signal Synthesis, Carrier Modulation and Demodulation	302
	6.5.1 The OFDM Transmitter	306
	6.5.2 The OFDM Receiver	307
6.6	OFDM for High-Speed PLC—Summary	308

Chapter 7 Conclusions and Further Work — **311**

Chapter 8 Reading List and Bibliography by Topics — **315**

8.1	Carrier-Frequency Modulation in High-Voltage Lines	315
8.2	Ripple Carrier Signaling	315
8.3	Standards and Regulation Issues	316
8.4	Spread-Spectrum Techniques	316

8.5	Fundamentals of Communications, Systems Theory, and RF Technology	317
8.6	PLC-Related Electronic Circuit Design, Application Notes	318
8.7	Communication over the Electric Power Distribution Grid	318
8.7.1	Channel Analysis and Modeling at Low Frequencies	318
8.7.2	System Concepts, Experimental Modems, and Test Results for Low Bit-Rate PLC	320
8.7.3	Doctoral and Habilitation Dissertations Concerning PLC	323
8.7.4	Channel Analysis, Modeling, and System Concepts for High-Speed PLC	323
8.8	Conference Proceedings	327
8.9	Books on Powerline Communication	328
8.10	WWW Links Related to PLC	329

Foreword

The term "powerline" has splashed across the media more or less spectacularly for quite some time. Many of these representations are not sound enough to convey a realistic view of this technology and its possibilities to the public. Even experts are not fully familiar with the subject, because powerline is a unique medium for message transmission. This may help explain why powerline research and development has had a shadowy existence until recently. The only worthwhile applications had been seen in niches like building automation or remote meter reading. Extremely restrictive regulations with regard to the usable frequency ranges and admissible transmission levels have also discouraged powerline developments. There has not been massive economic interest to drive the formation of a powerline lobby like those in other technological fields, which would be able to exert influence on standards and regulations.

This situation changed radically with the deregulation of the telecommunications and energy markets. One immediate consequence is the demand for alternative fast data connections on the local loop level. There is still a de-facto monopoly on this level, because all the copper in the telephone network still belongs to one source. Powerline communication could offer ideal solutions, mainly because the power supply infrastructure is denser than any other communication network. The full infrastructure from the provider to the home wall plug is there, ready for use without any additional installation cost. This means, for example, that fast Internet access from the wall plug will not remain just a vision but will gradually become reality. Development will advance at a pace so rapid that future users will have to be provided quickly with comprehensive information in easily understandable form, so that they can properly appreciate the new options at their fingertips.

Unfortunately, suitable literature has been lacking. The publications currently available can be divided roughly into two classes:

1. Scientific papers addressing highly specialized experts and describing specific technical detail.
2. Press articles made up in spectacular presentations to draw attention to the "powerline" novelty, without conveying useful technical background information.

Drawn from a large number of sources, this is the first book aimed at providing future users with comprehensive information about this new technology. It deals in an easily understandable way with all relevant aspects, from the historical roots to the possibilities and limits of powerline communication, taking the most recent results from research and development into account. In addition to physical and technical interactions, this book also discusses economic and regulatory aspects. The introductory chapter presents the most important information about the motivation for powerline communication and describes the properties of energy supply systems from the view of telecommunications. Subsequent chapters describe the possible uses of powerline communication based on existing standards, communication at very high data rates, and the physical limits. Another chapter discusses how problems regarding electromagnetic compatibility can be solved. Subsequently, promising transmission and access methods as well as the state-of-the-art in device and system development are presented. Finally, the last chapter discusses further developments and the future significance of powerline communication.

This book is intended to alert all potential users to the fascinating possibilities offered by powerline communication. Considering that its capacity allows theoretical data rates of over 100 Mbits/s, one can easily imagine the enormous innovation potential, which could be used to create considerable economic values. Meanwhile, the massive commitment of leading companies and a large number of research institutions certainly suggests that worldwide communication over powerlines may become as commonplace and omnipresent as the use of electric energy. The author sees this book as a mediating link between science and practice to put the complex powerline issue in the right light for the public. Hopefully, it will not only awaken interest and technical curiosity, but also lead to immediate practical application of the new possibilities.

The contributors to this book are too numerous to name them all. It will be easy for the readers to find them in the bibliography. We would like to thank all of them for their help.

The author wishes especially to thank his wife for her support and patience throughout the writing of this book, and to thank his editor and all the people at Prentice Hall who helped shepherd him through the book-writing process.

KLAUS DOSTERT

Krickenbach, December 2000

CHAPTER 1

Introduction

Connecting to the Internet may some day be as easy as plugging into the same wall outlet that serves your stereo. Powerline telecommunication uses existing electric lines to transmit broadband communications in home networking environments and to deliver telecommunication services to homes and businesses.

Despite the positives that powerline telecommunication can offer, questions remain about whether the regulatory and economic problems that have kept it from truly taking off will hinder new developments. Unless proper standards and regulations are developed globally, powerline runs the risk of living as a niche technology at best.

The beginning of communication over supply mains dates back about eight decades. Although electric networks were basically designed for lossless energy transmission, without considering telecommunication requirements, it was achieved early to bring together lossless energy transmission and reliable data transmission at a satisfactory level. At first, only the power supply utilities (PSUs) had been able to make profitable use, but this situation has changed recently.

Deregulation of the telecommunications and energy markets was initiated in 1998. PSUs have to face future competition in the electric power market, and they want to open up new business fields with growth potential in the deregulated telecommunications market. The "electricity" product can be expanded by special value-added services, such as automatic remote meter reading, various and transparent tariff plans, or other services in the field of "home automation" to eventually strengthen their customer base. The use of electric networks within the local loop to bridge the so-called "last mile" for telecommunication services is even more interesting. This could provide a real alternative to the existing national telecommunication networks, such as Deutsche Telekom in Germany, for all types of voice, fax, and data services, particularly for fast and low-cost Internet access for all homes. Furthermore, work is under way to exploit indoor powerlines as fast local area networks carrying digital audio and video information besides other data. This topic is of world-wide interest and is currently being pushed forward by the HomePlug Alliance in the United States and within the Information Society Technologies (IST) program of the European Union.

The possibilities and consequences will be so fundamental and comprehensive that they are hard to predict. The value of the new communication paths within the local loop in the form of an electric power distribution system has been recognized by PSUs and the industry. Numerous studies and field tests have shown that the channel capacity of typical distribution networks on the medium-voltage and low-voltage levels allows data rates of up to several hundred Mbits/s, given a frequency range of about 20 MHz. There is an enormous innovation potential that will create considerable economic values—for example, Internet access from the wall plug. The world's largest knowledge and information base will thus be available to everybody anywhere in the world over a ubiquitous infrastructure. High access costs that have been a massive obstacle to Internet use for many users may change dramatically.

In contrast to radio broadcasting, for example, Internet users can select the topics themselves and open up actively a valuable information medium, which will gradually become commonplace, because access over the regular domestic wall outlet is easy and relatively cheap. Worldwide communication, procurement of information, purchasing, and trade will become as ubiquitous as the consumption of electric power from the wall outlet.

The uses of the electric networks are not unlimited, of course, because when occupying a frequency band of about 9 kHz to over 20 MHz, existing services such as long-wave, medium-wave, and short-wave radio and amateur radio bands will be overlaid. For frequency allocation and the determination of level limits, tradeoff solutions will have to be worked out. Since communication over power networks is basically wire-borne, suitable measures have to be found to prevent inadmissibly high signal radiation. Major efforts are currently being made to work out solutions to ensure electromagnetic compatibility (EMC). This and other challenges of powerline telecommunication will be described in later sections of this book.

Chapter 2 describes the power supply system and its properties. Chapter 3 describes the historical development of powerline data communication. Chapter 4 discusses new possibilities for using the low-voltage level on the basis of European standards, in particular the European CENELEC standard EN 50065. Chapter 5 deals with the innovation potential from the recent deregulation of the power market, while Chapter 6 presents communication system concepts and hardware implementations. A brief summary and outlook of powerline telecommunication is given in Chapter 7. Finally, Chapter 8 provides an extensive bibliography structured by issue.

CHAPTER 2

The Electric Power Supply System and Its Properties

2.1 Topological and Electrical Structures

The starting point for the following studies is an analysis of typical network structures on the various voltage levels of the electric energy supply system (see Figure 2-1). We distinguish three levels: the high-voltage level (110–380 kV), the medium-voltage level (10–30 kV), and the low-voltage level (0.4 kV), each adapted to the bridging of certain distances. The voltage levels are interconnected by transformers, designed in such a

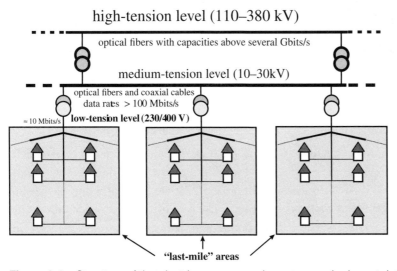

Figure 2-1 Structure of the electric energy supply system and relevant data flows.

5

way that the energy loss is as low as possible at the power frequency (50 or 60 Hz). For the high carrier frequencies typically used for data communication, the transformers are "natural" obstacles, which cause a perfect separation. This suggests a corresponding hierarchical structure for planning the communication system.

2.1.1 The High-Voltage Level

The high-voltage level serves for the long-distance transport of electric energy from the power station to the consumer, bridging distances from several dozen to several hundred kilometers. High-voltage networks are implemented almost exclusively in the form of three-phase current overhead lines. Compared to other multiphase systems, the three-phase system has the most favorable number of wires and thus the lowest investment costs. The aim is to build a symmetric three-phase system by

- using three voltages of the same amplitude with 120 degrees phase shift each for feed.
- building the lines symmetrically. This can be achieved by using the same material for all three wires, using the same wire geometry, and attempting to arrange the wires in an equilateral triangle at the same distances from the ground. This is often opposed by economic considerations with regard to the construction and foundation of the poles, so that one selects an asymmetric structure—for example, using an equal-sided triangle and swapping the wires about once every 20 poles cyclically to achieve balancing.
- attempting to equally distribute both the true and the reactive power load.

Assuming that there is appropriate symmetry, it will be sufficient to consider a one-phase substitute system in the calculation of a three-phase system.

From the physical perspective, there is no fundamental difference between the transmission of high-frequency signals over high-voltage lines and the transmission of energy with the technical AC frequency of 50 or 60 Hz. If we think of a one-phase substitute system simplified as a transmission line consisting of two parallel wires in free space, then both the current and the voltage on the wire pair cause a transverse electromagnetic field, which is the carrier of the transported energy. When the wire distance is much smaller than the wavelength of the transmitted signals, then the field energy concentrates mainly between the wires, and there is virtually no radiation of electromagnetic waves. Although energy transmission and the transmission of high-frequency signals on lines are similar physical processes, there is a considerable difference in the properties of the lines at the different frequencies. In particular, the loss along the lines increases considerably with higher frequency.

With the technical AC frequency of 50 or 60 Hz, the corresponding wavelength on high-voltage overhead lines is about 6000 or 5000 km, respectively. This is why we have to consider wave propagation effects in extended networks, i.e., we cannot assume DC-type conditions. The longest high-voltage lines in Western Europe extend over about 500 km.

In principle, high-voltage networks are designed for optimum energy transmission aimed at minimizing energy loss over long distances. Losses in energy transmission are essentially composed of (Joule's) heat loss due to the ohmic resistance of the wire material and leakage losses, e.g., losses due to leak currents along insulators. Although an increase of the nominal voltage reduces the heat losses, the leakage losses increase concurrently, so that an optimum tradeoff has to be found. Suitable dimensioning of the wire cross sections and selection of suitable material can help keep the heat losses low. The ohmic resistance responsible for heat losses can no longer be determined at 50 or 60 Hz, as in the case of DC, because the current distribution over the wire cross section is no longer constant due to the skin effect [T1, T2]. There is a current displacement toward the wire surface, and the effective resistance

$$R_{\text{eff}} = \frac{P_v}{I^2} \tag{2.1}$$

is larger than in the DC case. Equation (2.1) is a somewhat unusual definition used in energy technology, in which the effective ohmic resistance, R_{eff}, of a wire results from the ratio of the effective power P_v consumed longitudinally to the square of the effective value, I^2, of the traversing current.

The leakage losses depend on the quality and design of the insulators. They are basically smaller by several powers of ten than the heat losses in operation at normal rating. The leakage losses can be calculated only very roughly and are subject to strong climatic fluctuations.

At high voltages there are additional corona losses, originating from discharge activities in the environmental air, due to the high electric field strengths. Discharges start forming in dry air when the electric field strength exceeds approximately 15 kV/cm. In damp air or fog, much lower field strengths are sufficient. The thinner a wire carrying high voltage, the stronger the discharge effects, because very high marginal field strengths occur then on the wire surface. Corona discharges not only cause energy losses but also are sources of intensive high-frequency interference, which impairs radio reception, mainly in the long-wave and medium-wave bands, in the environment of high-voltage lines and the "carrier transmission over powerlines (CTP)" (see Chapter 3).

A suitable geometric arrangement of the high-voltage wires can help reduce the corona losses and the related interference considerably, without the need to increase the wire cross section to impractical or uneconomic dimensions. The arrangement of wires in trunks of four, where one wire is placed in each corner of a square, is one of the methods very successfully applied all over the world to limit corona losses. With a wire radius of approximately 1 cm and a square lateral length of 40 cm, for example, we can obtain an effective substitute wire radius of 22 cm (see, for example, [T1, T2]) with regard to the corona effect.

At the higher frequencies relevant for CTP, the resistance of a wire is much larger than in the DC case, due to the skin effect, and depends on frequency, because the skin depth, a, decreases with increasing frequency. For aluminum, which is the material mostly used for high-voltage lines, the following applies:

$$\frac{a}{\text{cm}} \approx 8.768 \cdot \frac{1}{\sqrt{f/\text{Hz}}} \qquad (2.2)$$

With a wire radius of approximately 1 cm (standard 380-kV line with four-wire trunk arrangement), the effective resistance for a two-wire system with a length of 500 km is $R_{50\text{eff}} \approx 30\ \Omega$ at 50 Hz [T1]. With the same line and an operating frequency of 500 kHz, we would have to expect $R_{500\text{k}} \approx 1023\ \Omega$. This example shows clearly that we have to expect much higher attenuation values in the transmission of high-frequency signals, compared to transmission at power frequency. In addition, the configuration of a high-voltage line with energy transmission is different from that for data communication. While the energy transmission always involves equal load on all three wires of a three-phase system, data communication uses generally only two wires. This means that we will obtain different results when studying wave propagation processes in energy transmission as opposed to data communication.

2.1.1.1 Energy Transmission over High-Voltage Overhead Lines

In this area, we first consider wave propagation processes for energy transmission at 50 or 60 Hz. We will base the derivation of the line properties for higher frequencies on this later.

High-voltage overhead lines can be described in the one-phase model like two-wire lines with the primary "line constants" R' (resistance per length), G' (leakage conductance per length), L' (inductance per length), and C' (capacitance per length). It should be noted that the L', C', and R' values depend on the frequency. To better understand, we use the example of a 380-kV line with four-wire trunks, arranged in a square

Topological and Electrical Structures

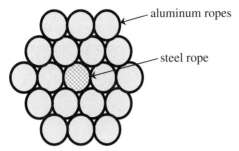

Figure 2-2 Cross section of a typical high-voltage wire.

with a lateral length of $s = 40$, where a single wire consists of a steel core in the form of a rope with a large number of aluminum wires twisted around it. Figure 2-2 shows a typical wire cross section. We see a highly flexible rope-type structure. An aluminum/steel standard cross-section ratio is $240\,\text{mm}^2/40\,\text{mm}^2$ with a total wire diameter of approximately 2 cm. The steel core increases the wire's resistance. Note that the steel core is not really significant for the current transport, because its resistance is approximately 25 times higher than that of aluminum. The individual aluminum wires are twisted around the steel core in two saucer-type levels, where the twisting direction is opposite in the two levels to achieve mechanical stabilization.

The twisting causes a minor extension of the wire length and thus an increase of the resistance, which can be 1–2%, depending on the wire type [T1]. It is difficult to calculate the implications of the skin effect at the power frequency with a wire structure like the one shown in Figure 2-2. This is why manufacturers of high-voltage lines indicate measured values of the effective resistance per length R'_{eff} at 50 or 60 Hz in tables. [T4] gives $R'_{\text{eff}} \approx 0.03\,\Omega/\text{km}$ for the 380-kV line selected in our example above. Typical values of the resistance increase due to the skin effect are in the range of 2–4% [T2]. An increase of the line temperature can also increase the resistance, in addition to the skin effect.

There is almost no way to calculate the primary line constant, G'. Measured values for G' for practical three-phase current overhead line configurations are given in tables. For instance, the value is $G' \approx 3.5 \times 10^{-8}$ S/km for the 380-kV line used in the above example.

With some simplifications, the line constant L' can be calculated with sufficient accuracy at an acceptable effort. With a two-wire line made of massively metallic wires (e.g., aluminum) with wire radius r and distance $d \gg r$, we obtain

$$L' = \frac{\mu_0}{\pi} \cdot \frac{d}{r \cdot e^{-1/4}} \tag{2.3}$$

where the factor $e^{-1/4}$ in the denominator of the argument of the natural logarithm considers the internal inductivity of the wire [T2]. With a 380-kV three-phase current overhead line as described above, the calculation of the inductance per length L' for the one-phase substitute model is more difficult due to the following circumstances:

- The wires are not massive, but stranded (with steel core).
- One phase conductor is formed from four individual wires, of which each one conducts 1/4 of the external wire current.
- The three phase conductors are arranged on the edges of a equal-sided triangle and inductively coupled.

The impact from soil, poles, and other conductive structures, such as guard wires, should be neglected from the outset. Nevertheless, the calculation required to obtain a one-phase equivalent circuit

$$L' = 0.813 \text{ mH/km} \tag{2.4}$$

for the inductance per length for the symmetric 380-kV three-phase current overhead line is relatively complex [D3]. Similar to the line constant L', the capacitance per length C' can also be determined at an acceptable cost under certain simplifications. After an equally extensive calculation, we obtain [D3]

$$C' \approx 13.89 \frac{\text{nF}}{\text{km}} \tag{2.5}$$

in the single-phase equivalent circuit of a typical symmetric 380-kV three-phase current overhead line.

The characteristic impedance Z_L and the propagation constant γ for the 380-kV three-phase overhead line we selected as an example can now be stated, initially at an operating frequency of 50 Hz:

$$Z_L = \sqrt{\frac{R' + j\omega L'}{G' + j\omega C'}} \approx 242.77 \, \Omega \cdot e^{-j3.12°} \tag{2.6}$$

and

$$\gamma = \alpha + j\beta = \sqrt{(R' + j\omega L') \cdot (G' + j\omega C')} \approx 1.059 \cdot 10^{-3} \cdot e^{j86.42°} (\text{km})^{-1} \tag{2.7}$$

The values α and β in (2.7) are called *attenuation constant* and *phase constant*, respectively. A line is said to be slightly lossy if

$$R' \ll j\omega L' \text{ and } G' \ll j\omega C' \tag{2.8}$$

In this case, the characteristic impedance can be calculated from L' and C' with a good approximation, and it is real-valued. With (2.4) and (2.5) it is

$$Z_L = \sqrt{\frac{L'}{C'}} \approx 241.9 \, \Omega \tag{2.9}$$

The attenuation constant α and the phase constant β can be calculated from (2.7), taking the relationship for $x \ll 1$ approximately into account. We obtain

$$\sqrt{1+x} \approx 1 + x/2$$

$$\alpha \approx \frac{R'}{2 \cdot Z_L} + \frac{G' \cdot Z_L}{2} \text{ and } \beta \approx \omega\sqrt{L'C'} \tag{2.10}$$

Equation (2.10) supplies

$$\alpha \approx (6.2 \cdot 10^{-5} + 4.233 \cdot 10^{-6})/\text{km} = 6.62 \cdot 10^{-5}/\text{km} \tag{2.11}$$
$$\beta \approx (1.047 \cdot 10^{-3}) \cdot 1.01/\text{km} = 1.055 \cdot 10^{-3}/\text{km} \equiv 6°/100 \text{ km}$$

for the 380-kV line selected as example.

A comparison of (2.6) and (2.9) shows that the characteristic impedance Z_L can be easily calculated as a real value with good approximation from the L'/C' relationship. We understand from (2.11) that the leakage conductance per length G' flows into the attenuation constant α only with approximately 6.4%. Therefore, it is sufficient for rough calculations of α to consider only the resistance per length R'.

The determined line values allow a mathematical description of the progress of the complex effective values of current $I(\ell)$ and voltage $U(\ell)$ along a high-voltage line as functions of the complex effective values of the current I_E and the voltage U_E at the line end and the location coordinate ℓ, which is counted starting from the line end. The following apply:

$$U(\ell) = U_E \cdot \cosh(\gamma \cdot \ell) + I_E \cdot Z_L \cdot \sinh(\gamma \cdot \ell) \tag{2.12}$$

and

$$I(\ell) = \frac{U_E}{Z_L} \cdot \sinh(\gamma \cdot \ell) + I_E \cdot \cosh(\gamma \cdot \ell) \tag{2.13}$$

In the area of energy transmission with the technical alternate current frequency of 50 or 60 Hz, the line equations (2.12) and (2.13) are used for exact determination of the line states. This is necessary here because, when operating a high-voltage line, for example, we have to avoid overvoltage states. When a long line runs idle at the end or is operated in light-load condition, with nominal voltage U_N at the beginning, the voltage is higher than U_N at the line end. This can be dangerous for insulators, transformers, and switches and therefore has to be prevented quickly and reliably by suitable compensation measures [T4].

A very special load condition of a high-voltage line is given if the line is terminated with its characteristic impedance Z_L at the end. In this case, the so-called natural power

$$S_{nat} = \frac{U_N^2}{Z_L^*} \approx P_{nat} = \frac{U_N^2}{Z_L} \tag{2.14}$$

is transmitted. This means that overvoltage can no longer occur, because the line operation is free from reflections. The effective value $U(\ell)$ of the voltage along the line decreases from the generator to the consumer with $e^{\alpha\ell}$. Equation (2.12) becomes then

$$U(\ell) = U_E \cdot e^{\alpha\ell} \cdot e^{j\beta\ell} \tag{2.15}$$

S_{nat} is a secondary characteristic line value. For the 380-kV three-phase current overhead line selected as an example, we obtain

$$S_{nat} = 594.8 \cdot e^{-j3.12°} \text{ MVA} \approx P_{nat} = 596.86 \text{ MW} \tag{2.16}$$

In practice, it is hardly possible to operate a high-voltage line with S_{nat} permanently, because the load changes, which the line operator can control only to a very limited extent, so that this factor determines the operating conditions. The operation of a line with S_{nat} is favorable with regard to undesired reactive power, because no reactive power appears in this case. It is always desirable to compensate the reactive power of the line, regardless of the current load [T4]. For natural power S_{nat}, it should also be noted that this is not the limiting power of a line. S_{nat} cannot be achieved due to the heat losses, in particular for lines with relatively low nominal voltage and thus also small characteristic impedance, e.g., cables. In contrast, high-voltage overhead lines can also transmit powers larger than S_{nat}.

Based on the above considerations, we should be careful not to draw the wrong conclusion that, with the adaptation at the line end, we would get equal situations for energy transmission and data communications, and thus accept an energy loss of 50%. A high-voltage line is terminated in operation with S_{nat} only at its end with its characteristic impedance. On its feed side, the generator is never equipped with an internal resistance that equals the characteristic impedance, but the generator's internal resistance is kept as small as possible.

If we try to draw a rough image of the operating condition of a high-voltage line, we calculate with the model of a lossless line, i.e., $R' = 0$ and $G' = 0$. This simplifies Equations (2.12) and (2.13) to

$$U(\ell) = U_E \cdot \cos(\beta \cdot \ell) + jI_E \cdot \sin(\beta \cdot \ell) \qquad (2.17)$$

and

$$I(\ell) = I_E \cdot \cos(\beta \cdot \ell) + j\frac{U_E}{Z_L} \cdot \sin(\beta \cdot \ell) \qquad (2.18)$$

The amplitude decrease due to the attenuation is considered separately by using the factor $e^{\alpha \ell}$. The simplification described above is also used for the design of CTP systems (see Chapter 3).

2.1.1.2 High-Voltage Overhead Lines for HF Signal Transmission

For high-voltage networks, particular measures are normally undertaken to keep the attenuation of high-frequency signals as low as possible. At the beginning and end points as well as on junctions, carrier-frequency blocking filters in the form of inductors with inductivity values of 0.5 to 2 mH are used, normally configured as cylinder-shaped air-core coils (see Figure 2-3). It should be noted that the usual carrier transmission over powerlines (CTP), which will be described in more detail in the following chapter, takes place in the range below 500 kHz, which results in relatively large inductance values. For adaptation to the line's characteristic impedance, special input and output coupling units are used. In terms of high-frequency signal transmission, we are not talking about a symmetric three-phase current system, but about an approximation toward a two-wire system. The influence of the soil and neighboring wires, poles, and guard wires can normally be neglected in the calculation of the line constants. The line constants L' and C' give now different values than in (2.4) and (2.5). The resistance per length R' will increase dramatically due to the skin effect, depending on the frequency. The leak conductance per length G', which is effective at 50 or 60 Hz, does not have any more significance here. Instead, the state of the medium air around the wires, depending in the weather, gains considerable influence on the high-frequency signal attenuation. In particular frost, but also fog, rain, and snow, show a clear impact.

The inductance per length L'_H has to be calculated according to (2.3), where the wire structure (e.g., four-wire trunks) must be taken into account. The capacitance per length C'_H according to [D3] results in

$$C'_H = \frac{\pi \varepsilon_0}{\ln \frac{d}{r}} \qquad (2.19)$$

Figure 2-3 High-voltage switching station with carrier-frequency locks.

where d is the wire distance and r is an effective radius depending on the wire structure. For the 380-kV three-phase current overhead line selected as an example with four-wire trunks (with a 40-cm distance between each wire, trunk distance 9.5 m), we obtain the following:

Topological and Electrical Structures

$$L'_H \approx \frac{\mu_0}{\pi} \cdot \ln\frac{950}{16.29} = 1.626\,\frac{\text{mH}}{\text{km}} \qquad (2.20)$$

and

$$C'_H \approx \frac{\pi \cdot \varepsilon_0}{\ln\frac{950}{17.34}} = 6.944\,\frac{\text{nF}}{\text{km}} \qquad (2.21)$$

When neglecting the losses, we can now state the real-valued characteristic impedance

$$Z_{LH} = \sqrt{\frac{L'_H}{C'_H}} = 483.9\,\Omega \qquad (2.22)$$

for two-wire coupling. For example, [T1] states a planning value of 500 Ω from practical experience.

To determine the high-frequency signal attenuation, we first need to determine the attenuation constant α. It can be calculated from R' and Z_{LH}, while neglecting G' with a good approximation based on (2.10). The particular attenuation effects caused by fog, frost, heavy rain, or snow, as mentioned above, cannot be meaningfully included in a calculation. The resistance per length R' depends on the frequency due to the skin effect. Using (2.2), we obtain the tailored value equation

$$R'(f) \approx 1.45 \cdot 10^{-3} \cdot \sqrt{\frac{f}{\text{Hz}}}\,\frac{\Omega}{\text{km}} \qquad (2.23)$$

The attenuation of high-frequency signals on high-voltage lines depends thus on frequency according to a square-root function, and we obtain

$$\alpha(f) \approx 1.479 \cdot 10^{-6} \cdot \sqrt{\frac{f}{\text{Hz}}}\,(\text{km})^{-1} \qquad (2.24)$$

If we look at a 380-kV three-phase current overhead line with a length of 500 km, for example, which is to be used for the transmission of signals with a carrier frequency of 1 MHz, then the attenuation experienced by this line due to ohmic losses is

$$e^{\alpha(1\,\text{MHz}) \cdot 500\,\text{km}} \approx 2 \equiv 6\,\text{dB} \qquad (2.25)$$

This very low value may surprise; unfortunately, practice has proven that this value cannot be used as a basis for system design, because a series of additional physical influences may lead to an increased attenuation of high-frequency signals. Frost, fog, heavy snow, and rain were already mentioned above. The formation of frost and ice on the wire surfaces is considered a particularly strong attenuation-increasing effect. Theoretical studies explain this with high dielectric losses in ice. In extreme cases of line icing, with frequencies of only around 100 kHz, attenuation values increased by 15 times versus normal dry weather have been measured [T1]. In Central Europe, one can assume that only relatively short lengths of a high-voltage overhead line are exposed to extreme weather impact, so that one does not have to use the measured maximum values of attenuation in the design.

But even in normal dry weather, high-voltage overhead lines with two-wire coupling show considerably higher attenuation values than one would expect, based only on (2.25), due to the ohmic losses. The main reason for this is losses in the soil, because the distance of the two high-frequency conducting wires to the soil is only about three to four times larger than the distance between the wires. While it is true that the metallic wires neighboring the high-frequency line pair—the third wire of the three-phase current system, additional three-phase current systems in the same line stretch, guard wires and poles—influence the characteristic impedance, their attenuation effect can normally be neglected [T1]. In contrast, the soil causes noticeable losses due to poor conductivity and thus high skin depth of the high-frequency electromagnetic fields. At high frequencies, there are also dielectric losses. The losses in the soil are considerably higher than in the wires themselves. The attenuation caused by the soil increases with higher frequency and is determined essentially by the ratio between wire distance d and the height H of the wires above ground. Table 2-1 shows a few practical design guidelines for attenuation D per 100 km as a function of frequency for a ratio of $d/H = 3.5$.

Table 2-1   Frequency-dependent attenuation by soil impact.

D per 100 km	f (kHz)
6.12 dB	200
9.25 dB	300
12.39 dB	400

A comparison of these values with (2.25) shows that the attenuation due to ohmic line losses can normally be neglected.

When it is a matter of studying the quality of an information transmission, the sole consideration of the attenuation properties is not sufficient. The interference scenario also has to be considered, because for the received signal quality, the ratio of signal power (or useful power) S to noise power N at the receiver, which is within the receiver bandwidth, is decisive. This means that the following relation is important: signal-to-noise ratio

$$\Gamma = S/N \qquad (2.26)$$

or in logarithmic representation, in dB,

$$a = 10 \log_{10}(S/N) \qquad (2.27)$$

The following section studies the typical interferences on high-voltage lines to produce a calculation basis for Γ.

2.1.1.3   High-Frequency Interference on High-Voltage Overhead Lines

With regard to the original cause, we can distinguish between two types of high-frequency interference on high-voltage overhead lines:

1. Short-time impulsive interference, which occurs aperiodically. Cause: switching events or atmospheric discharges (arcs).
2. Broadband, permanently present interference with relatively high level. Cause: coronary discharges, glow discharges, e.g., on isolators.

Interference of the first kind has broadband spectra which are, however, present only a short time. However, one can observe up to 300 interference impulses per second [T1]. The impulses reach considerable amplitudes, and they cause dangerous voltage peaks at the receiver devices due to their high-frequency portion, so that special protection measures are required for the receiver input levels. However, for the CTP signals, the impairment is low because CTP waveforms—as will be described in more detail in Chapter 3—are narrowband and long, compared to the typical impulse duration [T1].

Interference of the second kind can be modeled for a frequency of up to approximately 1 MHz as white Gaussian noise (see Figure 2-4). Their impact is much more critical than interference of the first kind, because they are constantly present at relatively high levels. In particular the power spectral density (PSD) depends strongly on weather. It increases considerably during rain, frost, or fog; the variation is approx. 20 dB. Figure 2-4 shows a typical PSD spectrum, where the level information refers to a bandwidth of 2.5 kHz. This value corresponds to the CTP channel basic grid. The fluctuation of the noise power between $N_{min} = -30$ dBm (i.e., 1 µW) and $N_{max} = -10$ dBm (i.e., 100 µW) results essentially from the climatic dependencies.

The minimum value is typical in dry, clear weather, while the interference level increases with increasing air humidity or line icing to eventually reach the maximum limit value under massive frost. In practice, however, the upper limit is extremely rare for

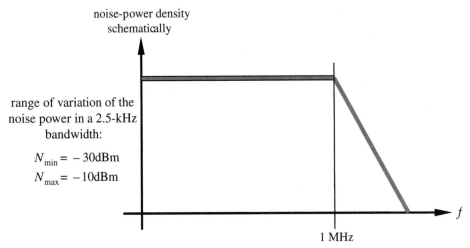

Figure 2-4 Typical PSD of noise power in high-voltage lines based on information from [T1].

very long distances. In general, distances influenced by heavy frost over a lengthy period will be less than 100 km.

2.1.2 The Medium- and Low-Voltage Levels

In recent years a large number of energy distribution networks have been modeled and developed as communication channels for frequencies of up to about 150 kHz; see the publications in Section 8.7.1 in the Bibliography. The transmission properties of the networks for frequencies up to the megahertz range have not been studied in the past, either in theory or in practice. Due to the wide range of different network topologies, which use different line or cable types, a general theoretical description is difficult, if at all possible. Technical accompanying studies can therefore not be avoided. A meaningful goal of theoretical studies is the description of the properties of individual components, which can help us gain gradual insight into the behavior of a complex network.

Medium- and low-voltage networks are built from overhead lines and cables, where the cables are generally laid underground. Medium-voltage overhead lines have nominal voltages below 110 kV; the typical values are 10–20 kV. Medium-voltage overhead lines supply normally electric energy to rural areas, small towns, or individual industrial companies or plants. Typical lengths of medium-voltage overhead lines range between 5 and 25 km. There are no more overhead lines in densely populated areas. The power to these areas is supplied exclusively from ground cables. On the low-voltage level, overhead lines are still found in many small towns and in areas with relatively old buildings. In many cases, the overhead lines are being replaced by aerial cables, e.g., where an underground installation would be impossible or uneconomic. Otherwise, the supply on the low-voltage level is generally via ground cables.

Medium-voltage overhead lines have relatively small line cross sections of maximum 95/15 Al/St, i.e., 95 mm^2 aluminum and 15 mm^2 steel rope cross sections, compared to high-voltage overhead lines. Due to the low voltage—typically in the range of 10–20 kV—the required pole heights are small, and consequently so are the wind forces they are exposed to. The pole constructions for medium-voltage overhead lines are very light, compared to high-voltage overhead line poles. Cables on the medium- and low-voltage levels are in use in a large number of variants. The wire materials used include copper and aluminum. With regard to the wire cross sections, we distinguish between round, sector-shaped, or oval shapes with single-wire or multiwire structure. The insulation material in modern cables is usually either polyvinyl chloride (PVC) or vulcanized polyethylene (VPE). Oil or paper is used for higher voltages. The entire multitude of cables with their special fields of application are not described here in detail, but Chapter 6 [T2] provides a good overview.

2.1.2.1 Properties of Lines and Cables on Medium- and Low-Voltage Levels
The following discussion refers to three-phase current networks. On the medium- and low-voltage levels, we distinguish between transmission networks and distribution networks. The borders between these two basic forms are blurred. While we can undoubtedly speak of a pure transmission network for electric energy on the high-voltage level, this is only partly the case for the medium-voltage level. The general rule is that the distribution function dominates with decreasing network voltage. The supply radius of medium-voltage networks varies between 5 and 25 km. Large radiuses are found in rural supply areas. For low-voltage networks, the typical supply radiuses of a feed point (low-voltage transformer station) are from 100 to 500 m [R3].

This section discusses the line properties on the medium- and low-voltage levels in two different frequency ranges with practical relevance: first, in a range near the power frequency, which reaches a maximum of 3 kHz, and second in a range of up to approximately 30 MHz, which is far above the power frequency.

Line Properties near the Power Frequency When using a power supply network additionally for the transmission of information, the impedance that exists, for instance, between two wires of the network plays an important role. As a general rule, the smaller the impedance, the more power has to be fed in for information transmission. Near the power frequency, the impedance of a power supply network is determined significantly by the current load. Therefore, the impedance is not constant, but varies over time. It is influenced by the voltage level, the pertaining load density, and the behavior of the consumers collective. The variation range of the impedance can be determined from the ratio between peak load and minimum load. For a rough classification, Table 2-2 shows indicative values of the peak load to minimum load ratio for the three voltage levels [R3]:

Table 2-2 Peak load to minimum load ratio on various voltage levels.

	Low-voltage level	Medium-voltage level	High-voltage level
Peak load : minimum load	7:1–19:1	3:1–8:1	2:1–3:1

We can see that the highest impedance fluctuations are on the low-voltage level, and that there is a considerable fluctuation width on the medium-voltage level, too. For a signal transmission, for example, in ripple control systems (see Chapter 3) with frequencies near the power frequency, this means that communication with matched system components (source, line, and sink) is excluded. The required transmit power for ripple control has to be oriented to the network load from the connected consumers.

Line Properties at Higher Frequencies

The Overhead Lines of the Distribution Network In connection with the previous studies on the high-voltage level, we start this section with the consideration of low-voltage overhead lines, which occur mainly in scarcely populated areas, but which are not meaningless today, if we want to open up new communication channels. This section describes "classical" low-voltage overhead lines, configured in the form of four wires, arranged as the edges of a square. For high-frequency signal transmission, the most interesting case from the technical view of the two-wire coupling is used as a basis, where the signal to be transmitted is injected between two phases. The neutral conductor should generally not be used. The reason is to avoid electromagnetic compatibility problems, which will be described later.

Overhead lines begin normally at a transformer station and, when they lead consistently all the way to all consumers connected to the supply network, they represent fairly good high-frequency waveguides with low attenuation. With regard to the characteristic impedance and the losses to be expected, we can see a strong similarity with the high-voltage networks described earlier, although the geometry is totally different. For signal transmission on high-voltage networks, particular measures are taken in order to keep the attenuation of the high-frequency signals as low as possible. At the start and the end points as well as on junctions, for example, carrier-frequency barriers in the form of inductors with inductivity values from 0.5 to 2 mH are used. As the typical applications (CTP) occur here below 500 kHz, relatively high inductivity values are required. Special input and output coupling units are used for matching to the line's characteristic impedance. With regard to high-frequency signal transmission, there is a good approximation to a two-wire system. The impact of the soil and neighboring wires or poles can be neglected in the calculation of the line constants.

For a low-voltage overhead line, when using the coupling to two phases as mentioned above, we can also use a two-wire model as basis. Assuming a wire distance of $s = 25$ cm and a wire diameter of about $d = 10$ mm, we obtain approximately the characteristic impedance

$$Z_{LF} = 120 \ \Omega \cdot \ln\frac{s}{d/2} = 470 \ \Omega \qquad (2.28)$$

Even papers published more than 25 years ago contain information about the properties of distribution networks, which match the considerations described here. As an example, we use Table 2-3 from [R3], which summarizes typical line values for medium- and low-voltage networks and the resulting propagation speed and characteristic impedance areas.

Table 2-3 Standard values for lines on the medium- and low-voltage levels.

Line Type	L' (mH/km)	C' (µF/km)	v (km/s)	Z_L (Ω)
Overhead line	0.9–1.57	7–13·10⁻³	2.92·10⁵–2.95·10⁵	347–358
Cable up to 1 kV	0.25–0.31	0.34–0.8	0.6·10⁵–1.06·10⁵	19.6–27
Medium-voltage cable	0.22–0.4	0.2–0.5	1–1.5·10⁵	28–33

To estimate the high-frequency attenuation, studies done for high-voltage overhead lines can also be used here. In an overhead system, there are losses virtually only because of the skin effect. The resulting attenuation values are surprisingly low, e.g., only 6 dB at 1 MHz over 500 km on the high-voltage level. Unfortunately, a series of other physical influences cause increased attenuation, such as frost, fog, heavy snow and rain. On the other hand, problems of this kind occur only on very long distances of over 100 km, so that they are insignificant in the low-voltage area and often also in the medium-voltage area.

The analyses conducted so far look at overhead lines as "isolated" waveguides, which are separated from their environment for the usable frequency band at the beginning and at the end and on junctions—for example, by inductors. Both the technical and the economic aspects of this type of measure in low-voltage networks will surely have to be checked in the individual case. In addition, there are other circumstances when the consideration of an overhead line as a "clean" HF waveguide is not admissible. Sometimes an overhead line does not start in a transformer station, but for instance it is fed over a ground cable in a length of up to several hundred meters. In rare cases, there is another transition from an overhead line to a ground cable. While such transitions cause no problem whatsoever at the power frequency (50 or 60 Hz), there are strong reflections at high frequencies, so that a considerable part of the HF signal power does not pass the transition, but wanders back toward the source. The reflection factor r states the ratio of the amplitude of the backward wave to the amplitude of the forward wave in a line. When two lines with a different characteristic impedance (Z_{L1}, Z_{L2}) are coupled, and if each line is terminated with its characteristic impedance at its end, then the resulting reflection factor at the coupling point is

$$r = \frac{Z_{L2} - Z_{L1}}{Z_{L2} + Z_{L1}} \qquad (2.29)$$

This gives the reflection factor

$$r_{EF} = \frac{50 - 470}{520} \approx -80.7\% \qquad (2.30)$$

for a ground-cable/overhead-line transition with $Z_{LE} \approx 50\ \Omega$ and $Z_{LF} \approx 470\ \Omega$. As will be derived later, the ground-cable characteristic impedance is near $50\ \Omega$.

Analogously, we obtain

$$r_{FE} = \frac{470 - 50}{520} \approx 80.7\% \qquad (2.31)$$

for an overhead-line/ground-cable transition. From the amount of the reflection factor, we can now determine the transmitted power and the reflected power at the transition. The ratio of transmitted P_T to the injected power P_E is

$$\frac{P_T}{P_E} = 1 - |r|^2 \qquad (2.32)$$

and the reflected power P_R is given by

$$\frac{P_R}{P_E} = |r|^2 \qquad (2.33)$$

This means that we obtain a reflection of 80.7% of the power arriving at the ground-cable/overhead-line transition, i.e., only approximately 19% pass into the overhead line. Assuming that there is a second overhead-line/ground-cable transition, then only about 3.6% of the original power that arrives at the first transition will remain, and no line attenuation has been taken into account yet. A simple ground-cable/overhead-

Topological and Electrical Structures

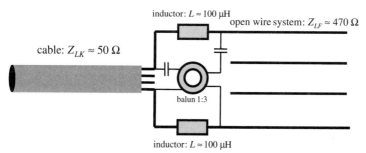

Figure 2-5 Low-reflection transition between ground cable and overhead line.

line transition or vice versa has thus an additional attenuation of approx. 7.2 dB, while two transitions in one line stretch would result in an attenuation of 14.4 dB. These values themselves are actually not critical, but they could cause a problem in conjunction with the entire path attenuation, so that appropriate countermeasures may be worthwhile.

Figure 2-5 shows the principle of a matching circuit, which creates an HF transition free of reflections, without impairing the safety of the energy supply. The inductors could be configured as cylinder-shaped air-core coils with the following dimensions:

Coil length: 20 cm
Coil diameter: 10 cm
Number of turns: 45

By use of magnetic materials, e.g., in the form of properly designed ferrite cores, the dimensions of the inductors can be considerably reduced. Note, however, that the passing 50-Hz or 60-Hz current can cause saturation and thus deteriorate the magnetic properties. In Figure 2-5, a ferrite ring core could be used, however, as HF balun without any problem. In this case, a turns ratio of

$$\sqrt{\frac{470}{50}} = 3$$

would have to be realized to match the characteristic impedances.

The Cables of the Distribution Network   Undoubtedly the cable is the most important component of distribution network. Both the mechanical and the electrical parameters of low-voltage cables are subject to wide dispersions, which means that the line parameters of interest here, such as attenuation or characteristic impedance, also have an accordingly high tolerance. To allow a simple start, analytical approximation calculations are done, which are sufficiently exact for an estimate of the parameters to be

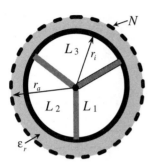

Figure 2-6 Schematic structure of a three-wire cable with concentric outer conductor.

determined. Figure 2-6 shows a schematic structure of a three-wire cable with sector conductors. The sector conductors form the phases L_1, L_2, and L_3, while the outer conductor functions as PEN (protective earth and neutral).

With symmetric feed, the voltage differences between the sector conductors of the cable are zero. Due to the symmetric arrangement and the symmetric termination, the voltage difference between the sector conductors along the line cannot change. As a consequence, the space between the sector conductors is field-free, and a radial electric field forms approximately toward the outer conductor. Here, very small gaps between the conductors in comparison to the radiuses r_a and r_i are assumed. This means that the field is very similar to that of a coaxial cable. This assumption has been proven for the frequency range up to 150 kHz in [D1] by numeric field calculation, and it can also be applied to much higher frequencies, because the field behavior is essentially determined by the geometry of the conductor arrangement. According to [D1], the capacitance per length, C', and the inductance per length, L', are then calculated as follows:

$$C' = \frac{2\pi\varepsilon_0\varepsilon_r}{\ln \frac{r_a}{r_i}} \quad (2.34)$$

and

$$L' = \frac{\mu_0}{2\pi} \ln \frac{r_a}{r_i} \quad (2.35)$$

In (2.34) and (2.35), r_a is the inner radius of the outer conductor and r_i is the radius of the sector conductor. ε_0 is the dielectric constant, μ_0 the permeability of free space, and ε_r the relative dielectric constant.

At high frequencies the resistance per length, R', is essentially determined by the skin effect. The reason is that the current flows now only in a relatively thin layer on the bent outer surfaces of the sector wires and on the inner surface of the outer conductor. The skin depth, a, depends on the frequency, f, and the specific conductivity, κ, of the wire, and can be calculated as follows [see also Equation (2.2) for aluminum]:

$$a = \frac{1}{\sqrt{\pi f \kappa \mu_0}} \tag{2.36}$$

Then we obtain the following for the resistance per length, R':

$$R' = \frac{1}{\kappa \cdot a}\left(\frac{1}{2\pi r_a} + \frac{1}{2\pi r_i}\right) \Rightarrow R' = \frac{1}{2}\sqrt{\frac{f\mu_0}{\kappa\pi}}\left(\frac{1}{r_a} + \frac{1}{r_i}\right) \tag{2.37}$$

The leakage conductance per length, G', of the line can be calculated by means of loss factor $\tan\delta$ of the dielectric from the capacitance per length, C':

$$G' = 2\pi f C' \cdot \tan\delta \tag{2.38}$$

The expression for the complex characteristic impedance Z_{LK} [see also Equation (2.6)] is then

$$Z_{LK} = \sqrt{\frac{R' + j\omega L'}{G' + j\omega C'}} \tag{2.39}$$

and for the propagation constant, $\gamma_K = \alpha + j\beta$, with the attenuation constant α and the phase constant β [see also Equation (2.7)], we obtain

$$\gamma_K = \sqrt{(R' + j\omega L')(G' + j\omega C')} \tag{2.40}$$

We obtain $R' \ll \omega L'$ for high frequencies, because R' grows proportionally with the square root of the frequency f. In addition, the loss factor $\tan\delta \ll 1$ for the most frequently used insulation material, polyvinyl chloride (PVC), at not too high frequencies, so that we can assume that $G' \ll \omega C'$. This means that a typical ground cable can generally be modeled as slightly lossy. The characteristic impedance Z_{LK}, the attenuation constant α, and the phase constant β can then be described with the well-known approximations (2.9) and (2.10):

$$Z_{LK} = \sqrt{\frac{L'}{C'}} \text{ (real)}, \quad \alpha(f) \approx \alpha_R + \alpha_G = \frac{R'}{2 \cdot Z_L} + \frac{G' \cdot Z_L}{2}, \quad \beta \approx \omega\sqrt{L' \cdot C'} \tag{2.41}$$

The attenuation $D(f, \ell)$ can be calculated from the attenuation factor $\alpha(f)$, depending on length ℓ in dB:

$$\frac{D(f, \ell)}{\text{dB}} = 20 \cdot \log_{10}(e^{\alpha(f) \cdot \ell}) = 8.686 \cdot \alpha(f) \cdot \ell \qquad (2.42)$$

The previous equations are now used as examples for a cable of the type NYCY70SM/35 to determine the values for characteristic impedance and attenuation, depending on the frequency, in order to get an idea of the conditions. The insulation of the cable is made of PVC, so that the dielectric properties of this material have to be considered in the calculation. Unfortunately, it will not help us gain general findings, because the dielectric properties of PVC depend heavily on factors like temperature and the addition of "softeners." In a first approximation, the relative dielectric constant, ε_r, applied over a logarithmically divided frequency axis, has a straight curve with the value $\varepsilon_r \approx 3.8$ at 1 MHz, which then drops to $\varepsilon_r \approx 2.9$ at 20 MHz. The loss factor tan δ of PVC shows also an approximately linear curve over a logarithmic frequency axis with a value of 0.05 at 1 MHz, which decreases even to approximately 0.01 at 20 MHz in line with increasing frequency. Figures 2.7 and 2.8 show the resulting curves of the characteristic impedance Z_{LK} and the attenuation $D(f)$ for $\ell = 1$ km in dependence on the frequency. Note that the characteristic impedance Z_{LK} remains constant in the 7–8-Ω range. This confirms the correctness of the assumption that the medium is in fact slightly lossy. As expected, the attenuation $D(f)$ increases as the frequency increases, reaching a value of about 50 dB at 20 MHz.

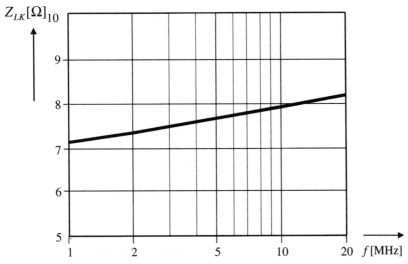

Figure 2-7 Characteristic impedance depending on the frequency for cable type NYCY70SM/35.

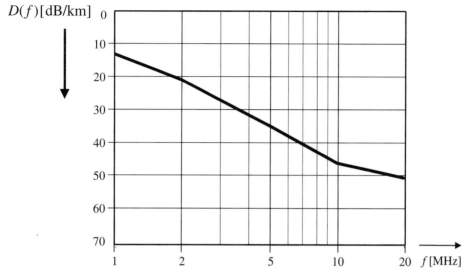

Figure 2-8 Attenuation depending on the frequency for cable type NYCY70SM/35 at a length of 1 km.

The cable studied in Figure 2-6 is not very commonly used in distribution networks in Europe. The most frequently used cable has a cross-section geometry of Figure 2-9. This structure differs strongly from the studied coaxial structure. We will

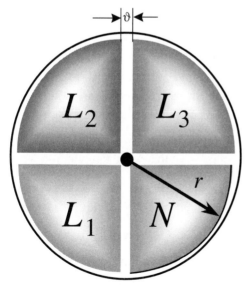

Figure 2-9 Schematic representation of the structure of a four-wire cable without concentric outer conductor.

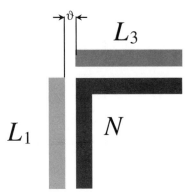

Figure 2-10 Strip-line model of the four-sector cable without concentric outer conductor.

see later that this is not a disadvantage worth mentioning for the transmission of high-frequency signals. At first sight, this does not seem logical, because one would assume clean conduct and shielding of the high-frequency electromagnetic fields in the coaxial structure. More detailed studies, which will be described later, will show that the operation of an energy cable in "coaxial mode" leads to considerable disadvantages with regard to the electromagnetic compatibility.

As for the coaxial cable in Figure 2-6, we may also assume for the cable built from four sector wires in Figure 2-9 that the gap, ϑ, between the wires versus the radius, r, is small. This means, for example, that the capacitive coupling between wires L_1 and N, L_3 and N, L_1 and L_2, and L_2 and L_3 is very high. In contrast, the coupling between the wires L_2, N, and L_1 and L_3 is obviously much smaller. With symmetric feed, the voltage of wires 1 through 3 versus the neutral conductor (N) is identical. Due to the symmetry, a potential difference between the wires 1 through 3 and N cannot arise in any other point of the line, either. Due to the condition $\vartheta \ll r$, the electric field can be approximated through the field of two strip lines, which are positioned at a 90-degree angle to each other, as shown in Figure 2-10.

The electric field in the strip lines is homogeneous, so that we obtain the following for the capacitance per length C' under the approximation $\vartheta \ll r$:

$$C' = 2 \cdot \varepsilon_0 \cdot \varepsilon_r \cdot \frac{r}{\vartheta} \qquad (2.43)$$

Due to the 90-degree angle between the two strip lines, the magnetic coupling of these two lines can be neglected. As a consequence, the inductance per length L' of the entire arrangement results from the parallel connection of the inductance of the two strip lines:

$$L' = \mu_0 \cdot \frac{\vartheta}{2r} \qquad (2.44)$$

The current flows essentially on the inner surfaces between the wires L_1 and N or L_3 and N, respectively. The current on conductor L_2 flows mainly in the peak of this sector due to the skin effect. As we have a parallel circuit, the resistance of L_2 versus the essentially smaller resistance values of L_1 and L_3 can be neglected. This condition results then in a resistance per length of the line of

$$R' = \frac{1}{2} \cdot \frac{1}{\kappa \cdot a} \cdot \frac{2}{r}, \qquad \text{thus} \qquad R' = \sqrt{\frac{\pi \cdot f \cdot \mu_0}{\kappa}} \cdot \frac{1}{r} \qquad (2.45)$$

Similar to the coaxial structure, the leakage conductance per length, G', results from $G' = 2\pi f C' \cdot \tan\delta$. The above equations apply equally to the characteristic impedance, to the phase constant, the attenuation factor, and the attenuation per length. We can now use this basis to determine the values of characteristic impedance Z_{LK} and attenuation $D(f, \ell)$ for the most frequently used cable types with four-sector structure, NAYY50SE and NAYY150SE.

First, it is useful to look at a typical distribution network structure, showing the fields of application of the cable types. Figure 2-11 shows the topology of a typical residential district supply. In this example, up to ten supply cables can lead out from the transformer station, feeding up to 40 house connections each. A supply cable consists generally of cable type NAYY150SE, while the junctions to the customer premises are implemented with cable type NAYY50SE. The two cable types differ basically only in their cross-section dimensions. PVC is generally the preferred insulation material.

Figure 2-12 shows the characteristic impedance values of the two cable types, depending on the frequency. The values are 22 to 25 Ω and 28 to 32 Ω, respectively, which means that they are significantly higher compared to the coaxial structure, but smaller depending on the frequency, which justifies the assumption of low losses.

Figure 2-13 shows each the attenuation in dB/km of the two four-sector cables. It is interesting to note that the values are very similar to those of the above-described coaxial structure; i.e., by no means significantly higher values result, as one would expect at first sight. Due to the fact that there is no symmetry and no shielding, there are less ideal field conditions in four-sector cables. This means that fields could enter into the soil around the cable, which would generally lead to higher losses. However, an exact analysis of the strip-line model in Figure 2-10 shows that the properties to be expected are not considerably worse than with the coaxial structure, because only relatively weak marginal fields will exist, while the main part of the high-frequency power is transported between the wires. These considerations make it plausible that all studied

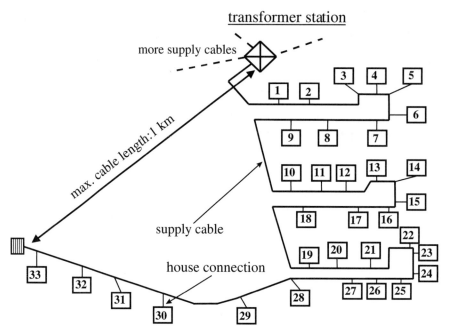

Figure 2-11 Topology of a typical residential district supply.

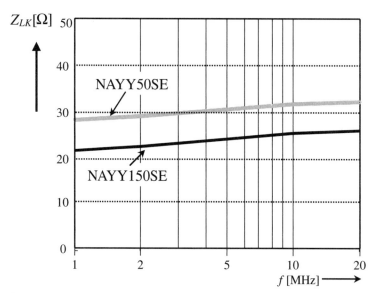

Figure 2-12 Characteristic impedance values of the most commonly used four-sector cables.

cables have very similar attenuation values. The slightly increased attenuation of the thinner cable (NAYY50SE) can be ascribed to the impact of the skin effect, because

Topological and Electrical Structures

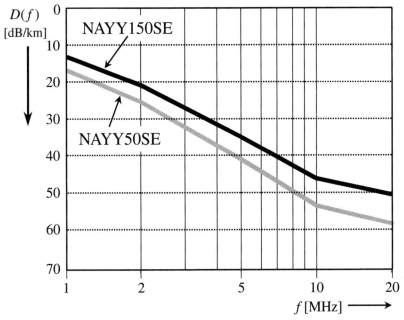

Figure 2-13 Attenuation per km for the most common four-sector cables.

opposing wire surfaces remaining for the transport of the high-frequency currents are smaller in this case.

2.1.2.2   Line Properties—Summary

The high-frequency properties up to 20 MHz of various line types used in distribution networks have been studied. The studies showed that the attenuation in cables is essentially determined by the properties of the dielectric. It is impossible to make precise and universal statements because there is a large number of various insulation materials in practice. The examples represented here apply actually only to a certain PVC type, which is often used as cable insulation, and for which data are available. The properties of PVC can depend on various factors, such as temperature. A cable temperature of 20 degrees C is assumed; this is much higher than the temperature of the soil, but it helps take some of the heating into account that occurs when the cable is loaded by 50-Hz or 60-Hz currents. Studies have shown that attenuation figures in energy cables do not reach critical values for high-frequency signals up to the 20-MHz range, even for lengths of 1 km.

In addition, we saw that overhead lines generally have a very low attenuation, and that even "ground cable—overhead line—ground cable" transitions altogether cause additional attenuation of only about 14 dB. This means that overhead lines, even when transitions are included, do not represent insurmountable obstacles for high-frequency signals, so that such unusual structures can also be used for communication purposes. In some cases, simple measures may be sufficient to achieve reflection-free matching to heavily reduce the attenuation at the transitions.

2.1.3 The Interference Scenario

The last part of this chapter deals with the typical interference environment found in the sections of electric distribution networks of interest for upcoming communication purposes. More specifically, these are the areas described as "last mile" in Figure 2-1. Of course, installation networks inside buildings are included. However, we will see later that it is meaningful to clearly separate signal transmission on house-internal networks and external distribution systems.

The significance of the medium-voltage network for a wide and general use for communication purposes is currently estimated to be low. Nevertheless, it may be useful for power supply utilities (PSUs) with regard to the required information transmission of future energy-specific value-added services. Various studies and research projects currently investigate the possibilities. Medium-voltage networks are generally fed from a high-voltage network over a transformer station. In turn, they feed normally a number of low-voltage transformers, apart from individual supply islands, such as remote small towns or industrial plants with high power demand. In the frequency range above 20 kHz, these transformers are almost perfect barriers, so that from medium-voltage lines on the one hand there is no signal coupling into the high- and low-voltage networks, and on the other hand interference coming from these networks is heavily reduced. Where overhead lines are concerned, the transmission properties of medium-voltage networks are similar to those of high-voltage lines, also with regard to the interference load. Information transmission tasks that involve medium-voltage networks—both overhead lines and cables—require a coupling from the medium-voltage to the low-voltage level. This means that the good properties of the medium-voltage lines are lost almost entirely, because the properties are now essentially determined by the low-voltage network, which also means that the interference will also be almost fully effective.

Hardly any other electric network shows a larger variety of signals than the energy distribution network on the "last mile" and within the supplied buildings. A large part of the interference is caused in electric machinery and devices during normal operation. In addition, there are many different narrow voltage or current peaks (transients) due to a wide range of switching events. In particular, building installation networks are electromagnetic open structures, so that there are numerous signals originating from spurious irradiation of various radio services. The irradiation of radio stations, mainly in the long-wave and medium-wave ranges, causes considerable voltages. In the normal case, they do not exhibit any significant interference effects; to the contrary: mains-supplied radio devices can even benefit from them. However, if telecommunication systems with high data rates have to operate in electric power networks in the future, then collisions will be unavoidable, unless special measures are taken, because the frequency ranges used are close or even overlapping. This concerns mainly the use of frequencies up to about 30 MHz, i.e., including the entire short-wave range. Chapters 5 and 6 describe the resulting EMC problems and suitable solutions. The following section provides an initial overview on the dominant interference types found in buildings and the connected distribution networks.

When analyzing the amplitude spectrum of interference at a wall outlet in a building, for example, three different classes can generally be identified: colored background noise, narrowband interference, and impulsive noise. Figure 2-14 shows a general approach to the description of interference and contains the following components that, together, have an impact at the location of a receiver:

Colored background noise: This kind of interference is of stochastic nature and can be described by a power spectral density (PSD). High PSD values are characteristic for networks, starting from the power frequency (50 or 60 Hz) to frequencies of about 20 kHz. A continuous decrease of the power spectral density is observed above these values as the frequency rises. At 150 kHz, there is often only 1/1000 of the value found at 20 kHz. Toward even higher frequencies, there is only background noise with very

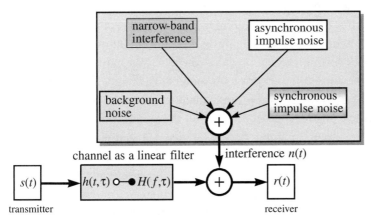

Figure 2-14 Approach for a general "impulsive interference model."

low power spectral density, which is normally called white noise. Of course, this excludes ranges subject to radio service irradiation.

Narrowband noise: The occurrence of clear needle-shaped elevations in the spectrum indicates narrowband interference, which occurs only within a narrow limited frequency range, but with high PSD. Narrowband interference at frequencies below 150 kHz, i.e., below the radio-frequency bands, can originate from switching power supplies, frequency converters, fluorescent lamps, or television sets and computer screens. At the higher frequencies, most narrowband interference comes from radio stations. These issues will be described in Chapters 4 and 5.

Impulsive noise: Impulsive noise is characterized by short voltage peaks that last about 10–100 μs and can reach beyond 2 kV. Basically, these are rare single events, which are caused by on and off switching events, and they are different from frequently occurring periodic events. Periodic impulses are mostly caused by phase controls (e.g., dimmers) and occur during each zero-crossing of the network voltage, i.e., with a frequency of 100 Hz. Depending on the duration of this interference, one or several bits in a data transmission can be corrupted. Such errors have to be prevented by suitable channel coding measures.

The following sections describe examples of interference scenarios in various network access points to introduce the issue. The field of interference modeling is currently the subject of scientific research. Projects include a model that, with the corresponding hardware support, will soon allow us to simulate all relevant interference signal constellations in the laboratory environment in order to test conveniently the behavior of communication systems under almost real conditions. These issues will be described in more detail in Chapters 4 and 5.

At very low frequencies, i.e., near the power frequency of 50 or 60 Hz, the interference level inherent there is primarily caused by harmonics and nonharmonics of the power supply voltage. In addition, there is impulse-type interference. Nonharmonics have an arbitrary frequency position to the power frequency. The main source of nonharmonics that are effective over lengthy periods is motors. This is the reason why this interference is most frequently found on the low-voltage level. Overtones form in the motors, and their frequency position is determined by the number of grooves in the slider and the stator and the motor's number of revolutions. A spectral emphasis taking these influences into account cannot be given due to the large number of commonly used motor types. With high-performance motors (10–60 kVA), overtones reach considerable amplitudes of several percent of the network voltage and can impair the reliability of ripple control (see Chapter 3). Although ripple control systems basically avoid harmonics of the power frequency, there is no workaround for nonharmonics. In practice, however, disturbances are infrequent, because motors with the highest potential interference effect are high-performing asynchronous motors. Such motors are normally operated with a more or less constant number of revolutions and low slippage, so that they generate a very narrow interference spectrum around a powerline voltage harmonic, so that ripple control can avoid this problem. In some cases, e.g., in extensive distribution networks in rural areas with low load density, there are occasional malfunctions, e.g., during the operation of a thresher, where heavily fluctuating numbers of revolutions are unavoidable.

Another source of nonharmonics is switching events. Switching events cause a continuous spectrum with amplitudes decreasing inversely proportional to the frequency. In a power supply network, interference voltages originating from many switching events that are statistically independent overlap at any given time, forming a noise level that decreases as the frequency increases. The sidelobes of such interference spectra can be observed up to very high frequencies.

Harmonics of the network voltage can arise on both the medium-voltage and the low-voltage level. We distinguish between harmonics of the voltage and harmonics of the current. The former arise mainly in generators and depend essentially on the load; they are normally passed on from the high-voltage level to the medium-voltage and low-voltage levels. Current harmonics form mainly on the medium- and low-voltage levels, e.g., in transformers and power converters. In transformers, overtones arise with load-independent amplitudes due to the bent magnetization curves. Power converters in the medium-voltage network, for example, cause load-dependent current harmonics. On the low-voltage level, consumers with nonlinear impedance characteristics are sources of load-dependent current harmonics. The number of consumers of this type has been increasing to an alarming number, mainly due to the enormous proliferation of switching power supplies.

The interfering effect of overtones of the network voltage is generally negligible for communications. Ripple control systems can avoid them in a targeted way, and toward the higher frequencies the amplitudes of the overtones decrease to a level such that an interference effect can be excluded. However, the situation is totally different when the operation of electric devices in the network generates very specific high-frequency oscillations and when they are transferred into the network.

Based on Figure 2-15, the following sections study a few typical disturber types found in buildings.

Curve 3 reflects the colored background noise found, for example, in the "quiet" network of a one-family home. Background noise does not represent a critical interference factor for all types of information transmission. The level is relatively low, so that

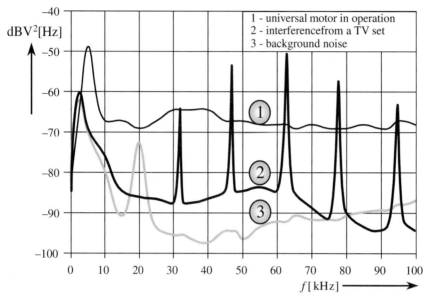

Figure 2-15 Possible interference scenarios in building installation networks; see, for example, [N4, N5].

in many cases one has to go to the limit of sensitivity of usual measurement instruments to be able to record it at all.

In contrast, the interference caused by so-called universal motors, shown in curve 1, are much more critical. Such motors are found in many household appliances, such as mixers, hair dryers, vacuum cleaners, and hand drills. Within the frequency range represented in this example, the interference is mostly white, and the power spectral density is clearly above the background noise with more than 20 dB. When such an interference source is very close to a receiver, it can develop a very hefty interference effect on the information transmission. Curve 2 is an impressive example of the massive narrowband interference that a television set can cause. We can clearly see the harmonics of the line frequency at 15,734 Hz. While the basic oscillation does not appear, probably due to an adapted filter in the device, the harmonics are very strong. A transmission system that relied, for example, on the spectral ranges, which are around these harmonics, could no longer function reliably when the TV was on. With standard television sets, the harmonics' spectral positions are fixed, so that one could basically exclude these ranges from information transmission. In contrast, most modern personal computers (PCs) use color monitors, which are configured as so-called multifrequency or multisync devices with changeable line deflection frequency. The currently selected operating mode of the PC graphics card defines the frequency. The interference effect of these monitor types is comparable with that of TV sets, with the important difference that the interference spectrum can no longer be avoided in a targeted way.

So far, we have been looking at stationary interference. However, a large number of interferences observed in power supply networks have impulsive and stochastic character. This may not come as a surprise when we consider that switching events, which are the main triggers of such interference, occur constantly in large numbers within electric networks. When recording the interference spectrum by use of a spectrum analyzer, the "stationary" curves often show needle-shaped elevations, and their height and position change from recording to recording. In contrast to what one would suspect at first sight, these peaks are by no means narrowband disturbers. Instead, the disturbers are mostly broadband, but appear only for a short time. Due to the very long recording time of the spectrum analyzer, compared to the interference duration, impulsive noise appears as needles on the screen. In reality, each needle means that, for a moment, there was an elevated power spectral density over a broad spectral range. From the literature only a few approaches to impulsive noise analysis and modeling are known. A basic approach from [D3] is presented here before we go on to new concepts.

The interference impulses mostly observed in low-voltage networks have an approximately triangular shape. System theory describes a triangular impulse with height 1 and basic width 2 by the symbol $\Lambda(t)$ [F10]. In the frequency range, $\Lambda(t)$ corresponds to a spectrum with a $[\sin(f)/f]^2$-shaped curve. With the symbol ⊶ for Fourier transform and the abbreviation $\text{si}(f) = \sin(f)/f$, we obtain [F10]:

$$\Lambda(t) \mathrel{\circ\!\!-\!\!\bullet} \text{si}^2(\pi f) \tag{2.46}$$

Using a triangular impulse with height A_p and basic width T_p, the following applies when using the relation of similarity of the Fourier transform [G9]:

$$A_p \cdot \Lambda\left(\frac{2t}{T_p}\right) \mathrel{\circ\!\!-\!\!\bullet} \frac{1}{2} \cdot A_p \cdot T_p \cdot \text{si}^2\left(\pi f \frac{T_p}{2}\right) = S_\Lambda(f) \tag{2.47}$$

We then obtain the energy density spectrum of a triangular impulse with height A_p and basic width T_p from (2.46) as

$$|S_\Lambda(f)|^2 = \frac{1}{4} \cdot A_p^2 \cdot T_p^2 \cdot \text{si}^4\left(\pi f \frac{T_p}{2}\right) \tag{2.48}$$

Introducing impulse height A_p in volts and T_p in seconds in (2.47), then $|S_\Lambda(f)|^2$ has the dimension V²s/Hz, i.e., the energy density in Ws/Hz results at a resistance of 1 Ω. The energy density according to (2.47) calculated so far belongs to an individual noise impulse. This result is not of much use yet. The impact of single impulses can also be studied better in the time range than in the spectrum.

System theory describes single events by use of energy signals. Energy signals have an energy density spectrum; see (2.47). When an energy signal occurs repeatedly or frequently, for example in the form of a single noise impulse, which does not mean that it is necessarily periodic, then it is referred to in system theory as a power signal [F10, F12], with a pertaining power spectral density. In the case of periodic, nonoverlapping noise impulses, we obtain a line spectrum with an envelope that has the form of the power spectral density of a single impulse. The line spectrum has spectral lines with the gap of the pulse repetition rate. This fine structure of noise-impulse power spectral densities is normally of no interest in practice, because on the one hand the line density is normally very high, and on the other hand there are seldom exact periodic noise impulses. This means that the lines in the power spectral density spectrum do not have a fixed frequency position, but are "smeared." This has been proven with a number of records taken by use of a spectrum analyzer. We generally obtain a continuous interference power spectral density. This applies in particular to the interference spectrum generated by a general-purpose motor (see curve 1 in Figure 2-15). We observe a more or less strong spark formation, the so-called "commutator sparking," strewing noise impulses in different forms and heights in quick but generally not equidistant sequence during dynamic operation of the device driven by that motor.

To consider the entire interference power of impulsive noise, we need to look at the square of the spectral effective value, which we will call A_{eff}^2 in the following. It can be determined by integrating the power density over the frequency, for example, from the interference power spectral densities shown in Figure 2-15. If we use curve 1 (general-purpose motor), we obtain the following estimate, because the interference power spectral density is relatively constant with a mean value of -65 dB V^2/Hz:

$$\frac{A_{eff}^2}{dBV^2} \approx -65 + 10 \cdot \log_{10}(100,000) = -15 \tag{2.49}$$

If we now look at noise impulses with a height of 20 dB above the broadband effective value and with a duration of 65 μs, we obtain approximately

$$\frac{|S_\Lambda(f)|^2}{dBV^2 s/Hz} \approx -15 + 40 + 20 \cdot \log_{10}\left(\frac{65 \ \mu s}{2 \ s}\right) + 10 \cdot \log_{10}\left[si^4\left(\pi \cdot f \cdot \frac{65 \ \mu s}{2}\right)\right] \tag{2.50}$$

$$= -64.7 + 10 \cdot \log_{10}\left[si^4\left(0.0325\pi \cdot \frac{f}{kHz}\right)\right]$$

for the relevant energy power spectral density from (2.47) and (2.48).

When using an impulse repetition rate of 100 s^{-1}, corresponding to the zero-crossings in a 50-Hz power supply network, with (2.49), we obtain the relevant interference power spectral density of

$$\frac{|L_\Lambda(f)|^2}{\text{dBV}^2/\text{Hz}} \approx -64.7 + 10 \cdot \log_{10}(100) + 10 \cdot \log_{10}\left[\text{si}^4\left(0.0325\pi \cdot \frac{f}{\text{kHz}}\right)\right] \quad (2.51)$$

At the frequency $f_1 = 20$ kHz and based on (2.50), we obtain the interference power spectral density

$$|L_\Lambda(f_1)|^2 \approx -58 \text{ dBV}^2/\text{Hz} \quad (2.52)$$

When comparing this result with Figure 2-15 (curve 1), we can observe an astonishingly good match, because at 20 kHz the value of approximately -67 dBV^2/Hz measured for the interference power spectral density is of the same order of magnitude.

Table 2-4, taken from [D3, N8], shows important description values for impulsive noise in the form of mean values. The classification used here is not arbitrary; it is based on a large number of measurement results. In simplified form, the same value A_{eff}^2 based on (2.48) is assumed for the three impulsive noise classes.

Table 2-4 Classification of impulsive noise in three classes.

Impulse Energy:	High	Medium	Low
Class:	1	2	3
Impulse duration	65 µs	20 µs	5 µs
Impulse height above the effective value	20 dB	15 dB	10 dB
Repetition rate	120 s^{-1}	200 s^{-1}	400 s^{-1}

Class 1 impulses with high energy have a large amplitude and long duration. They occur quasi-periodically, which is normal for a large number of impulsive disturbers. We then obtain a high interference power spectral density at low frequencies, which drops according to (2.49) with an si$^4(f)$ function with increasing frequency. This kind of impulsive noise is frequently found at building power supply networks. The repetition rate of the impulses rich in energy corresponds often to double the power frequency; i.e., the noise impulses occur in connection with zero-crossings of the mains voltage.

When studying impulse noise of class 2 in Table 2-4, we obtain a resulting power spectral density, which is clearly below that of class 1 disturbers. This is due to several mutually influencing factors:

- Power supply networks have a lowpass character, so that shorter impulses, compared to class 1, face higher attenuation.
- The impulse amplitude is generally lower.
- The width-repetition rate product is lower than in class 1, so that there is altogether less power.

Class 3 impulses with very short duration must have relatively low amplitudes due to the lowpass character of the mains network. Measurements have shown that their repetition rate is often a multiple of the power frequency. This is why such disturbers normally represent almost white noise up to frequencies of several hundred kilohertz. The commutator sparking in universal motors is a typical example of sources for class 3 impulsive noise.

The above discussion allows us to derive only very rudimentary findings for the modeling of real impulse noise of all the types occurring in power supply networks. To the author's knowledge, for high frequencies (approximately 500 kHz to 30 MHz) no findings have been published, but several studies have been started.

In general, an analysis in the time domain has proven to be meaningful for a detailed universal description of impulse noise in power supply networks. Three parameters are of particular interest here: the impulse amplitude, the impulse width, and the "interarrival" time between impulses. This discussion concerns stochastic parameters, which means that they cannot be predicted or directly analyzed, so that a statistical description is required. This approach involves high equipment and time costs, because only a very large number of recordings and their detailed analysis will supply general statistically reliable values. Initial studies have been done with standard instruments, such as a remotely operable digital storage, oscilloscope, and a spectrum analyzer. The enormous volume of data involved here has to be stored on large fixed disks for later evaluation by special PC software. Most of the required software programs have to be written exclusively for this purpose, because commercial tools are useful only to a limited extent.

The methodology described is not suitable for a parallel approach at different locations and in different network types, as would be required for sufficient statistical reliability of the results. For this reason, university institutes and industrial organizations are cooperating to develop a general-purpose PC-based measurement system, including the computing power of a very fast digital signal processor (see Chapters 5 and 6): iPLATO ≡ IAD Power Line Analyzing Tool. In addition to recording and analyzing impulsive noise, this measurement system will also allow appropriate interference synthesis for simulation and test purposes.

CHAPTER 3

Historical Development of Data Communication over Powerlines

3.1 Possibilities and Limits of Classical Usage Types

Power supply networks began to be used additionally for data transmission soon after full-coverage electrification. The carrier-frequency transmission of voice over high-voltage lines began in 1920. High-voltage networks are attractive for data transmission mainly for historical reasons. Important tasks to maintain the function of extensive high-voltage networks are operations management, monitoring, and troubleshooting. These tasks require a fast bidirectional flow of messages, e.g., between power plants, transformer stations, switching equipment, and coupling points to neighboring networks. There was no full-coverage telephone network at the beginning of electrification. Even today, the availability of a telephone connection cannot be taken for granted at every point of a high-voltage network where data transmission is required. In addition, active connections for telemetering and monitoring tasks over telephone exchanges are not suitable, because interruptions could be dangerous, and because they are economical only over short distances. This chapter provides a short overview of the classical use of powerlines for data transmission, which has been almost entirely by the PSUs for operations management and optimum energy distribution. There are important differences at the various voltage levels. While the information flow is relatively large and bidirectional on the high-voltage level by using CTP and working with low transmit power, the RCS on the medium- and low-voltage levels allow only a very low unidirectional information flow from a PSU to the consumer, requiring enormous transmit powers.

3.1.1 Carrier Transmission over Powerlines (CTP)

Together with the use of the carrier-frequency technology over postal lines around 1920, one began to apply this technology for data transmission over high-voltage networks [T1]. Considerable technical input and investment, supported by the fast progress in the development of refined data communication methods, made it possible to unify low-loss energy transportation and reliable information transmission over power supply networks. The need for a comprehensive information network for electric supply companies, as a medium for operations management and other purposes, resulted from the fast growth of full-coverage high-voltage networks to supply large areas with electric energy. The tasks for the data networks of electric supply companies can be grouped into three main classes:

- Operations management
- Monitoring
- Limitation and removal of failures

Operations management is intended to ensure optimum energy distribution, so that on the one hand no unused capacities are maintained and on the other there are sufficient reserves for peak loads. The economic fulfillment of these tasks requires reliable and fast data transmission. Monitoring concerns making the operations state of the network transparent. Status parameters regarding energy requirement, voltage, and frequency are of major interest in this respect. In case of failures in the high-voltage area, fast exchange of information between power plants and transformer stations as well as distribution stations on the cause of failure is important so that the impact can be minimized. In this respect, the transmission of information has to work reliably over the large distances inherent to high-voltage networks, even under failure conditions.

A data network within the supply company's serving area connects power stations, transformer plants, switching equipment, and coupling points to neighboring networks on the high-voltage level. In addition, a national grid, including a number of power supply companies, requires a data remote network. In the past, the largest portions of data have been transmitted analogously to voice data. In the course of time, automatic telemetering and remote monitoring of status parameters without the use of operators have become increasingly important, so that telephony plays a rather subordinate role today, while digital data communications become more and more dominant.

The PSUs planned from the very beginning to build their own data networks for their information transmission. The general telephone network was found to be unsuitable, particularly for telemetering and remote monitoring, because it is not available everywhere, and even very short interruptions can have dangerous consequences. In addition, the use of leased lines would not be economical for large distances. Most power supply companies have always seen the high-voltage network as a "natural" medium to transmit their operational information, because it connects all important stations. The necessity of data networks for the PSUs themselves led to the quick development of the CTP introduced in Chapter 2. This technology has gained increasing significance during the course of its development, not least because of the expansion of national grids. This transmission technology did not lose any of its actuality in the age of omnipresent automation and computer deployment. It still forms the backbone of a refined and thus economical energy supply management.

As mentioned in Section 2.1.1.2, special channels are created for high-frequency signals on the high-voltage level by the use of carrier-frequency blocking filters in the form of inductors at the start points, end points, and junctions. In addition, one will use special input and output coupling units to match the impedance of CTP transmit and receive devices to the characteristic impedance of the high-voltage lines. Typically, broadband coupling filters constructed as shown in Figure 3-1 are used. The high-voltage capacitors C_1 and C_2, together with the grounded coil L_1, provide a secure separation of the high-voltage from the CTP devices connected to coil L_2. The attenuation of a coupling filter as shown in Figure 3-1 is very low, i.e., normally smaller than 1 dB.

Although high-voltage overhead lines were not designed for data communication, they represent good waveguides (see Chapter 2). The reliable bidirectional transmission of data is possible at low transmit power. A relatively broad spectrum is available,

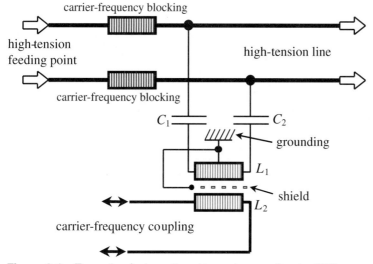

Figure 3-1 Example of a broadband two-wire coupling for CTP.

which can be almost fully used thanks to the good transmission properties, and it can be easily divided into individual channels. This division of the available transmission spectrum is called *frequency multiplexing*. The transmission technologies used here are based on narrowband conventional modulation methods, which can be easily implemented, and which ensure sufficient failure safety due to the good channel properties. At the beginning of CTP, there was only the transmission of voice by means of amplitude modulation. Later, telemetering and telecontrolling tasks were added more and more, for which digital transmission technologies are of interest. Today, voice transmission plays a subordinate role. The following sections describe the historical analog CTP technology.

The frequency range usable for CTP is between 15 and 500 kHz. The lower limit is given by the cost for coupling equipment. Below 15 kHz, the required capacitors with large capacity with the high-voltage stability required are too expensive to be economical. The upper limit is essentially given by the attenuation, which rises heavily as the frequency increases. The frequency range from 300 to 500 kHz is not tolerant to frost and should be avoided over long distances in northern countries.

High-voltage overhead lines are relatively good waveguides in the frequency range for CTP. There is a relatively well-defined characteristic impedance, so that sources, sinks, and junctions can be matched. Reliable information transmission is therefore possible at low transmit power over large distances. A value of 10 W, corresponding to 40 dBm, has proven to be a suitable transmit power in practice, because it allows the bridging of sufficiently large distances without disturbance. On the other hand, the interference for long-wave and medium-wave radio caused by CTP is limited to an acceptable size. When planning the frequency allocation, care should be taken not to use those frequency ranges occupied by neighboring radio stations. In addition to a potential interference in radio reception caused by CTP, this would also represent a potential impairment of CTP itself by irradiation from strong broadcasting stations. In contrast to lines which are exclusively used for data communication, so that they can be well protected against external interference impact, high-voltage lines have to handle primarily the transport of electric energy at high voltage. Thereby, discharge effects (corona losses) are unavoidable, causing special disturbances with high power spectral density and large fluctuations, depending on the weather. To ensure a sufficiently interference-free reception, CTP requires a signal-to-noise ratio of $\gamma > 20 \equiv 13$ dB at the receiver input. Table 3-1, taken from [T1], gives the distances that can be bridged under most favorable and unfavorable conditions for three transmission frequencies: 200, 300, and 400 kHz.

Table 3-1 Examples of CTP distances at various frequencies and interference powers.

	Frequency		
	200 kHz	**300 kHz**	**400 kHz**
Distance with heavy corona discharge	571 km	376 km	283 km
Distance with weak corona discharge	900 km	594 km	445 km
Distance difference	329 km	218 km	162 km

We see clearly that CTP under favorable circumstances can bridge the enormous distance of 900 km with a transmit power of only 10 W. Even under unfavorable circumstances, it still reaches almost 300 km. These brief studies show that CTP is a mature technology and well adapted to the medium, leaving hardly any room for significant improvements.

In the beginning, there was the task of handling operations management of the power supply on the high-voltage level by means of voice. The voice-frequency band of 300 to 2400 Hz was to be transmitted at lowest possible cost for both the transmit and the receive devices. Under the given marginal conditions, only amplitude modulation (AM) was suitable to feed the data onto the high-frequency carrier. AM with its variants offers not only simple solutions, but solutions that today are still optimal for voice transmission problems within the scope of CTP. We will study this interesting aspect in more detail. The theoretical basics used here are described in [F3, F10, and F12].

We will use the following symbols:

$\hat{u}_T$ Amplitude of a sinusoidal unmodulated carrier.
ω_T Angular frequency of the carrier.
$\hat{u}_N$ Amplitude of a sinusoidal message signal used to modulate the carrier.
ω_N Angular frequency of the message signal.
$\hat{u}_T(t)$ Amplitude of the modulated carrier (transmitted signal).
$s_{AM}(t)$ Transmitted AM signal.

We distinguish three variants of amplitude modulation (AM):

Linear AM This is a double-sideband AM (DSB-AM) with suppressed carrier. This modulation can be implemented by use of a ring mixer, for example. The amplitude of the modulated carrier is

$$\hat{u}_T(t) = k_1 \cdot \hat{u}_N \cdot \sin(\omega_N t) \tag{3.1}$$

where k_1 is a positive real number between 0 and 1. Equation (3.1) says that the amplitude of the transmitted signal depends linearly on the amplitude of the message signal; hence the name linear AM. With a linear DSB-AM, we obtain the transmit signal

$$s_{AM}(t) = \frac{k_1 \hat{u}_N}{2} \cdot [\cos(\omega_T - \omega_N)t - \cos(\omega_T + \omega_N)t] \tag{3.2}$$

Equation (3.2) shows that no carrier signal is transmitted. Instead, two spectra are carried that are symmetric to the carrier frequency; they are called sidebands, and each contain the same message. The upper sideband is used to map high message frequencies onto high transmitted frequencies, and in the lower sideband, high message frequencies are mapped onto low transmitted frequencies. Therefore, the message is said to have normal position in the upper sideband and inverted position in the lower sideband. The demodulation of the linear AM requires some effort, because the suppressed carrier has to be added at normal phase. Linear AM occupies a transmission spectrum with twice the message bandwidth. Due to the high receiver cost, it is not considered for use in CTP, although the missing carrier would be an advantage, particularly with multi-channel carrier-frequency devices, due to the low intermodulation risk.

Usual Double-Sideband AM This modulation can be achieved, for example, with active analog multipliers, but also with simple passive components, such as diodes with a nonlinear characteristic that partly obeys a square function. The requirements for the quality of the modulator components are considerably less than for the linear modulation, which needs maximum symmetry, e.g., within the ring mixer to ensure that the carrier is really neatly suppressed. The usual DSB-AM occupies a spectrum with twice the message bandwidth, similar to linear AM, but where the carrier does not disappear. With the modulation depth $m = \hat{u}_N / \hat{u}_T$, the following applies to the transmitted usual double-sideband AM signal:

$$s_{AM}(t) = \hat{u}_T \cdot \sin(\omega_T t) + \frac{m \cdot \hat{u}_T}{2} \cdot [\cos(\omega_T - \omega_N)t - \cos(\omega_T + \omega_N)t] \tag{3.3}$$

The carrier is transmitted, allowing a very simple receiver structure, because the demodulation can be done by simple rectification and lowpass filtering. This is the main reason for the use of the usual double-sideband AM in long-wave and medium-wave broadcasting. However, a major drawback of this modulation is that a sinusoidal oscillation with large amplitude, namely the carrier, has to be transmitted, but does not contain any information. The only purpose for the carrier transmission is to enable simple demodulation.

The amplitude of the transmitted AM signal based on (3.3) achieves the value $2\hat{u}_T$ with modulation depth $m = 1$. This has undesirable consequences, because the operating limits of the transmitter's power stages are reached through the carrier that does not transport information, so that the useful transmitted power is reduced, compared to a transmission without carrier. Since the beginning of the CTP development, the usual DSB-AM nevertheless played an important role for the transmission of the voice band of 300 to 2400 Hz. The modulated signal according to (3.3) can be transmitted in a frequency band of 5 kHz. In Germany, the spectrum on high-voltage lines used for voice transmission was therefore equipped with a 5-kHz channel grid. In several other countries, an 8-kHz grid taken from the postal telegraphy over multiplex was introduced. Until about 1940, voice-transmission devices with DSB-AM with carrier were used almost exclusively. They haven't entirely disappeared yet in places where neither a lack of spectral resources, nor unfavorable signal-to-noise ratios, justifies the use of the single-sideband AM described below.

Single-Sideband AM (SSB-AM) Both in the linear and in the usual DSB-AM, the message to be transmitted is contained twice in the transmitted signal, namely in the upper sideband (in normal position) and in the lower sideband (in inverted position). Therefore, it is possible to save bandwidth, for example, by transmitting only one sideband without carrier. This is offered by SSB-AM. For instance, a SSB-AM transmit signal can be yielded from a linear modulated DSB-AM signal according to (3.2), by cutting the undesired sideband through a filter. Particularly for voice signals, as they have to be transmitted by means of CTP, this can be easily implemented, because there are no spectral components in the range from 0 to 300 Hz. The frequency gap of 600 Hz between the sidebands allows the use of simple filters for the entire CTP range from 30 to 490 kHz, when implementing the modulation with an intermediate filter stage. This means that first the message is modulated onto a 12-kHz intermediate carrier with the usual DSB-AM. The desired sideband can then be simply filtered out and shifted to the final frequency position by mixing.

On the receiver side, the demodulation of SSB-AM is simple, if a signal with the exact carrier frequency is available in the receiver. This signal has to be added to the received signal with sufficiently large amplitude. Then, an envelope demodulation will be sufficient, e.g., by use of a rectifier followed by a lowpass filter, in order to recover the message. In contrast to the linear AM, this modulation does not require the addition of a normal-phase carrier; but the frequency must match exactly. This can be achieved by sending pilot frequencies with the transmitted messages. For example, one such pilot frequency can be divided simply to yield the required carrier signals for many SSB-AM channels.

From about the year 1940, due to the increasing lack of spectral resources, SSB-AM began to be used for voice transmission over high-voltage lines. The 5-kHz channel grid introduced in Germany was modified to a 2.5-kHz grid for SSB-AM. Accordingly, there was a 4-kHz grid in other countries, resulting from the 8-kHz basic grid. Eventually, Germany also introduced 4-kHz channels for the transmission of digital information at high data rates.

In addition to amplitude modulation, frequency modulation (FM) was also used to a lesser extent for voice transmission in CTP. Both the modulation and the demodulation are simpler in FM, compared to SSB-AM. However, FM is not suitable to reduce the requirement of spectral resources, compared to the usual DSB-AM. Also, the signal-to-noise gain remains relatively low in FM, because the frequency deviation has to be limited to 2 kHz to allow transmission of a voice signal with an upper limit frequency of 2.4 kHz within a 5-kHz-wide channel. This narrowband FM allows voice transmission over high-voltage lines with the same bandwidth requirements as the usual DSB-AM.

We will now study the signal-to-noise gain for a final quantitative comparison of the three modulation methods, SSB-AM, DSB-AM with carrier, and narrowband FM, used for voice transmission within CTP. For this purpose, we will consider the useful signal-to-interference power ratio before the demodulation, γ_D, and after the demodulation, γ_E. The useful power influencing γ_E corresponds to the power of the received unmodulated carrier in these considerations for FM and DSB-AM with carrier. For SSB-AM, it is the power of the transmitted sideband, when the transmitter is fully modulated by a signal of the angular frequency ω_N contained in the spectrum. The expected interference is white Gaussian noise caused by corona losses.

For SSB-AM, all useful voltages and interference voltages existing in the high-frequency position of a channel will be equally transferred into the low-frequency position. This means that the signal-to-noise ratio will remain unchanged. It is

$$\gamma_D = \gamma_E \quad \text{for SSB-AM} \tag{3.4}$$

In DSB-AM with carrier, the demodulation through normal-phase addition of the sidebands (3.3) produces a useful signal voltage with the amplitude $m\hat{u}_T$, i.e., we obtain the received carrier amplitude multiplied by the modulation depth m, while the interference voltage reaches the $\sqrt{2}$ – fold compared to SSB-AM, due to the two sidebands. With the same γ_E as in SSB-AM, we obtain

$$\gamma_D = \frac{m^2}{2}\gamma_E \qquad (3.5)$$

for DSB-AM with carrier. Comparing (3.4) and (3.5), we can see that the SSB-AM versus the DSB-AM with carrier is always better by at least 3 dB. This means that, with the same transmit power, SSB-AM bridges larger distances than DSB-AM.

For narrowband FM, the signal-to-noise gain depends on the frequency deviation, h_{FM}, following a square law. Using the upper limit frequency, f_g = 2.4 kHz, of the voice signal to be transmitted, we obtain approximately

$$\gamma_D \approx \frac{2}{3} \cdot \left(\frac{h_{FM}}{f_g}\right)^2 \cdot \gamma_E \qquad (3.6)$$

for narrowband FM. With (3.6) for h_{FM} = 2 kHz, we obtain $\gamma_D \approx \gamma_E$, as for SSB-AM. However, narrowband FM requires twice the bandwidth of SSB-AM.

Soon after the introduction of CTP, it was found that the major part of the tasks of operations management of high-voltage networks cannot be handled optimally with voice connections alone. The remote transmission of measurement values (telemetering) and switching events can be triggered much easier, faster, and more reliably over long distances (telecontrolling) by transmitting digital information. For example, voice transmission would not support a timely full-coverage remote monitoring of critical measurement values. In telemetering and telecontrolling, higher requirements are placed upon transmission safety, compared to telephony, because a natural redundancy as in human voice is not inherent in the signals. The corruption of a single data bit could lead to critical errors. Suitable coding can help add the required redundancy to digital data for sufficient error safety. Such coding methods are state of the art today and have been generally used in the transmission of digital information. At the advent of CTP, however, the fundamental theoretical background on which coding methods are constructed today was not known. But even with such knowledge, the building of coding and decoding equipment would have been impossible due to the lack of digital devices. However, another simple way was found: CTP allows bidirectional data traffic, so that high security can be realized, for example, for critical switching events, by using feedback. Where a feedback channel is impractical for cost reasons, then the same measurement value or telecontrolling command can be transmitted several times to obtain high safety.

When the first telemetering and telecontrolling tasks in high-voltage networks had to be performed, the technology of single-sideband amplitude modulation was already known. The transmission band from 30 to 490 kHz accommodated up to 184 voice channels in the 2.5-kHz grid, if needed. Therefore, it was logical to also insert the telemetering and telecontrolling signal transmission into this channel grid. The relevant data rates were low at the beginning, starting with 50 bits/s, and increasing later to 100 and 200 bits/s. Amplitude or frequency shift keying of a carrier out of the voice-frequency band from 300 to 2400 Hz was used. As the required bandwidth for a telemetering or telecontrolling channel, the 1.6-fold of the data rate was used, i.e., 80 Hz at 50 bits/s. With an additional "safety gap" of 40 Hz between the channels to allow separation with simple means, one obtains a 120-Hz grid at 50 bits/s. This means that 18 telemetering or telecontrolling channels with 50 bits/s fit into one 2.5-kHz SSB-AM voice channel. For 100 bits/s, the channel grid is 240 Hz, so that nine channels with this data rate fit into 2.5 kHz. With 200 bits/s and a 480-Hz grid, only four channels fit into 2.5 kHz. With the advent of automation, the demand for extensive data transmission increased, e.g., to transmit the entire operating state of complex systems. Higher-rate digital transmission channels with 600, 1200, 2400 bits/s and more had to be used. A full 2.5-kHz channel is required each for 600 and 1200 bits/s. To transmit data at 2400 bits/s, the 4-kHz channels mentioned above are used.

With recent developments and the use of mostly digital modulation methods, it is possible today to implement a data rate of 64 kbits/s in two 4-kHz channels, i.e., in a bandwidth of 8 kHz. In this case, QAM (quadrature amplitude modulation) is generally used (see Chapters 5 and 6), so that a spectral efficiency of 8 bits/s/Hz can be achieved. We will now look at the most important requirements for such a channel and come back to more details in Chapter 5. A general theoretical limit in the form of channel capacity C[1] in bits/s was described by Shannon [F1] in 1947 for channels with a bandwidth B and affected by white Gaussian noise as follows:

$$C = B \cdot \operatorname{ld}(1+\gamma) \tag{3.7}$$

where γ is the signal-to-noise ratio. For 8 bits/s/Hz, this means that the logarithm to the base two of the expression $(1 + \gamma)$ has to be 8; i.e., a signal-to-noise ratio of $\gamma \approx 256 \equiv 24$ dB is required. This is not a particularly high value, and one can assume that there are much better values in numerous CTP channels, so that the technical implementation of 64 kbits/s is possible with a sufficiently low bit error rate.

---

[1] Theoretically, the data rate C can be transmitted over this channel without errors.

A particularly interference-resistant transmission technology was developed for cases where maximum reliability and highest speed in remote release of switching events are required, and where a feedback or readback of commands is not feasible for time reasons. The approach of this technology is astonishingly similar to modern spread spectrum techniques. It is a method with "four-frequency shift keying," which is similar to frequency hopping (FH). One of the four frequencies is used as "idle frequency" for permanent monitoring of the connection and the readiness of the transmit and receive devices. The other three frequencies each trigger a special switching event, when they are detected in the receiver. The receiver evaluates only the frequencies, but not the incoming amplitudes. Impulsive noise with large amplitudes can be effectively limited, and receive-level bounces due to sudden attenuation change on the transmission path impair the information transmission only to a minor extent. The fact that only one frequency out of four is sent is known to the receiver and is used to ensure high safety against unintended switching due to false alarms.

In closing, this section describes a few transmit power issues. The power of the unmodulated carrier is specified as nominal power P_N of a CTP transmitter for frequency modulation and DSB-AM with carrier. In SSB-AM, the nominal power is the power of the sent sideband, when the transmitter is fully modulated from the message spectrum with a signal with fixed frequency. In Germany, $P_N = 10$ W is defined as the maximum value for nominal power for CTP operation. The peak power is another important value in the design of transmitters, particularly for the output stages in CTP. For DSB-AM with carrier and a modulation depth of $m = 1$, the transmit amplitude reaches twice the nominal value [see Equation (3.3)], so that the peak power is $4P_N = 40$ W. This value had been defined as peak power that a single transmit device can output in CTP for historical reasons. In devices used for telemetering and telecontrolling, a number of channels can be accommodated in a 2.5-kHz frequency band. Two cases are distinguished here with regard to allowable transmit power: When 1 to 4 channels are accommodated in 2.5 kHz, the 10-W nominal power limit applies. With more than 4 channels, the 40-W peak power limit applies. The entire nominal power in a 2.5-kHz band is calculated as the sum of powers of the unmodulated carriers from the individual telecontrolling and/or telemetering channels. Assuming the same carrier power in the individual channels, then the expected peak power equals the square of the number of channels, multiplied by the single-channel power. As an example, Table 3-2 shows the described relationships for a few practical cases.

We see from Table 3-2 that the peak power of 40 W is never exceeded. Up to a channel number of 4, the individual power per channel decreases in inverse proportion to the number of channels. This means that the sent summated power reaches the maxi-

Table 3-2 Nominal powers and peak powers for multichannel CTP devices [T1].

	Number of Channels					
	1	2	4	6	12	18
Single-channel power in watts	10	5	2.5	1.1	0.28	0.12
Summated power in watts ($\equiv$ nominal power)	10	10	10	6.6	3.4	2.2
Peak power in watts	10	20	40	40	40	40

mum admissible value of 10 W, creating optimum prerequisites for a high signal-to-noise ratio on the receiver side for each individual channel. With more than 4 channels, the peak power limit of 40 W becomes effective. The number of channels determines the peak power according to a square law, so that the summated power of 10 W can no longer be utilized from a number of channels of 5 and more. As shown in Table 3-2, only 8 W can be achieved with 5 channels, and only 2.2 W with 18 channels. This means that, when using 5 channels, a signal-to-noise ratio loss of approximately 1 dB has to be expected, and this loss increases even to approximately 6.6 dB with 18 channels. This discussion shows that the distance covered will decrease when occupying a 2.5-kHz band with a large number of telecontrolling and/or telemetering channels.

3.1.2 Ripple Carrier Signaling (RCS)

In contrast to the high-voltage level, the medium- and low-voltage levels are used mainly for other operations management tasks. These concern mainly load distribution, i.e., the avoidance of extreme load peaks and the smoothing of the load curve. It was found very early that, here too, the power supply network could be used for additional data transmission together with the conventional tasks. Ripple carrier signaling has been used since about 1930. In contrast to overhead lines on the high-voltage level, medium- and low-voltage networks are poor message transmission media. The information has to be fed in with a very high transmit power, dimensioned according to the network's peak load, because RCS works at low frequencies near the power frequency. Low frequencies used as carriers for the messages allow information to flow over the transformers between the medium-voltage and the low-voltage level without particular and generally costly coupling measures. On the other hand, the data rate is low, and the information can flow from the power supply company to the consumers only in one direction (unidirectional). For the tasks involved in load distribution, low information quantities are normally sufficient, because often only enabling or disabling commands have to be issued.

More recently, the power supply companies have been reaching the technical limits of RCS increasingly. Beyond the pure load distribution tasks, a large number of telecontrolling and telemetering processes are of interest for the current deregulation of the electrical energy market. Telemetering with RCS is not possible due to the unidirectional functionality. In addition, RCS works at full coverage, which means that all consumers in a large number of low-voltage networks, connected to the same medium-voltage network, are addressed in parallel. For a selective addressing of many individual consumers, the transmission capacity offered by RCS is not sufficient. This is, therefore, a field for new technologies, like those described in Chapter 4.

Initial RCS concepts were developed relatively early. The oldest known document on RCS is the German Patent number 118717 from 1898 (inventor: Loubery, Paris). However, the patent could not be transformed into a technically usable method at the time when it was granted. One of the first practical applications was the so-called Telenerg project of Siemens, conducted in Potsdam in 1930, a multifrequency method in which the mains voltage was overlaid by carrier-frequency impulses in the 280–600 Hz frequency range [R1]. In 1935, AEG introduced their so-called Transkommando scheme for practical use in Magdeburg and Stuttgart, Germany [R2]. This scheme was based on the amplitude modulation of the mains voltage and the transmission of information in time-division multiplexing (TDM).

After 1935, a single-frequency method with time-division multiplexing was developed, where an audiofrequency is serially injected into the mains as carrier [R3]. The current methods are based on this single-frequency time-division multiplexing technology.

As mentioned above, the RCS frequency range is near the power frequency. Therefore, the signals can be easily transmitted over transformers and experience only minor attenuation in the whole supply network. Medium- and low-voltage networks normally have a large number of junctions, in contrast to high-voltage lines. They are mainly implemented in the form of cables. A transmission technology similar to CTP is excluded in these networks, because a formation of matched high-frequency lines with constant characteristic impedance and low attenuation is impossible. The required high-frequency transmit power would lead to problems of electromagnetic compatibility. In contrast, audio-frequency signals in the frequency range of approximately 125–3000 Hz can be injected on the medium- and low-voltage levels at the high power required, and they propagate with low loss all the way to the end consumer. Interference effects are negligible here, although the transmit power is high—in practice around 0.1–0.5% of the maximum apparent power transmitted by the network [R3]. This high transmit power is necessary, because in this frequency range, all active consumers take in considerable parts of the audio-frequency power. A heavily loaded network represents a low load resistance for the audio-frequency source. Feeding is therefore always serial.

RCS requires an enormous effort for the transmitter, so that its use is always unidirectional, i.e., from the power supply company to the consumer. The very desirable opposite direction is inconceivable with this technology. The required transmit power has to be oriented to the network load from the connected consumers. In contrast, the impact of the number of RCS receivers in a given network on the impedance can be neglected. The required audio-frequency signal voltage at an RCS receiver input is determined by the interference level present at that input. The interference level effective in RCS is caused primarily by harmonics and nonharmonics of the mains voltage, and some degree of impulsive noise. By far the most critical interference is given by harmonics of the power mains frequency, because they have numerous high-power sources, which are in operation over long periods or even permanently.

Nonharmonics have an arbitrary frequency position to the mains frequency. For this reason, they cannot be avoided in a targeted way by simply selecting the RCS frequencies appropriately. The main source for long-lasting nonharmonics are motors, and their interference is found mainly on the low-voltage level. Overtones that form in motors have a frequency position determined by the number of grooves of slider and stator and the motor's number of revolutions. A spectral emphasis concerned mainly with these influences cannot be stated due to the large number of motor types in practical use. High-performance motors (10–60 kVA) achieve overtone amplitudes with several percent of the power voltage and can put the RCS reliability at stake. In practice, however, such impairments are rare; they normally occur only when several unfavorable circumstances coincide. High-performance asynchronous motors are definitely the devices with the largest potential interference effect. Such motors are preferably operated with an approximately constant number of revolutions and minor slippage, so that they generate a very narrow interference spectrum around a harmonic of the mains voltage. RCS is able to avoid such cases in a targeted way by defining its frequency appropriately. If a motor operation with heavily fluctuating number of revolutions cannot be avoided, as is the case with a thresher, for example, then occasional interference in ripple control receivers can occur. This may be the case particularly in extensive distribution networks in rural areas with low load density.

Switching events are another source of nonharmonic overtones. The continuous spectrum that results from switching events has amplitudes that decrease in inverse proportion to the frequency. In a supply network, interference voltages overlay at any given point in time, caused by many statistically independent switching events, and they form a noise level that decreases as the frequency increases. We can observe the spectral sidelobes of such interference all the way to very high frequencies (see also Section 2.1.3). The interference effect of nonharmonics to ripple control receivers is generally low. This is mainly due to the narrowband character of the receivers.

Harmonics of the mains voltage can be generated both on the medium-voltage and the low-voltage levels. We distinguish between harmonics of the voltage and harmonics of the current. Voltage harmonics form mainly in generators and are almost *independent from the load*. They are usually passed on from the high-voltage to the medium- and low-voltage levels. Current harmonics are generated mainly on the medium- and low-voltage levels, e.g., in transformers and power converters. Transformers may generate overtones with amplitudes that are *independent from the load* because of the bent magnetization characteristic of their iron core. Power converters in a medium-voltage network can cause current harmonics which *depend on the load*. On the low-voltage level, consumers with nonlinear characteristics are sources of current harmonics which *depend on the load*. This includes, for example, all devices equipped with switching power supplies.

The potential interference effect caused by harmonics of the mains voltage is determined by the interaction between voltage harmonics and current harmonics. The interference voltage existing in one location, x_e, is the overlay of the voltage harmonic existing there with the voltage produced by the current harmonic there. The latter depends for each current harmonic on the impedance $Z_{HN}(x_e)$, inherent in the network at location x_e for this harmonic, and on the source impedance $Z_{HQ}(x_e)$ of this current harmonic at x_e. This means that the interference effect from overtones of the mains voltage varies strongly in terms of location and time. The low-voltage network can be modeled in any place by a complex two-terminal network composed of resistive, inductive, and capacitive elements. In general, this two-terminal network has numerous parallel and series resonance characteristics, which can vary in location and time. The interference effect of a current harmonic is obviously highest when a parallel resonance is present, and when the source impedance, Z_{HQ}, for this harmonic is very high at the same time. The reason is that the quality of the parallel resonance is high in this case, and a high voltage with the frequency of the current harmonic can occur. Up to 10% of the mains voltage has been observed in extreme cases [R3].

The direct interference effect on a ripple control receiver is generally hard to predict. It may depend both on the spectral position and on the time behavior of the interference. This means that interference caused by switching events is normally less critical, despite broad spectrum and large amplitudes, because the corresponding peaks are very short, compared to the duration of a typical ripple control waveform. In contrast, long-lasting interference as normally caused by motors during their startup phase or with a heavily fluctuating number of revolutions is critical. From a spectral view, the situation is comparable to a relatively slow chirp[2] signal, affecting a ripple carrier frequency over a rather long period of time.

Permanent harmonics of the mains voltage are generally the largest interference potential, unless they can be avoided by an RCS system in a targeted way. Unfortunately, they can normally not be suppressed at their origins. For example, the load-independent overtones transmitted from the medium-voltage network into the low-voltage network exist permanently. The same holds true for overtones caused by transformers between the medium- and the low-voltage levels. Although load-dependent overtones are effective only during the startup phase of the disturber, they still have to be considered as permanent interference with regard to the ripple control signals.

A good basis for quantitative acquisition of the total interference effect from overtones of the mains voltage in a ripple control receiver is the definition of an effective value, U_{interf}, of the equivalent to the sum of all partial interference voltage effective values, $U(f)$, in the transmission band, $\Delta f_e = f_2 - f_1$, of the receiver. When the ripple control receiver filter has the transfer function $H_E(f)$, then

$$U_{\text{interf}} = \frac{1}{|H_E(f_e)|} \sqrt{\sum_{\Delta f_e} [|H_E(f)| \cdot U(f)]^2} \qquad (3.8)$$

We can now derive a measurement procedure from (3.8). In the event that we have to deal with n discrete interference spectral lines within Δf_e, we first have to measure the effective values $U(f_i)$ for $i = 1, ..., n$ selectively, and then weigh them with the transfer function $|H_E(f_i)|$ and perform a geometric addition. Next, we normalize to the transfer function value $|H_E(f_e)|$ at the desired receive frequency, f_e, and finally obtain U_{interf}.

To ensure that impulsive noise is also recorded correctly, the measurement time for the effective values, $U(f_i)$, has to correspond to the duration of the expected ripple control waveforms. For continuous interference spectra, the described measurement method is not practical. One of the better solutions would be the use of a spectrum analyzer. As mentioned above, the interference effect of continuous interference spectra can usually be neglected in practice.

For a functional RCS operation, a voltage effective value of $U_{\text{RCS}} > U_{\text{interf}}$ is required at the RCS receiver in any case. It has been observed that the occurrence of

---

[2] Linear frequency modulation.

faulty switching is low if the ratio [R3] is

$$\frac{U_{\text{RCS}}}{U_{\text{interf}}} = 1.5\text{--}2 \qquad (3.9)$$

The voltage U_{interf} according to (3.8) cannot be stated exactly, because it depends mostly on the network utilization. Extreme values are found when comparing the power supply network of a residential area with that of a commercial or industrial area. When the nominal mains voltage (effective value) is U_N, then one can observe a ratio of

$$\frac{U_{\text{interf}}}{U_N} \approx 0.001\text{--}0.0025 \qquad (3.10)$$

in purely residential areas and

$$\frac{U_{\text{interf}}}{U_N} \approx 0.0014\text{--}0.0035 \qquad (3.11)$$

in industrial areas.

Using (3.9) and (3.11), we can state indicators for the values of U_{RCS} required in ripple control receivers:

$$\frac{U_{\text{RCS}}}{U_N} \approx 0.005\text{--}0.007 \qquad (3.12)$$

This means that U_{RCS} should be between 1.15 and 1.61 V for $U_N = 230$ V. U_{RCS} is the voltage required at the ripple control receiver, which is by no means identical with the voltage, U_{SRCS}, to be injected by the transmitter. The latter has to be higher by a multiple. An upper limit for U_{SRCS} is given by two marginal conditions. First, large signal amplitudes with audiofrequencies above the mains frequency can cause disturbing interference in different appliances that may even impair their operation. Second, the cost for the generation and coupling of large audio-frequency signal amplitudes into a distribution network increases approximately with the square root of the injected power [R3]. The most important cost factor here is not the additional energy consumption, but the cost for transmitter and coupling equipment. RCS signals are generally injected on the medium-voltage level to achieve a large network coverage with only few feeding points. The power to be injected is determined by the current impedance of the network to be supplied. Table 2-2 gives an overview of the fluctuations to be expected.

Dimensioning of the transmit power for RCS tends to high functional safety and accompanying power reserve. Exact general power specifications cannot be given due to the large number of different network configurations and different consumer collectives. Transmitted powers in the 0.1–0.5% range of the respective network peak power are used as rough planning values. For practice, this means powers between 10 and 100 kW. In exceptional cases, for example when the network experiences a high capacitive load, the transmit powers can be more than 1 MW. The power supply companies generally have extensive experience with RCS and a mature measuring technology. This means that they can easily optimize the required transmit powers for their distribution networks.

Quasi-stationary operating conditions can be used as a basis when transmitting information on the medium- and low-voltage levels with carrier frequencies near the mains frequency. This means, for example, that the zero-phase angle of the carrier signal at the feed location has changed only to a minor extent during propagation to the most distant consumer. Both the impedance and the transfer function of a power supply network are subject to persistent fluctuations in terms of amplitude and phase due to load changes. With high-frequency carrier signals like those traversing several periods during the propagation time, the coherence is lost in power supply networks. This leads to drawbacks for signal detection at the receiver side. In contrast, low-frequency carrier signals allow the beneficial coherent transmission despite unexpected changes in the transmission properties of power supply networks. The assumption of quasi-stationary conditions is justified, when the wavelength, λ, of the carrier signal is larger than the 10-fold line length between the information source and the sink. The propagation speed of electromagnetic waves in medium- and low-voltage networks depends on the respective line constants, L' and C'. It is generally lower than the speed of light and significantly lower in cables than in overhead lines. Table 3-3 provides a summary for typical ranges of line characteristics for medium- and low-voltage networks and the resulting propagation speed and characteristic impedance ranges. This means that we can now state "supply limit lengths," l_{vg}, for various RCS frequencies. We assume that the respective wavelength, λ, is at least the 10-fold of the maximum distance between the RCS feeding point and the RCS receiver. When these limit lengths are maintained, then we can assume quasi-stationary conditions, so that a coherent transmission is ensured.

Table 3-3 Supply limit lengths l_{vg} of lines on the medium- and low-voltage levels for coherent RCS operation.

	f = 200 Hz	f = 500 Hz	f = 1 kHz	f = 2 kHz
Overhead line: l_{vg} in km	150	60	30	15
Cable: l_{vg} in km	50–75	20–30	10–15	5–7.5

In practice, the distances are far below the limit values given in Table 3-3.

3.1.2.1 Possibilities and Limitations of RCS

There is a high degree of branching in power supply networks on the medium- and low-voltage levels to enable fine distribution of electric energy upon demand. Such networks do not have a fixed characteristic impedance or even an approximately constant input and output impedance, as would be desirable for feeding and receiving of signals carrying information. Each consumer that is switched onto or off the low-voltage network influences the impedance for a transmitting device. At frequencies far above the mains frequency, the transmission properties become finally so unfavorable that a transmission of messages with conventional, simple modulation technologies, as used for CTP or RCS, is excluded. The transmit powers that would have to be used here would lead to problems of electromagnetic compatibility and would also be uneconomical.

When the first RCS systems of practical use were introduced at the beginning of the 1930s, complex modulation schemes were virtually unknown. The state of the technology at that time would not have allowed an implementation anyway. RCS may well have survived a period of almost 70 years only because its most important task has remained virtually unchanged, i.e., optimization of system functions for energy distribution. This includes the extensive balancing of the load curve in power supply to economize on power-plant peak power, the availability of which is uneconomical due to the typically brief demand. After all, RCS concerns the problem of compensating for poor storage capability of electric energy by targeted interventions across all areas of the low-voltage network. RCS is primarily designed to solve the problem of a simple but full-coverage telecontrolling, consisting of the transmission of on/off information in networks with supply radiuses between 5 and 25 km. During periods of peak load, major consumers, such as night-storage heating systems with a functionality not related to time, have to be switched off in a targeted way and switched on again in slack periods. In the course of time, RCS was given other tasks such as tariff switching or even the time-specific operation of equipment, e.g., street lamps. Telephone or radio networks had not been taken into consideration for such tasks from the very beginning. The telephone network reached a coverage comparable to that of the electric energy supply network only much later. Even today with a high telephone coverage, it is by no means sure that there is a telephone connection near each electric house connection or distributor. It would have been uneconomical to build and operate radio networks at the advent of RCS, not to mention the lack of reliability and the high proneness to failures. Even today with modern radio technology, it would be difficult to achieve profitability for several reasons, including lack of spectral resources and problem cases caused by unfavorable terrain situations or shadowing by buildings.

This means that the use of powerlines for the transmission of information is logical and at the same time the most economic solution. RCS is used worldwide and has reached a high degree of maturity in Europe, particularly in Germany. However, note that this holds true only for the west. RCS hardly exists in the eastern part of Germany and will most likely not be used in the future. In view of the current deregulation of the electrical energy market, these federal states focus more on the required new technologies (see Chapter 4). But even in the western federal states, further expansion of RCS is not planned; however, the use of existing equipment appears meaningful over a certain period of time.

RCS is used to transmit little information, according to its original tasks. However, the transmission has to be highly reliable, because the correct receipt of a telegram cannot be confirmed. Low-voltage transformers are easily surmountable in view of the low RCS frequencies. This means that extensive low-voltage networks can be served from few feed points on the medium-voltage level. Transmitting powers between 10 and 100 kW and in some cases even up to 1 MW may be required to reach a U_{RCS}/U_N ratio of at least 0.005 (i.e., 1.15 V with 230 V) needed for the safe triggering of an RCS receiver at each consumer location. As can be seen from Table 3-3, the theoretical supply radiuses of RCS are considerable, and low frequencies allow reaching the largest distances. This is why the frequency range below 1500 Hz is preferred.

In view of the fact that there is no feedback option, RCS has to be designed with a high degree of failure safety from the very beginning. The so-called "ripple control telegrams" were developed for this purpose; they are used to map highly redundant simple switching information. Initially, multifrequency methods had been used for some time, where a sum from several signals of different frequencies was transmitted. The receiver side had to detect all sent frequencies in parallel before the function allocated to a frequency combination was triggered. In 1940, the multifrequency methods were replaced by single-frequency time-division multiplexing methods. This change was mainly motivated by the experience that generating and injecting multifrequency signals into the mains network was much more difficult and costly, compared to the monofrequency alternative [R3].

From the very beginning, RCS was used to basically transmit digital information, so that corresponding digital modulation methods were of primary interest. With single-frequency methods, it is relatively easy to map information by on/off switching (amplitude shift keying—ASK) to the carrier. This type of modulation has become widely used because it is simple, and it is today almost the only type used. In addition, frequency shift keying (FSK) methods, which map the H and L information bits each onto a different frequency, have been tested. However, FSK has not gained wide use for similar reasons as the multifrequency methods. ASK is the dominating technology today, and its superiority is mainly ascribed to the simple transmitter construction. Transmitter cost is one of the highest costs due to the required high power. Any simplification of equipment for signal generation, modulation, amplification, and coupling is most welcome, as long as it does not reduce transmission safety. A major drawback of ASK is its general proneness to false alarms in the carrier pauses. On the other hand, the RCS system design by its special waveforms with regard to frequency, amplitude, and duration offers a priori a high degree of safety against such false alarms, mainly by the following measures:

1. One selects carrier frequencies from the 125 to 3000 Hz range, which do not coincide with harmonics of the mains frequency. Before the final frequency definition, one checks the interference spectrum in the environment of the planned frequency over a long time to ensure sufficient noninterference.

2. The transmitting power is selected with a generously dimensioned safety gap, depending on the network load, so that an RCS signal in the amount of 5 per thousand of the network nominal voltage (approximately 1.15 V at 230 V) is guaranteed at the receiver side. The threshold of RCS receivers is reached by signal amplitudes with 4 per thousand of the nominal mains voltage.

3. The duration of an RCS telegram is between 28 and 180 s, with very few exceptions. Ten to 60 impulses with intermediate pauses, which are equal or have the same length as the impulses, are sent during this period. This means that one telegram includes between 20 and 120 bits. The impulse duration is generally within the range from 0.1 s to maximum 8 s. Signals with this length offer high resistance against transient interferers. High interference amplitudes can be effectively suppressed by limiters at the receiver input.

Practical implementations have proven that RCS ensures a high degree of safety against faulty switching or lost telegrams, even without special coding methods, which are normally used in modern communication technology. On the one hand, this may be thanks to the seemingly coarse solution of the attenuation problem with enormous transmit power as well as to the long duration of the waveforms used. The high transmit power allows surmounting almost all disturbers, and the long waveform duration allows high selectivity in the receivers, i.e., very narrowband filtering. To increase switching safety, the telegram design is redundant, using binary codes which are not fully exploited. This means that there are error-control and error-correction options, the full utilization of which has been made possible only more recently by digital technology.

In cases where maximum switching safety is required, the telegrams are simply repeated at intervals of several minutes [R3]. A typical example is when night-storage heating systems are switched off early in the morning, particularly in winter. If these heating systems were not switched off due to lost telegrams, then the supply network could collapse, e.g., from the peak load at the beginning of the industry's working time.

Actually, RCS waveforms have a long duration for historical reasons. At first, only mechanical systems were available for modulation in the transmitter and signal processing in the receiver. An example is given further below.

Although ripple control telegrams generally trigger simple processes, such as on/off switching, there is a large number of various addressees, which have to be served selectively, and for which a series of different switching events have to be handled. Ripple control telegrams include 20–120 bits of information. This is a relatively large amount of information in view of the fact that they concern simple switching events. In the course of time, changing and additional tasks have led to a multitude of ripple control telegrams. We consider here a usual variant as a representative example. The example concerns a two-step process in which some of the information bits are used for an object or group addressing (step 1) and the rest contain the switching information (step 2). For example, object addressing means that a single major consumer in an industrial plant is addressed for the purpose of sudden release. An example for group addressing is the addressing of the street lights in a city.

RCS uses a simple coding procedure. This may also be explained with historical reasons. It uses either complete binary codes, in which all states are occupied, or m-out-of-n codes [R3]. The two-step method mentioned above uses m-out-of-n codes, both for object and group addressing and for the switching information. Normally, $m = 1$ is set for the switching information to achieve high redundancy that can be used for error detection and error correction, while $m > 1$ is generally selected for object and group addressing, which reduces the safety against transmission errors. Of course, the use of redundant codes is meaningful only if there are ripple control receivers which include appropriate error-detection and error-correction options. Electronic ripple control receivers with this capability have been introduced more recently; they have built-in high-performing digital equipment, e.g., in the form of a microprocessor.

The following typical example describes the telegram structure and, related thereto, the RCS possibilities. Assume that we are using a relatively small telegram length of 20 bits. The first 8 bits are used with a "3-out-of-8" code for object and group addressing, and the remaining 12 bits are implemented with a "1-out-of-12" code for the switching information. This telegram allows the transmission of 56 addresses and

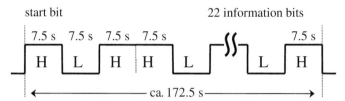
Figure 3-2 Transmitter telegram with the A/Z1 method.

12 switching events each; i.e., a total of 672 different bit combinations is allowable. The redundancy guarantees high safety against faulty switching.

Numerous variants of the technology have been developed in the course of RCS's 70-year history, including a simple start/stop method that had been in use by the name of A/Z1 until about 1954. We will describe this technology briefly as a representative historical example. The typical mechanical receiver function is rather interesting from today's viewpoint. In a waveform transmission with the start/stop method, both the transmitter and the receiver require time references that are stable enough so that, after an initial synchronization (start impulse), no significant time shift can occur between the received waveforms and the receiver's "evaluation organs" for the duration of a telegram (see Figure 3-2).

Figure 3-2 shows the impulse scheme of the A/Z1 method. The entire telegram takes about 172.5 s. The impulses and the pauses have a length of 7.5 s each. The first impulse is the start impulse, which activates the receiver to detect the telegram. Next come 22 steps, in which either a carrier is transmitted to send an H bit, or no carrier for an L bit. This means that the A/Z1 telegram comprises 22 bits. During the transmit process, an audio-frequency generator with relatively high output power is switched on and off according to the impulse pattern. Generators for RCS will be described further below.

The ripple control receiver for the A/Z1 method works on the basis of the start/stop principle. The functionality is described on the basis of Figure 3-3 (see also [R3]).

The audio-frequency signals are filtered from the mains voltage with a series-resonant circuit, which is formed from the tapped coil with inductivity L and the capacitor with capacitance C. Diode 22 is used to rectify the voltage present at the coil of the resonant circuit and applied to the charging capacitor, C_s. The impulse receive relay 14 is connected to this charging capacitor over a glow lamp 21. The glow lamp switches the relay on as soon as it reached its ignition voltage. This means that it has the function of a threshold switch and ensures that faulty switching due to noise is prevented. In idle state, which is shown in this example, the turning arm 13 of the synchronous selector 12 maintains the start/stop contact 18 open, so that the synchronous

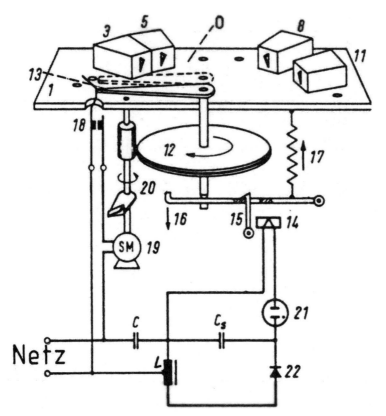

Figure 3-3 Receiver for the A/Z1 method.

motor 19 stands still. At the same time, the anchor 15 of the relay 14 is hooked in lever 16.

The effect of the start impulse of a telegram is that the receive relay 14 pulls the anchor 15. The lever 16 is unlocked, and the spring 17 lifts the lever 16 and thus the synchronous selector 12 and its switching arm 13 up, so that switch 18 is closed. The synchronous motor 19 is started and pulls the lever 16 down by means of cam 20 in each cycle. Cam 20 makes exactly one cycle during the time of one impulse, i.e., in 7.5 s. The use of a synchronous motor ensures the time stability required for safe telegram detection. The received telegram can be stored mechanically on ring 0 of the synchronous selector 12 by means of up to 22 flip switches. Figure 3-3 shows four such flip switches labeled 3, 5, 8, and 11. When the switching arm 13 bypasses a flip switch in lifted state, then this flip switch is switched on. In nonlifted state (anchor 15 is not pulled), the switching arm causes the bypassed flip switch to switch off. The synchronous selector 12 make exactly one cycle during the duration of the entire telegram, and the switching arm 13 assumes the position shown in the figure again; it then switch 18, so that the synchronous motor is stopped. Subsequently, the receiver goes back into idle state. This example should provide a rough overview of the RCS receiver technology; other variants are described in [R3, R5].

Modern RCS receivers do not require any mechanical parts for their functionality, except for the power switching relay at their outputs. More recently, high-performing digital circuits, e.g., in the form of microcontrollers and ASICs (application-specific integrated circuits) have contributed considerably to improve the functional safety. It is now possible to utilize the redundancy in the transmitted telegrams entirely for error detection and error correction. In addition, signal filtering is virtually fully digital—usually implemented in ASICs. These modern technologies allow the low-cost, exactly reproducible, long-lived steep-sloped filters, because the components neither drift nor age. Moreover, switching safety can be further increased, thanks to the possibility of storing several telegrams and additionally with the use of a "background clock." This clock allows extensive control of lost telegrams. For example, purely time-controlled emergency switching programs can be activated upon demand, as after a total failure of a connection.

While the technology of ripple control receivers has changed radically due to microelectronics, hardly anything "revolutionary" occurred on the transmitter side. The improved receiver technology ensured RCS's longevity. The most important change on the transmitter side was the gradual replacement of motor generators by static power converters for signal generation. In closing, this section provides a short insight into the generation and injection of ripple control signals.

A ripple control transmitter (RCT) includes three functions:

1. *Generator function*: Generates an audio-frequency carrier with sufficient power and frequency stability.
2. *Modulator function*: Modulates the carrier—generally by on/off switching—based on a time pattern determined by the message to be transmitted.
3. *Coupling unit*: Couples the modulated carrier signal into the power supply network, generally on the medium-voltage and occasionally also on the low-voltage level.

The power that an RCT has to make available is considerable. Audio-frequency powers between 10 and 100 kVA are requested, depending on the peak load of the supply network. Audio-frequency generators for RCT can be roughly divided into alternate-current and three-phase current generators. With alternate-current generators, the power is fed into the neutral conductor circuit of the network, and with three-phase gen-

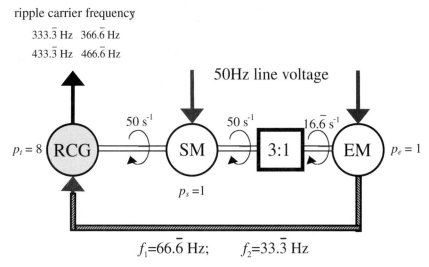

Figure 3-4 Line-commutated rotating converter for the generation of high-audio-frequency power (no mains harmonics) according to [R3].

erators, coupling is done symmetrically onto the three phases of the network. Basically, each generator of an RCT transforms a part (0.1–0.5%) of the power supplied to the consumers at the mains frequency (50 or 60 Hz) into the audio-frequency position; hence the name "audio-frequency converters." There are two totally different constructions of audio-frequency converters: rotating (also called motor generators) and static. With regard to the frequency reference, we distinguish between line-commutated and externally commutated converters. In line with the historical development, we will first describe an example of a three-phase motor generator.

Figure 3-4 shows a rather complex example of a line-commutated rotating converter that can be used to generate audio-frequency signals, which contain no mains harmonics. The synchronous motor SM has the number of pairs of poles, $p_s = 1$, and its stator is fed with the mains voltage of frequency $f_0 = 50$ Hz. The number of revolutions of this motor is then $n_0 = 3000$ min^{-1} = 50 s^{-1}. A gear with a reduction of 3:1 is used to drive an exciter EM (also called dynamo) with a number of revolutions of $16^2/_3$ s^{-1}. The exciter EM has the number of pairs of poles, $p_e = 1$, and its stator is excited with the mains voltage (50 Hz). At the slider of the exciter, the two frequencies $f_1 = (50 + 16.666)$ Hz = 66.666 Hz and $f_2 = (50 - 16.666)$ Hz = 33.333 Hz can be taken over wipers and fed to the stator of the audio-frequency generator RCG, which is also implemented in the form of a synchronous engine. The audio-frequency generator RCG has the number of pairs of poles, $p_t = 8$, and is driven by the synchronous motor SM with a number of revolutions of $n_0 = 50 \text{ s}^{-1}$. The audio-frequency generator now supplies the following four frequencies at the wipers of its slider:

$$f_{t1,2} = p_t n_0 66.\overline{6} \text{ Hz and } f_{t3,4} = p_t n_0 33.\overline{3} \text{ Hz},$$

that is,

$$f_{t1} = 466.\overline{6} \text{ Hz}, f_{t2} = 333.\overline{3} \text{ Hz}, f_{t3} = 433.\overline{3} \text{ Hz}, f_{t4} = 366.\overline{6} \text{ Hz}.$$

The desired frequency is filtered in the narrowband coupling circuit described further below.

An externally commutated rotating converter can be developed from Figure 3-4 by replacing the exciter EM, for example, by a quartz oscillator or a tuning fork generator, each followed by a power amplifier. Externally commutated converters allow a more flexible selection of the frequency, because a tuning fork or a quartz crystal with appropriate frequency can be produced at less cost than the gear from Figure 3-4, and it can be changed when necessary. In general, externally commutated converters have a higher frequency stability, compared to line-commutated constructions. However, the difference is insignificant for RCS, because the mains frequency is held stable with high cost within a fluctuation range of ±2 per thousand. If extreme fluctuations occur due to failures in mains operation, then the RCS is stopped whenever the frequency change is more than +1% or –2%.

While only rotating converters in the form of motor generators were able to generate the high required powers in the audio-frequency range at the beginning of the RCS technology, new developments tended exclusively to static power converters. Core elements of static power converters are thyristors, used as electronically controllable power switches, and microprocessors to supply more or less complex control signals for these switches. More recent developments for the replacement demand, which are necessary to a certain extent to maintain existing ripple control systems, use also IGBTs (isolated gate bipolar transistors) as power switches. But even before there were thyristors on a semiconductor basis, static power converters were built with controllable rectifiers in the form of gas-filled thyratrons. Due to the inertia of thyratrons, their applicability is limited to frequencies below 250 Hz [R3].

Static power converters consist basically of two functional elements: rectifier and inverter. Both elements together form a converter. In a static power converter, a direct voltage is first yielded from the mains voltage (DC link). Second, the direct voltage (DC) is transformed into alternate voltage (AC) with the desired frequency by means of appropriate on/off switching by controllable rectifiers (thyristors) and interaction with appropriate reactive elements (inductances and capacitances). We will not provide a detailed description of the functionality of static power converters here, but rather refer to the literature [R3]. The most important benefits of static power converters, compared to the rotating configurations, are:

- Immediate readiness for operation
- Low maintenance effort
- High flexibility in the frequency selection
- Small space requirement and little installation cost
- Silent operation

The modulator in a ripple control transmitter switches the audio-frequency carrier signal on and off according to a time pattern determined by the message; see Figure 3-2. First, modulation can be executed directly in the power stages, i.e., on the input side at the feeding network, or on the output side at the controlled network, by means of a contactor. Second, modulation can be implemented over control circuits, in which low power has to be switched. This approach is the rule for static power converters. These designs allow the exclusive use of semiconductor components. For motor generators (see Figure 3-4), modulation is always implemented directly by a contactor in the power stage toward the controlled network. The coupling circuit of a ripple control transmitter has to selectively transmit the audio-frequency carrier power to the mains network and safely separate between mains and RCT. Customary are either simple series-resonant circuits or band filters. We use the symmetric series coupling for a three-phase system as an example. Three-phase series coupling can be applied on both the medium-voltage and the low-voltage level.

Figure 3-5 shows the principle. The primary sides of the three equally built coupling transformers T_1, T_2, and T_3, together with the three equal capacitors C_1, C_2, and C_3, each form a series-resonant circuit, with the audio-frequency to be sent as the resonance frequency. The audio-frequency voltage at the primary sides of the coupling transformers is generally high; 1000 V is a typical value in low-voltage networks. A transmitter then has to reduce this voltage to approximately 1.5% of 400 V, i.e., 6 V. Thus, the required turns ratio is approximately 167. This high turns ratio can be risky for RCTs, because it raises the mains voltage. In practical applications, suitable "burdens," for example in the form of series-resonant circuits, matched to the mains frequency at the primary side of the transformer have to be used to keep the transformed mains voltage low. Table 3-4 shows important characteristic values for RCS systems:

Possibilities and Limits of Classical Usage Types

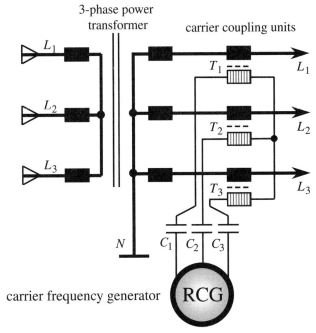

Figure 3-5 Example of a symmetrical three-phase series coupling of an audio-frequency generator [R3].

Table 3-4 Characteristic data of ripple control systems.

	Frequency range
	168–3000 Hz (preferably 200–1500 Hz)
Typical telegram size	20 bits–120 bits
Telegram duration	30 s –180 s
Single waveform duration	0.1 s–7.5 s
Typical pause duration	0.4 s–2.75 s, if the telegram contains pauses
Typical start impulse duration	0.4 s–5 s
Typical injected power	0.1%– 0.5% of the mains network peak power
Typical injected voltage	1.5% of the nominal mains voltage
Receiver threshold	880 mV (adjustable in modern receivers)

CHAPTER 4

New Usage Possibilities of the Low-Voltage Level Based on European Standards

4.1 The European CENELEC Standard EN 50065 for the Frequency Range Below the Long-Wave Broadcast Band

The European CENELEC Standard EN 50065, which rules the use of the frequency range from 3 to 148.5 kHz [NO3], has been in force since the end of 1991. Figure 4-1 gives a basic overview of the definitions of this standard. EN 50065 differs considerably from other regulations, e.g., those applicable in the United States or in Japan, where a

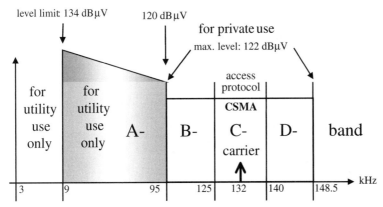

Figure 4-1 Frequency ranges and PLC signal level limits specified in EN 50065.

frequency range to approximately 500 kHz is available. In addition, signal injection between neutral conductor and protective ground is admissible in these countries. The use of the protective ground conductor means that the interference is very low. Due to the differences in standards, it is not possible to buy PLC systems in the United States or Japan and use them in Europe. All attempts to modify imported equipment to match the EN 50065 have failed.

The European Standard 50065 rules the frequency use and the admissible signal levels for communication over powerlines below the long-wave broadcast band. Figure 4-1 is divided into two parts, where frequencies below 95 kHz (A band) are reserved for use by PSUs, while the range from 95 to 148.5 kHz (B, C, and D bands) is for private use, mainly within buildings, e.g., for automation tasks. A maximum signal level of 122 dBµV (= 1.25 V) applies equally to the B, C, and D bands, while the level in the A band may be 134 dBµV (= 5 V) at 9 kHz and can be lowered to 120 dBµV (= 1 V) up to 95 kHz, as shown in Figure 4-1.

Due to the bridge circuit specified in the measurement standard (see Figure 4-2), the level actually present at the mains may be twice as high, i.e., 10 V at 9 kHz, decreasing to 2 V at 95 kHz. The measurement standard also defines the use of twice the value, i.e., 2.4 V, in the B, C, and D bands. While the standard does not specify anything with regard to the access protocol in the A, B, and D bands, it specifies CSMA (carrier sense multiple access) for the C band. The application of this method, which is well known from Ethernet technology, has proven to be a problem in practice. Due to

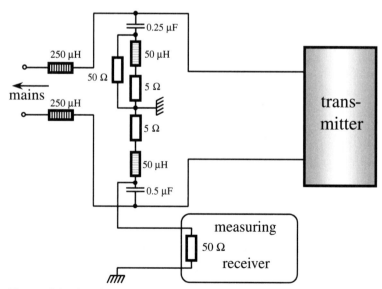

Figure 4-2 Level measurement according to EN 50065.

high attenuation fluctuations, the detection of carrier signals cannot be guaranteed when there is a high number of distributed participants, e.g., in a large building.

The most important section in Figure 4-1 available for the PSUs is the A band (9–95 kHz). Plans to implement value-added services in this section will be described in detail later. Moreover, the measurement regulations of EN 50065 specify that the transmit signal amplitudes have to be determined as peak values at the output of a band-pass filter in a width of 100 Hz. The consequence is mainly that modulation methods using the broadband transmit signals have advantages over narrowband methods.

The frequency range reserved for private users is divided into three bands, referred to as B, C, and D bands. These bands all have the common characteristic that a transmit level of 122 dBµV ($\approx$ 1.25 V) must not be exceeded, similar to the measurement regulation for the A band. The level is clearly lower, compared to the A band, which is attributed to the shorter distances which are typical within buildings.

Though the EN 50065 specifies a separation of PSU and private uses, in terms of frequency, a large number of meaningful applications entail a functional interplay of both areas.

Based on the EN 50065 there are technologies available today which allow bi-directional data traffic at rates of up to a few thousand bits/s. Often, several hundred households are connected directly over a distribution network supplied by the same transformer. The network is normally divided into several branches in star shape (see Figure 2-11). Figure 2-11 shows one out of up to 10 branches with a row of supplied customers. The maximum distance between two modems can be up to one kilometer. Much larger distances do not occur in practice, taking into account the applied optimization criteria, such as material and construction costs, and the observance of voltage drop and grounding conditions.

The following section briefly describes the consequences resulting from the individual regulations of the standard, before we give details of the transmission technology and the structure of corresponding devices for use in the frequency range specified in EN 50065. First, a brief summary of the most important statements:

- The output spectrum of a transmit device is determined by use of a spectrum analyzer with a peak value detector within a bandwidth of 100 Hz.
- The transmitter has to be operated in such a way that both the bandwidth and the output signal show the maximum admissible values according to manufacturer information.
- The spectral bandwidth (B in Hz) is determined by the width of the frequency section, in which all spectral lines are less than 20 dB below the maximum spectral line.

- The output level is measured over a time span of one minute by use of a peak-value detector. This measurement can be carried out by use of a spectrum analyzer with a transmission band equal to or larger than the bandwidth B of the transmitter output.

- The measured level must not exceed 134 dBµV at 9 kHz, and it decreases linearly to 120 dBµV at 95 kHz over a logarithmic frequency scale.

- For broadband signals, i.e., a signal bandwidth of $\geq$ 5 kHz, the measured level must not exceed 134 dBµV. Moreover, the signal spectrum, measured by use of a peak-value detector with a bandwidth of 200 Hz, must not exceed 120 dBµV in any range.

This means that, in addition to the signal amplitude for broadband signals, the EN 50065 limits also the spectrum of the transmitted signal. For the latter, a bandpass filter with a bandwidth of 200 Hz followed by a peak-value detector is used for testing. The allowed signal amplitude for a broadband signal can be higher than for a narrowband signal, which favors broadband signals. The spectrum limitation is normally decisive for broadband signals.

Based on EN 50065, the following section describes the impact of the limitation of both the signal level and the signal spectrum on the maximum transmission power of a transmission system in connection with the defined measurement principle.

4.1.1 The Impact of Limiting the Signal Level

In the following description, the measurement filter has a bandwidth larger than or equal to the signal bandwidth. The transmit signal $s(t)$ of a transmission system is thus not changed by the measurement filter. This means that the value measured by the peak value detector is

$$s_P = \max\{|s(t)|\} \tag{4.1}$$

With a peak power of

$$S_P = \max\{|s^2(t)|\} = s_P^2 \tag{4.2}$$

and a mean power of

$$S = E\{s^2(t)\} \tag{4.3}$$

we can define the peak-to-mean-power ratio

$$\chi = \frac{S_P}{S} \quad (4.4)$$

where $E\{.\}$ is the expected value of each. When using $\hat{s}$ to designate the allowed limit value at the output of the peak detector, then the following applies to the maximum possible transmission power:

$$S_{max} = \frac{s_{P,max}^2}{\chi} \quad (4.5)$$

Table 4-1 shows the peak-to-mean-power ratio, the maximum transmission power, and the mean transmit amplitude

$$A_m = \sqrt{2 \cdot S_{max}} \quad (4.6)$$

for three different modulation methods. The maximum signal level $\hat{s}$ was set to 134 dBμV according to EN 50065.

Table 4-1 Maximum transmission power with signal-level limitation; from [H21].

Modulation Scheme	χ	S_{max}	$\bar{A}$
Unipolar ASK	4	6.28 V^2	3.54 V
PSK	2	12.56 V^2	5 V
16-QAM	3.6	6.98 V^2	3.74 V

We can see that limiting the signal level represents a certain disadvantage for modulation schemes with a nonconstant transmit amplitude (ASK, QAM). The difference in the maximum transmission power is 3 dB between PSK and unipolar ASK, and 2.55 dB between PSK and 16-QAM.

4.1.2 The Impact of Limiting the Signal Spectrum

The relation between transmission power and peak value of the measurement filter output signal is relatively complex with broadband signals. [D8] contains a detailed description of "memoryless" modulation schemes. We are summarizing important results, which are needed.

Chapter 5 deals again with the determination of peak values in connection with capturing limit values of unintentionally radiated signals when telecommunication over powerlines is considered. Whether "quasi-peak values" or peak values should be measured is currently still the subject of discussion. If a peak-value measurement is specified, then this results in considerable disadvantages for powerline communication, because the modern digital transmission methods used normally transmit broadband signals with random character. The more sophisticated these methods, the more the amplitude distribution of the transmit signal approximates a Gaussian distribution, in which arbitrarily large amplitudes can occur. However, as the probability for peak values to occur is very small, acquisition by measurement technologies would be virtually impossible and, on the other hand, the interference effect would also be low.

First, it is important to look at the probability density function of the peak values at the output of a detector, following a narrow filter (B = 200 Hz), with typical broadband transmit signals of interest in powerline communication.

Figure 4-3 shows an accumulation in the range of medium peaks, while very large and very small values are infrequent. In practical measurements, this means that the acquisition of a large peak value depends on the measurement duration. The longer the measurement time, the higher the probability that a rarely occurring very high peak is acquired.

To better understand the essential consequences from the above peak-value distribution, the next two figures show the maximum transmission power and the maximum energy per bit of four different standard modulation schemes, depending on the bit rate r_b. Based on EN 50065, the following parameters were used:

measuring bandwidth: B_M = 200 Hz,
max. signal amplitude: $\hat{S}$ = 5 V $\triangleq$ 134 dBµV,
measuring time: 1 minute.

Figure 4-4 shows that the admissible transmission power of modulation schemes with carrier suppression (BPSK, QPSK, QAM) initially increases with increasing bit rate. Beyond a certain bit rate, which has a different value in each scheme, the transmission power can no longer be increased, because the signal-level limitation is now decisive, rather than the limitation of the signal spectrum. In ASK, the transmission power may be increased only slowly in line with the bit rate. It approximates a limit value asymptotically. The reason for this behavior, which deviates strongly from that in the other methods, is that, in ASK, only slight spreading of the spectrum occurs with increasing bit rate, so that a large portion of the transmission power is always falling into the measuring filter. This means that ASK cannot achieve a good transmission quality in general, because the resources allowed by EN 50065 are poorly utilized.

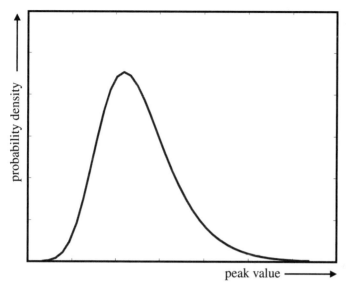

Figure 4-3 Probability density of the peak values of typical powerline transmit signals; from [D8].

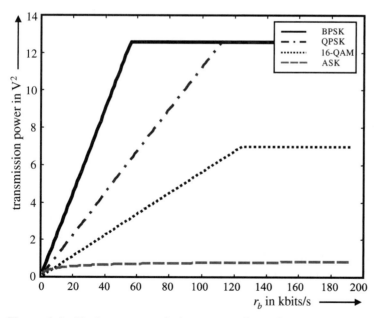

Figure 4-4 Maximum transmission power, depending on the bit rate, for various modulation schemes; from [D8].

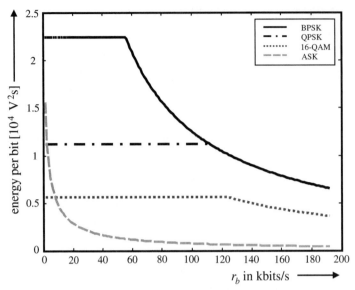

Figure 4-5 Maximum energy per bit, depending on the bit rate, for various modulation schemes; from [D8].

An important parameter for comparing the performance capability of transmission methods is the energy per bit, or the product of transmission power and duration for one data bit. The higher this value at the receiver in relation to the interference power density present there, the fewer errors will occur in the transmission. In comparisons, it is therefore customary to apply the bit error rate, e.g., as a function of the ratio of energy per bit E_b to the noise-power density N_0 [F12].

In conventional communication technology, a limited signal power is normally assumed. By utilizing this limit value, the energy per bit drops inversely proportional to the bit rate, because the duration of a data bit decreases as the bit rate increases. In contrast, the situation is totally different in communication over powerlines. To illustrate this difference, Figure 4-5 shows the energy per bit over the bit rate for four different standard transmission schemes. In the schemes without carrier transmission (BPSK, QPSK, QAM), the energy per bit remains constant until the limitation of the signal level becomes the controlling factor. This means that the curves kink exactly at the values of the bit rate at which the transmission-power limitation given in Figure 4-4 became effective. A comparison of the schemes shows that, up to a bit rate of 56 kbits/s, the energy per bit of BPSK is twice as high as that of QPSK and four times higher than that of 16-QAM. Above 110 kbits/s, only BPSK can transmit with the same energy per bit as QPSK. QAM always performs worse than BPSK and QPSK due to the unfavorable (higher) peak-to-mean-power ratio. With ASK, the energy per bit decreases rapidly as the bit rate increases. With larger bit rates above 10 kbits/s, ASK can transmit only with a considerably lower energy per bit, compared to the other methods, so that accordingly it will be more prone to interference.

Several important conclusions can be drawn from the above description:

- "Simple" modulation schemes, such as BPSK, can transmit with a higher transmission power and energy per bit up to a certain bit rate, compared to "more bandwidth-efficient" methods, such as QPSK or QAM. The reason is simply that the transmission spectrum of BPSK is spreading quickly as the bit rate increases, while there is only a minor spreading for QPSK and QAM; hence the attribute "bandwidth-efficient." This means that the share of the transmission power that falls into the measuring filter is almost unchanged under increasing bit rate in QPSK and QAM, while it decreases in BPSK.
- Carrier-suppressing modulation (BPSK, QPSK, QAM) can transmit more power at higher bit rates, compared to schemes where a carrier always remains, such as ASK.
- In carrier-suppressing schemes, the transmission power increases proportionally to the bit rate up to a certain value, so that the energy per bit is constant. This means that, in white noise with a constant noise-power density N_0, the E_b/N_0 ratio and thus the bit error rate is virtually independent of the bit rate.

4.2 Signal Coupling, Signal Attenuation, Access Impedance

4.2.1 Signal Coupling

To transmit information, signals with a frequency far above the mains frequency of 50 or 60 Hz have to be injected to the mains within the scope of a use specified in EN 50065. The sections describing the ripple control technology showed that very high powers are required to inject signals near the mains frequency, e.g., at several hundred Hz. This means that RCS is absolutely unsuitable for bidirectional transmission and thus for modem operation. As we will see further below, the mains access still has a very low impedance even at several kHz, depending on the local circumstances. Normally, impedance values allowing a "reasonable" design of transmission devices result only above frequencies of about 20 kHz.

When connecting a modem for data transmission to the mains, basically two message signal paths have to be made available. First, it is desirable to feed the transmission signal with the admissible upper limit value of the amplitude; second, the receiver signal should be "clean"; i.e., ideally it should be fed to the receiver input by separating undesired portions, such as those corrupted by interference. In both cases, it is necessary to provide a safe separation of the mains voltage. This task is more complex than one may at first assume, because not only does the mains experience a "smooth" sinusoidal voltage at 50 or 60 Hz, but there are also numerous harmonics far above this basic frequency, which can have considerable power. In addition, there are transients, i.e., short impulses with high amplitude, which can contain sufficient power to destroy the unprotected transmitter power stage or the receiver input of a modem. The modem will certainly not have a long life expectancy, unless proper protection measures against such transients are provided.

For coupling to the mains, a galvanic separation in the form of a transformer with highpass character has basically proven to be suitable. The highpass filter makes sure that the mains voltage and the major part of its harmonics are kept away from the modem. Coupling is normally implemented in parallel, i.e., within the building between neutral conductor and a phase (e.g., at a wall plug), and generally between two phases in the area between the transformer station and the house service connection. Here, the neutral conductor should ideally be avoided for EMC reasons, as will be shown later. A serial feed as in RCS would also be possible; but it has not proven to be suitable in practice. Serial feeding would mean high cost for disconnection of the wires in the transformer station. For serial feed within a building, a high-frequency short circuit on the modem side has to be applied to ensure that a usable HF current can form. This holds basically true also for a transformer station. There is another major drawback: the transmitter structure requires the use of magnetic materials in the form of ferrites. Such materials cannot carry arbitrarily large magnetic fields. The manufacturers normally state the value of the saturation induction B_{sat}, which ranges generally between 300 and 500 mT[1] for the materials discussed here. If B_{sat} is exceeded, then the material loses most of its magnetic properties; i.e., we eventually have a transformer with air-core coils. This occurs if a significant magnetic flux is caused by the 50-Hz or 60-Hz current for a transmitter on the mains side, which cannot be avoided in serial coupling. Basically, a transformer with air-core coils will not be able to fulfill the desired function of signal coupling to a satisfactory level at the frequencies discussed here.

Unfortunately, even parallel feed cannot prevent a certain amount of 50-Hz or 60-Hz current from flowing through the transformer's mains side, because the transmitter side requires a particularly close coupling. Fortunately, this current with parallel feeding can be controlled by appropriate component dimensioning, and in contrast to the serial case, it does not depend on the consumer current currently flowing in the

---

[1] 1 mT ≡ millitesla = 10^{-3} Vs/Am.

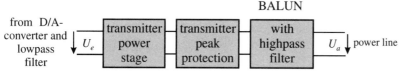

Figure 4-6 Signal coupling at the transmitter side.

lines. Weak coupling is basically sufficient at the receiver side, so that the mains-side premagnetization can be neglected. Though it is generally possible to use only one coupler both for the transmitter and for the receiver signal transformation, separating them has proven to be a good idea in practical applications. We have already mentioned the main reason: the transmitter side requires strong coupling in order for the transmission amplitude, which the transmitter output stage provides, to reach the network without attenuation, while weak coupling allows good filtering of interference that is desirable at the receiver side. The stronger the coupling, the more interference—being present mainly at low frequencies due to the power supply operation—is transported to the receiver input, where it will have to be suppressed at very high filtering cost. The following studies of typical coupling circuits explain these phenomena. First, an overview based on simple functional blocks is given.

Figure 4-6 shows the transmitter side. The transmitter's final stage receives the complete analog transmission signal, which has to be injected into the network over a lowpass filter, which is generally fed out of the A/D converter of a digital signal synthesizer. The coupler should try to transport this signal to the power supply network without losses and, at the same time, separate the mains voltage safely from the transmission's final stage. The coupling transformer has to have a strong highpass effect, because the ratio between mains voltage and transmission signal amplitude is generally higher than 30. This is not a big technical problem, because the frequencies differ by a factor of 400–3000. Nevertheless, as mentioned before, there is a threat to the transmitter's final stage from the mains side due to fast transients. It is, therefore, absolutely necessary to implement appropriate protection measures. So-called suppressor diodes have proven suitable for transmission systems which use the EN 50065 frequency range. Such diodes are characterized by fast response (a few nanoseconds) and the capability to quickly convert high quantities of energy into heat, without suffering any damage. Unfortunately, suppressor diodes have a relatively high capacity, so that they are not usable in high-frequency (several MHz) applications. This means that special protective circuits are required for the telecommunication systems presented in Chapter 6.

As mentioned above, the coupling in Figure 4-6 has to be strong to prevent significant signal-amplitude loss, so that high-energy transients can pass the transformer in

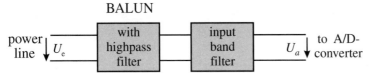

Figure 4-7 Signal coupling at the receiver side.

the modem direction. The transient protection has to provide enough attenuation to prevent the risk of destroying the transmitter's output stage. A successful practical circuit design will be shown later.

Figure 4-7 shows the receiver side. The functional block labeled "Transformer with highpass filter" provides, here too, for a safe separation of the 50-Hz or 60-Hz mains voltage, and it suppresses low-frequency spectral parts up to about 20 kHz, which contain interference mainly caused by the mains operation.

Although the basic structure of the coupling equipment (transformer with bandpass filter) in Figures 4-6 and 4-7 is the same, the dimensioning of the individual components has to be different. While the transmitter side requires a rigid coupling for known reasons, the receiver side requires loose coupling. This favors the filter effect (highpass), and there is no unwanted reaction on the network impedance. The latter is very important for the access methods described in Section 4.3 to function correctly. For energy-related value-added services, so-called master-slave protocols will most likely be used, where only one transmitter can be active within the corresponding network section. However, the number of receivers can be very high—up to several hundred. For this reason, the effects of receivers on the network properties, particularly on the network impedance and the transfer function, must be kept as small as possible. A large number of receivers rigidly coupled to the network would reduce the network impedance, thus unnecessarily burdening the transmitter.

Figure 4-8 shows the circuit of a coupling transformer with highpass effect. Capacitor C separates the mains voltage, i.e., about 230 V in Europe will drop off at this capacitor. It has to be suitable for permanent operation at 250 V AC and offer a DC endurance of at least 1000 V. On the transmitter side, C is normally selected between 1 and 2 µF, while a value of 100 nF is sufficient on the receiver side. Calculation examples will follow later. Resistance R is made as small as possible on the transmitter side (rigid coupling). It is essentially composed of the ohmic resistance of the transformer's winding on the mains side and the resistance of a fuse (not shown in Figure 4-8), which is always used in practical applications for safety reasons, particularly for signal feed at the crossbar system of a transformer station. On the transmitter side, R should be much smaller than 1 Ω.

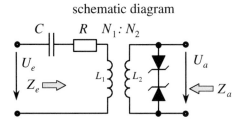

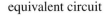

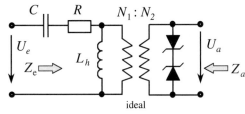

Figure 4-8 Coupling equipment schematic diagrams.

On the receiver side, R contributes essentially to the implementation of weak coupling. For this purpose, R is selected much larger than the mains impedance. In practice, values between 100 and 150 Ω have proven to be suitable.

When dimensioning the inductors L_1 and L_2, care should be taken that their impedance $(2\pi f_i L_{1,2})$ is always much higher than the connected resistances in all operating frequencies f_i. In the calculations, we can mainly concentrate on the mains side, i.e., on the correct dimensioning of L_h in the equivalent circuit shown in Figure 4-8. The reason is that it is desirable to have a small output resistance ($\approx 0\ \Omega$) on the transmitter side which is mainly influenced by the transmitter's output stage. By selecting the turns ratio $N_1:N_2$, the voltage supplied by the output stage is transformed appropriately, so that the desired signal amplitude occurs in the mains, which is generally the maximum admissible signal amplitude. $N_1:N_2$ is normally in the range 1/2–1/4.

On the receiver side, the transformer is usually followed by an active bandpass filter, the input resistance of which can be optionally selected within wide limits. In practice, values in the range from 5 to several hundred ohms have been successful. This means that the $N_1:N_2$ ratio can normally be set to one.

Ring cores made of ferrite material have generally proven suitable for the transmitter design. They offer an almost ideal guidance of the magnetic field with a very low stray field. This is an important prerequisite for the implementation of rigid coupling. A large number of trials with other core forms led to unsatisfactory results. Figure 4-9 shows three typical designs with corresponding windings as they have been used in

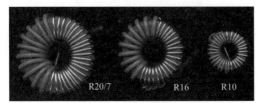

Figure 4-9 Transformer designs for energy-related value-added services.

many modems for energy-related value-added services in field trials. The labels R10, R16, and R20 give the diameters in millimeters.

We see that winding of the transformers can be relatively compact and simple. The figure also shows that it is possible to achieve a good insulation of the windings, so that modems can be operated on the mains without risk. Table 4-2 summarizes the most important technical data for the three cores shown above. This information will be used in the following sections for some important studies with regard to saturation effects and maximum transmittable signal amplitude.

Table 4-2 Important data on ferrite ring cores for powerline applications.

	Core Type		
	R10	R16	R20/7
Outer diameter	10 mm	16 mm	20 mm (height: 7 mm)
Material designation	N30	N27	N27
A_L value in nH/N^2	1760	1290	1930
Eff. magn. cross section	$A_{e10} = 7.83$ mm^2	$A_{e16} = 19.73$ mm^2	$A_{e20} = 33.63$ mm^2
Saturation induction	$B_1 = 380$ mT	$B_2 = 480$ mT	$B_3 = B_2 = 480$ mT

Core R10 is suitable for the receiver side, while R16 and R20 are both used on the transmitter side. Core R16 is generally usable, but it is operated close to its limits when utilizing the upper limit values of EN 50065. In these cases, it is recommended to use Type R20 or even a larger core.

For the mains-side number of turns, a value of $N_{1e} = 20$ for the receiver side and a value of $N_{1s} = 10$ for the transmitter side seem to be good practical choices. The reasons and possible tolerances for a meaningful definition of the numbers of turns are described in the following. First, we describe the receiver side. This concerns mainly the task to effectively suppress low-frequency spectral portions below about 20 kHz. It is therefore a good idea to design the coupling transformer choosing the R and C values (see Figure 4-8) so that we have a highpass filter with a 3-dB limiting frequency of 20 kHz. This is the case when $R = 120\ \Omega$, $C = 100$ nF, and the number of turns $N_{1e} = 20$. With an A_L value of 1760 for core R10, this results in an inductance $L_{1e} = 400 \cdot 1760$ nH $= 704\ \mu$H. At about 19 kHz, a series resonance results, at which the minimum input resistance, i.e., $R = 120\ \Omega$, appears. This value is generally higher by a multiple than the mains impedance at 20 kHz, which will be presented later based on measurement results. Toward higher frequencies, i.e., in the frequency range used for information transmission, the input impedance grows and takes inductive character. Toward lower frequencies, the impedance also grows, but with a capacitive character. At the mains frequency (50 or 60 Hz), virtually only the reactive impedance of the capacitor is decisive; it is 31.83 kΩ for 50 Hz, so that the mains voltage (230 V) drops almost completely across this capacitor. The amplitude of the current, which now flows at 50 Hz, is thus

$$\hat{I}_{1e} = 230\ \text{V} \cdot \sqrt{2} \cdot 2\pi \cdot 50\ \text{s}^{-1} \cdot 0.1 \cdot 10^{-6} \frac{A_s}{V} \approx 10.22\ \text{mA}$$

This value is required further below to calculate the premagnetization of the coupling transformer core. We need to ensure that the induction caused by the 50-Hz current remains far below the saturation induction given in Table 4-2. The major part of the admissible induction has to be available for the transmitter signal.

In this connection, the requirements are higher on the transmitter side—first, because it requires a rigid coupling, which demands a considerably larger coupling capacitor C on the mains side; second, because the transmitted amplitude is generally much larger than the received amplitude. The following example is a transmission coupler with core type R20 from Table 4-2. With a number of turns of $N_{1s} = 10$ on the mains side, we obtain inductance $L_{1s} = 100 \cdot 1930$ nH $= 193\ \mu$H. In the schematic diagram (according to Figure 4-8), resistance $R = 0$ and capacitor $C = 1\ \mu$F were selected. At 50 Hz, this capacitor has an impedance of 3183 Ω, which results in a current amplitude of

$$\hat{I}_{1s} = 230\ \text{V} \cdot \sqrt{2} \cdot 2\pi \cdot 50\ \text{s}^{-1} \cdot 10^{-6} \frac{A_s}{V} \approx 102.2\ \text{mA}$$

i.e., ten times the value of the receiver coupler. Although with 1 μF, the capacitor is pretty big and rather expensive due to the required high-voltage capability, it still has an impedance of almost 8 Ω at 20 kHz. Much smaller impedance values can, however, occur at many points of the mains, so that another increase of the coupling capacitance—generally by connecting several capacitors in parallel—should be considered in the individual case. However, the impact of the increase of the premagnetization caused by this should be exactly observed. Because of its general significance, this critical issue is discussed in detail below.

In general, the relationship

$$\hat{\Psi} = L \cdot \hat{I} \tag{4.7}$$

applies to any coil with inductance L, i.e., the amplitude of the magnetic trunk flux is proportional to the amplitude of the flowing current. With the ring cores of interest here, an ideal course of the magnetic field with good approximation can be assumed; i.e., the field lines follow the core form and the flux is equally distributed over the cross section. Consequently, the following relationships apply:

$$\hat{\Psi} = L \cdot \hat{I} = N^2 \cdot A_L \cdot \hat{I} = N \cdot \hat{B} \cdot A_e \tag{4.8}$$

This means that inductance L, calculated from the square of the number of turns N, multiplied by the A_L value, and the magnetic trunk flux are the N-fold of the flux Φ in the core cross section, which results from the product of induction B and effective magnetic cross section A_e, according to Table 4-2. Equation (4.8) can now be solved for induction B, and by inserting the corresponding units, we obtain the following tailored equation:

$$\frac{\hat{B}}{mT} = \frac{N \cdot A_L \cdot \frac{\hat{I}}{A}}{\frac{A_e}{mm^2}} \tag{4.9}$$

We can now use Equation (4.9) to calculate the premagnetization through the 50-Hz current. For the receiver coupler with core R10 in our example, we obtain $\hat{B}_{ve} \approx 46$ mT, and $\hat{B}_{vs} \approx 58.6$ mT results for the transmitter coupler with core type R20. We can see that these values remain sufficiently below the respective saturation induction so that only minor limitations result for the transmittable signal amplitude. This point will be described below.

The following applies to the voltage induced in a ring core coil:

$$\hat{U}_i = \frac{d\psi}{dt} = \frac{d}{dt}\{B(t) \cdot A_e \cdot N\} = 2\pi f \cdot \hat{B} \cdot A_e \cdot N \tag{4.10}$$

if the induction $B(t)$ changes sinusoidally with the frequency. Equation (4.10) can be used to directly calculate the maximum admissible voltage at a transformer winding, depending on the admissible maximum induction $\hat{B}$ as a function of frequency f. This shows that the voltage increases linearly with the frequency. This is a generally known fact: the higher the frequency, the smaller and more compact a transformer can be designed.

For an exact determination of the maximum transmittable voltage, the premagnetization through the 50-Hz current calculated above has to be considered; i.e., it has to be deducted from the saturation induction. The available induction for the R10 core is then

$$\hat{B}_{ve} = (380 - 46)\text{mT} = 334\text{mT} \qquad (4.11)$$

and for the R20 core of the transmitter coupler we obtain

$$\hat{B}_{ve} = (480 - 58.6)\text{mT} = 421.4\text{mT} \qquad (4.12)$$

The next two figures show the admissible transmission voltage (mains side) for the receiver and transmitter coupler, respectively, of our example as a function of frequency.

One can easily see that the admissible transmission amplitude varies within a large range of about 7–50 V. Of course, this is a consequence of the broad frequency range, extending from 20 to 150 kHz, more than seven octaves. The critical point is always the lowest frequency to be transmitted. Once it is ensured that no saturation can occur there, one is on the safe side for the rest of the transmission band. The coupler shown in Figure 4-10 is usable for all EN 50065 applications, except if one wants to transmit below 20 kHz.

As described in an earlier section, the very low frequency range is of little interest for two reasons. First, the mains impedance is extremely low here, that is, normally smaller than one Ω; second, the noise-power density present from the mains operation is very high.

The transformer considered in Figure 4-11 is intended for transmission purposes. Here, too, we can see the wide variation range of the usable transmission amplitude. In addition, we can see the existing reserves. At 20 kHz, the transmittable amplitude is about 18 V.

The admissible maximum amplitude is 10 V when the limit given by EN 50065 is fully utilized. At the highest frequency (about 150 kHz), even more than 130 V could be transmitted without threat of saturation. For numerous applications that do not go down to 20 kHz, smaller ring cores could be used on the transmitter side, e.g., type R16 from Figure 4-9.

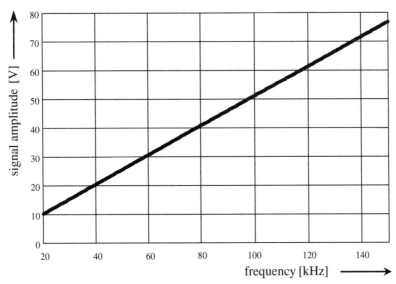

Figure 4-10 Signal amplitude over the frequency for a receiver coupler with ring core R10 and $N_{1e} = 20$ turns on mains side.

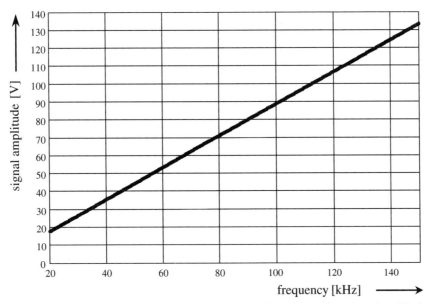

Figure 4-11 Signal amplitude over the frequency for a transmitter coupler with ring core R20 and $N_{1e} = 10$ turns on mains side.

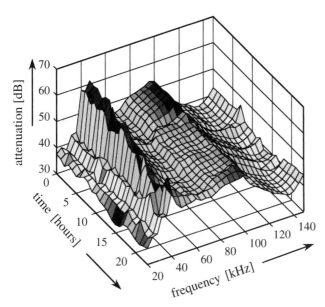

Figure 4-12 Long-term recording of attenuation.

4.2.2 Transmission Attenuation

Hundreds of courses of the transfer function were recorded and evaluated within the scope of extensive studies in various network structures (see Section 8.7.1 of the Bibliography). This section provides a summary of the characteristics based on Figure 4-12.

Figure 4-12 shows a typical three-dimensional graphical representation of the attenuation over the most important portion of the spectrum specified in EN 50065, over a period of 20 hours. The maximum attenuation values are at about 65 dB. Depending on the length of a path, attenuations of more than 80 dB can also occur. Figure 4-12 shows also that both the frequency and the time have a significant influence on attenuation, in addition to the path length. One can see frequency-dependent variations up to over 30 dB and time-dependent variations of about 10 dB. The extreme values measured at certain paths can be even more significant, i.e., there can either be very sharp notches,[2] or the transfer function can be "oblique." In this case, the attenuation generally increases toward higher frequencies, so that one has a channel with low-pass character. This issue will be discussed in Chapter 5.

For data transmission, the properties of the powerline channel described here mean that it is necessary to use modulation schemes resistant to various types of

---

[2] Strong attenuation over a narrow frequency range in the form of a gap.

frequency-selective attenuation, and which do not fail even when relatively broad ranges of the transmission spectrum are temporarily not usable.

4.2.3 Access Impedance

The transmitter of a modem has the task of injecting a voltage into the mains that reaches at most an amplitude as specified in EN 50065. The required transmission power can be calculated easily when the input impedance of the mains is known. The smaller the impedance, the more power is required. Of course, the most unfavorable conditions are found at the crossbar system of a transformer station. Among other factors, the parallel connection of a large number of outgoing trunks causes very low impedances. Figure 4-13 shows how the impedance depends on frequency. There is a variation range of about 0.2–2 Ω (see also Section 8.7.1 in the Bibliography). This means that we have to provide powers in the 25–250 W range if we want to achieve a transmission amplitude of 10 V.

This leads to very costly transmitter output stages in practical applications. Moreover, the design of the coupler components, in particular the capacitor, leads to higher costs. In some implementations, which will be presented in Section 4.4, the developers did not go beyond a maximum transmission power of 100 W for cost reasons and tried to select the frequency range used as far up as possible, i.e., generally above 50 kHz. The evaluation of an extensive database with impedance measurements confirmed that the mains access impedance increases always at a mean from 10 to 150 kHz. Figure 4-13 also shows the time dependence of the access impedance, in addition to its frequency dependence. As mentioned in an earlier section, the measurement location has a considerable influence on the value of the access impedance. In a transformer station, the value is much lower than it is at a wall plug inside a building. Also, it is fortunate that the typical house service connection has a much higher access impedance than the crossbar system of a transformer station, so that transmitter output stages, which can supply a maximum power of a few watts, will be sufficient for the operation of modems on the electricity customer side.

4.3 Modulation and Access Methods for Use under EN 50065

The transmission of digital information uses a process called "modulation," which transforms a data stream, consisting of logical ones and zeros, into a signal suitable for the transmission over a specific medium. This process involves a shift in the frequency range. The original data bits are lowpass signals,[3] which occupy a spectrum around the

---

[3] By definition, the spectrum has to reach down to the zero frequency for lowpass signals.

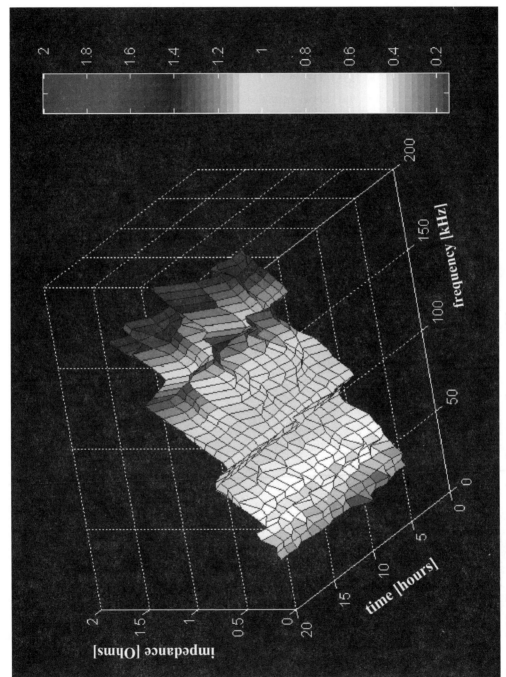

Figure 4-13 Course of the access impedance over frequency and time.

zero frequency that corresponds to approximately twice the data rate. Only bandpass signals are basically suitable for powerline communication, with the lowest spectral part being much higher than the zero frequency. In particular, a considerable gap to the mains frequency (50 or 60 Hz) has to be maintained. A suitable carrier with high frequency matching the data signal is modulated to generate a bandpass signal. The modulation results in a spectrum in the close neighborhood of the carrier, which transports the data in various forms. Some common modulation methods were briefly presented and described in Chapter 3. The result of a modulation process can be a spectrum which either is narrow, that is, the bandwidth corresponds approximately to twice the data rate, or is very broad, which means that the bandwidth is a multiple of the data rate. In the latter case, we speak of a band-spreading modulation (see Section 8.4 in the Bibliography).

Narrowband modulation is always the better choice when a "friendly" transmission channel is to be used with optimum spectral efficiency. Typical examples are radio relay systems or radio and television broadcasting, both terrestrial and over satellites. In the transmission channels used in these applications, the transfer function is virtually constant over the frequency, i.e., all ranges of the available spectrum are always equally well usable. And there are no interferers, except for background noise, which is very weak, compared to the transmission power, and which is generally white; i.e., it has the same power density at all frequencies.

Broadband modulation methods are characterized mainly by the fact that the information is transmitted at high spectral redundancy, i.e., with a much higher bandwidth than would be required by the data rate. This means that resistance against narrowband interferences and against selective attenuation effects can be achieved.

4.3.1 Narrowband Modulation Schemes and Their Properties

The modulation schemes described in this section are based on a sinusoidal carrier signal. Such a carrier signal is characterized by three parameters—amplitude A, frequency f, and zero phase φ:

$$s(t) = A(t) \cdot \sin[2\pi f(t) \cdot t + \varphi(t)] \qquad (4.13)$$

Equation (4.13) includes all basic possibilities of modulation. In the amplitude modulation (AM), a time-dependent change for the carrier amplitude $A(t)$ occurs according to the data signal. In the simplest case there are only the two values zero and one for $A(t)$, i.e., a zero data bit disables the carrier, while a one data bit enables it. This method is called *amplitude shift keying* (ASK).

If we describe the digital data stream to be transmitted by

$$d(t) = \sum_{i=-\infty}^{\infty} b_i \cdot \text{rect}\left(\frac{t - iT_b}{T_b}\right) \qquad (4.14)$$

where b_i is the "data vector" consisting of zeros and ones, and T_b is the bit duration, then ASK can be described by the following formula:

$$s_{\text{ASK}}(t) = d(t) \cdot A \cdot \sin(2\pi ft) \qquad (4.15)$$

ASK is a "historical" modulation method for data transmission over powerlines. It was used in early RCS methods [D3]. Moreover, integrated circuits were produced and marketed for ASK for bidirectional data transmission with low transmission powers in the EN 50065 frequency range [D3]. This means that low-cost and easy operation of in-house data transmission is possible, and with a little luck, links can be found that function more or less without problems over longer periods of time.

In general, however, ASK is not deemed suitable for low-power data transmission over powerlines. As soon as there are major disturbances, one has to expect failure of the transmission links, because added noise has a direct effect on the received signal amplitude, which contains the information. For example, interference can lead to a situation where the receiver gets a more or less large signal although zero bits are transmitted, for which the carrier is disabled. In other words, even at moderate attenuation, a clean separation of zero and one bits in the receiver is no longer possible. Figure 4-14 shows a simple signal example, where the upper rectangular function represents the digital data signal and the lower curve represents the modulated bandpass signal to be

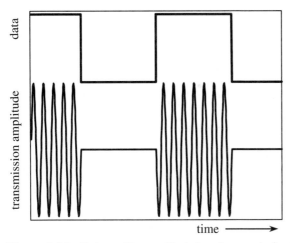

Figure 4-14 Data and transmitted signal example for ASK.

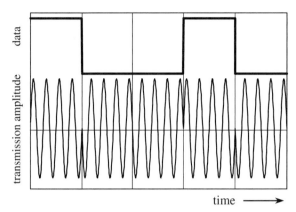

Figure 4-15 Data and transmitted signal example for BPSK.

coupled to the mains. Obviously, the major drawback of this simple modulation is the fact that nothing is sent during the zero bits. Figure 4-14 also shows that ASK is not a modulation with suppressed carrier. This is another drawback, because a certain part of the allowable transmission power is used for the transmission of a carrier signal portion that does not carry any information.

Phase-modulation methods achieve much better results. One of these methods, called *binary phase shift keying* (BPSK), is particularly simple and efficient. In BPSK, the binary data stream modulates the zero-phase angle of the carrier in such a way that the carrier phase is switched between 0 and 180 degrees, corresponding to the logical states of the data signal. Figure 4-15 shows a signal example.

BPSK is not particularly sensitive to amplitude or frequency influences, and it is the least sensitive to broadband noise among all narrowband modulation schemes discussed here. The reason is the large "Euclidean distance" between the two states within the signal space [G14]. On the other hand, BPSK could be sensitively disturbed, if fast phase hops were to affect the modulated carrier on its way from the transmitter to the receiver in the mains. Unfortunately, although such effects are rather infrequent in usual interference scenarios, they cannot be totally excluded. Using Equations (4.13) and (4.14), we can describe a BPSK signal as follows:

$$s_{\text{BPSK}}(t) = A \cdot \sin[2\pi f t + d(t) \cdot \pi] \qquad (4.16)$$

In Figure 4-15, the rectangular signal represents the data to be sent. The 180-degree phase hops at the bit borders are clearly visible in the transmitted signal. BPSK is a carrier-suppressing modulation, which means that the entire transmission power is available for the information-carrying spectral parts [F10, F12]. BPSK has hardly been used in simple systems that use the EN 50065 range. One reason may be the rather costly phase detection in the receiver. Most implementations prefer *frequency shift keying* (FSK), which will be described below. In future telecommunication systems, BPSK will certainly gain more significance.

One advantage of FSK is that transmitter and receiver systems can be easily implemented in both analog and digital technologies. Already in the eighties, a number of manufacturers offered chip sets for powerline applications (see [D3], for example). While analog technologies dominated initially, digital technologies changed the landscape more and more at the beginning of the nineties. This means that, initially, simple oscillators were used on the transmitter side and PLLs (phase-locked loops) were used for signal detection. Later, transmitters were increasingly implemented in the form of digital frequency synthesizers of various kinds, while receivers were equipped with digital correlators. Today, several transmission systems based on FSK (with certain modifications, which will be described later) are produced in series and used in large numbers, both for building automation and for energy-related value-added services offered by PSUs. Figure 4-16 shows the principle of FSK based on a data-signal and a transmitted-signal example.

FSK changes the carrier frequency (bottom signal) suddenly, corresponding to a logical zero or one in the data stream (top rectangular signal). This example sends a low frequency for zero and a high frequency for one. The distance between the two frequencies can be optionally selected within a wide range for a digital system implementation. The lowest possible frequency distance is the value of the data rate; i.e., the two frequencies must be 4.8 kHz apart at 4.8 kbits/s. The distance may also be an integer multiple of the data rate. This offers advantages (see Section 4.4) for certain applications. Frequency gaps that are not integer multiples of the data rate are useless, because this would mean a loss of the "orthogonality" (see Figure 4-17) and compromise the fault-

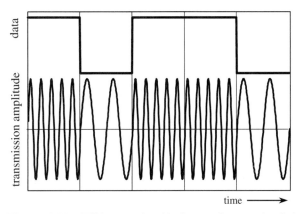

Figure 4-16 FSK example with data and transmitted signals.

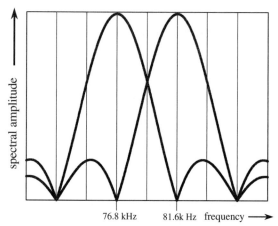
Figure 4-17 FSK example with orthogonality.

less detection in the receiver [F11]. If the distance between the two FSK frequencies corresponds to the data rate, then we obtain the lowest possible total bandwidth for the transmitter signal; it corresponds roughly to the triple data rate. Such a case is shown in Figure 4-17: data rate = 4800 bits/s, f_L = 76.8 kHz, f_H = 81.6 kHz. Orthogonality requires that all zero locations coincide within the spectrum. This example fills the transmission bandwidth relatively well, because half of the two spectra generated by modulation around the carriers overlap. We will return to this issue in Chapter 6, when we will discuss OFDM (orthogonal frequency division multiplexing). If the carrier distance is larger, the spectral efficiency drops, so that unused gaps form; but they can be desirable with regard to the interference resistance, as mentioned above.

It is usually a good idea to use the FSK modulation without phase hops at the data bit transitions, which means phase-continuous modulation (see Figure 4-16); otherwise, broadband spectral parts would form in the transmitted signal. Although these parts would hardly represent a drawback for the desired transmission, they could cause major problems with regard to electromagnetic compatibility due to out-of-band interferences. A particularly critical situation would be the operation of an FSK system in the D band. In compliance with EN 50065, a limit value of 66 dBµV (quasi-peak value in the 200-Hz bandwidth) has to be maintained, while use of the D band (up to 148.5 kHz) is admissible with a level of 122 dBµV (peak value in the 100-Hz bandwidth). Along this limit, there must be a sharp drop of the power spectral density of more than 55 dB, so that any measure taken to avoid undesirable spectral portions is welcome. The most expensive and worst idea would probably be the installation of filters at the transmitter output, rendering any type of implementation uneconomical. Phase-continuous FSK can be easily implemented with digital system concepts.

In general, FSK is a relatively robust transmission scheme, because corrupted frequencies on the transmission path are much less likely than corrupted amplitudes. Overlaid interference cannot affect the frequency of a transmitted carrier, as long as certain thresholds are not exceeded. Errors occur in FSK when one of the two frequencies is either extremely attenuated or overlaid by a strong narrowband interference. Such situations occur frequently in mains, as mentioned before. This is the reason why FSK cannot normally meet high requirements with regard to availability of a link and low bit error rate. Still, it was possible to implement relatively reliable data transmissions in building automation systems based on FSK. However, these implementations required modifications of the standard FSK signal scheme and additional measures, such as channel coding and special detection algorithms (see Section 4.4). Equation (4.13) and the introduced data vector b_i can be used to describe an FSK signal, consisting of zeros and ones, as follows:

$$s_{\text{FSK}}(t) = A \cdot \sum_{i=-\infty}^{\infty} \text{rect}\left(\frac{t - iT_b}{T_b}\right) \cdot [b_i \cdot \sin[2\pi f_1 t] + \bar{b}_i \cdot \sin[2\pi f_2 t]] \qquad (4.17)$$

Given by the data signal, i.e., at $b_i = 1$, a carrier at frequency f_1 is sent during bit duration T_b, while frequency f_2 is sent while $b_i = 0$.

4.3.2 Spread-Spectrum Techniques

Band-spreading modulation (see Section 8.4 in the Bibliography) was originally developed for military communication. In such applications, resistance against intentional or inadvertent interferers is achieved by high spectral redundancy. In addition, spread-spectrum techniques allow a transmitted signal to become almost "invisible" for outsiders. In the past, these techniques were characterized by extreme system costs, so that they were implemented only in special applications. Thanks to recent progresses in the field of microelectronic systems, spread-spectrum techniques have become feasible for almost any type of application and are being used now in applications like mobile radio, satellite communication, detection and ranging, and navigation. Their high resistance against all types of narrowband interference as well as selective attenuation effects make spread-spectrum techniques highly suitable for use in powerline communication. In addition, EN 50065 favors the use of broadband modulation. The following sections provide a brief overview of several spread-spectrum techniques used in mains networks, including their typical properties. Chapters 5 and 6 describe this issue in detail in connection with telecommunication systems.

4.3.3 Pseudonoise Direct Sequencing (PN-DS)

With this spreading technique, the sinusoidal carrier is generally modulated with the data signal using binary phase shift keying (BPSK), and then 0/180-degree phase hops based on a fast binary pseudonoise (PN) sequence are added. The latter generates a broad spectrum around the carrier, and the useful information contained in the data signal is distributed over this spectrum [S3, S4]. On the receiver side, there is initially a "despreading" with exactly the same synchronized binary pseudonoise sequence as was used in the transmitter. Subsequently, the data can be retrieved from the restored narrowband signal. It is important that all types of narrowband interferences, which overlaid the signal on its way from the transmitter to the receiver, are now subject to the spectral spreading process during "despreading" in the receiver, and can be filtered out before the narrowband demodulation process. These issues will be described in detail in Chapter 6 in our discussion of CDMA (code-division multiple access) for powerline communication.

An important parameter for the performance of a PN-DS system is the ratio of the bandwidth of the transmitted signal to the bandwidth of the data signal. This ratio is called processing gain, and it is normally above 1000 for military applications. In the mains, the maximum processing gain would be only 18 with a data rate of 2400 bits/s and a bandwidth of about 85 kHz (A band). It can be shown that a small processing gain like this is not useful in practice, because most of it will be used up by "implementation losses," so that the benefits over narrowband technologies would be lost. A detailed description of this issue would go beyond the scope of this book. We will content ourselves with a summary of the two most important points:

1. *Synchronization of the pseudonoise sequence on the receiver and transmitter sides.* This cannot be achieved without timing errors. In particular with heavily disturbed received signals, when the processing gain is to be fully utilized, both the acquisition and the tracking required to produce and maintain synchronism are very critical and error prone.

2. *Influence of the carrier phase from consumers and interference sources in the mains.* Assuming a broad frequency band, a highly distorted and time-variant phase response can be observed in some mains networks, leading the despreading in the receiver to be extremely poor, although the synchronization may be good. This can mean a much smaller effectively usable processing gain than theoretically expected. Suitable channel equalizers could be used to avoid this problem; unfortunately, such equalizers are very costly, and they would have to be adaptive

due to the time variance. Chapter 6 discusses this issue in connection with telecommunication systems.

4.3.4 Frequency Hopping (FH)

This section discusses *frequency hopping* (FH) in detail, as it forms the basis for a large number of system concepts for current and future powerline communication. The reasons for this FH preference will be explained gradually as the applications are discussed.

FH is a classical spread-spectrum method; it makes no use of a carrier signal with fixed frequency, but instead uses a large number of waveforms at various frequencies. The FH signal changes frequency in quick succession with the hop rate, h_r. The shorter the dwell time, $T_h = 1/h_r$, at one frequency, the less determinate is the FH signal, i.e., it appears more noise-like. The dwell time T_h is also called frequency validity interval or chip duration, depending on the context. An FH waveform with amplitude A and instantaneous frequency f_m is described by

$$s_{\text{FH}}(t) = A \cdot \text{rect}\left(\frac{t}{T_h}\right) \cdot \sin[2\pi f_m t] \quad (4.18)$$

The waveform in (4.18) occupies not only the spectral line f_m, but a continuous spectrum

$$s_{\text{FH}}(f) = A \cdot T_h \cdot \text{si}\left[\pi \cdot T_h \cdot (f - f_m)\right] \quad (4.19)$$

because frequency f_m is occupied only for a short time, T_h. $S_{\text{FH}}(f)$ is symmetric to f_m. As shown in Figure 4-17, this is basically also the case with FSK. In (4.19), si $(\cdot)$ is an abbreviation for sin $(\cdot)/(\cdot)$. Assuming matched filter reception, a transmission band with bandwidth B_g can be occupied by a maximum of

$$N_{\text{FH}} = \lceil B_g \cdot T_h \rceil - 1 \quad (4.20)$$

FH waveforms. $\lceil \cdot \rceil$ is the integer portion of $(\cdot)$. This means that the smallest admissible frequency distance equals the hop rate h_r. We speak of a set of orthogonal waveforms, because a receiver equipped with matched[4] filtering is perfectly capable of receiving any desired waveform from this set; i.e., it can supply the maximum of the autocorrelation function for the desired waveform, while all other waveforms of the set are perfectly suppressed. An orthogonal FH waveform set, which is grouped around a "center

---

[4] Explanations will follow in connection with matched-filter receivers in Section 4.4.

frequency" f_0, located in the center of the transmission band with bandwidth B_g, for example, with odd N_{FH} according to (4.20), is described by

$$s_{FHi}(t) = A \cdot \text{rect}\left(\frac{t}{T_h}\right) \cdot \sin\left(2\pi\left(f_0 + \left[i - \frac{N_{FH}+1}{2}\right] \cdot h_r\right)t\right),$$

with $i \in \{1, ..., N_{FH}\}$ (4.21)

For example, the waveform $s_{FH1}(t)$ has its spectrum symmetrical with the lowest frequency, $f_0 - [(N_{FH} - 1)/2]h_r = f_0 - (B - h_r)/2$, located at the lower band end. The waveform with the frequency $f_0 + [(N_{FH} - 1)/2]h_r = f_0 + (B - h_r)/2$ is located at the upper band end, accordingly with $i = N_{FH}$.

FH can be thought of as an expansion of the above-discussed FSK modulation. The following example explains the relationship. In this example, FH uses not only two frequencies, but five, to transmit a data bit. We can imagine that the information (one data bit) is present in five separate discrete spectral positions. The benefits are obvious: interference or complete deletion in one or two of these positions cannot corrupt the data transmission. The data bit can be easily reconstructed in the receiver with a simple majority decision (three out of five).

Figure 4-18 represents an example with an "H" and an "L" data bit (in the upper part) and the respectively allocated frequency sequences (in the lower part). In this example, the "H" bit is represented by a sequence of ascending frequencies, $f_1, f_2, f_3,$

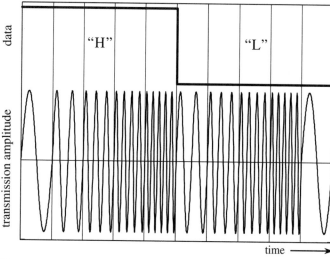

Figure 4-18 Data and transmission signal example for FH with five frequencies per bit.

f_4, and f_5, which are sent consecutively in fixed time slots (chip duration), corresponding to one-fifth of the bit duration.

The frequency changes abruptly at the chip limits, without a transient and without a phase hop.

To better understand, we have selected the frequency values for this example so that the number of oscillations fitting into a time slot increases by one from one time slot to the next. This means that the first time slot contains one oscillation and the fifth contains five. Identical frequencies were used for the "L" bit, but in another time sequence, i.e., f_2, f_3, f_4, f_5, and f_1, in the above example. Note that this is not the only option, because when thinking of the five frequencies as binary variables, we can compose $2^5 = 32$ different combinations, of which only two are used here. This high redundancy is eventually responsible for the high resistance to interference that can be achieved. When selecting the two combinations for the "L" and "H" bits, however, we have to observe that the frequencies in the respective time slots of the data bits are always different in order to achieve optimum resistance to interference. Regarding this side condition we still obtain more than two combinations, so that we could transmit more than one bit with the five frequencies, without compromising the resistance to interference. The following section studies this aspect in connection with the introduction of a new scheme called *modified frequency hopping* (MFH).

Figure 4-19 represents the spectrum of the sum of the five frequencies. This spectrum does not normally occur in practice, because the frequencies are sent consecutively and not concurrently (in contrast to OFDM, described in Chapter 6).

In principle, global synchronization allows five transmitters to share the frequency resources. We see in Figure 4-19 that the spectra at the individual frequencies are not falsified despite signal overlapping. The reason is the perfect orthogonality; i.e., all zero

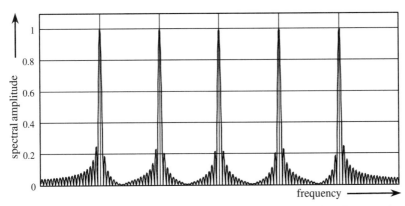

Figure 4-19 Magnitude spectrum of a sum of FH signals at five frequencies.

locations of the spectrum coincide. A receiver equipped with matched filters can separate and evaluate the transmitted waveforms optimally (see also Section 4.4).

The use of more than five carriers, or further spreading, can achieve an even higher resistance to interference. In general, FH offers the possibility of a very broad spectral spreading, where the spectrum does not necessarily have to be coherent, and where no pseudonoise (PN) sequence with high clock rate is required, in contrast to DS. This simplifies the synchronization problem enormously. Owing to this fact, mains networks allow even synchronization with the mains voltage as a globally available reference (see Section 4.4). The success of FH and derivative technologies for powerline communication is ascribed mainly to this possibility, among other advantages.

FH can be thought of as a method with fair frequency economy despite the high redundancy, as shown in the following example. Figure 4-19 shows the total spectrum under full utilization of a certain bandwidth B_g. If a hop rate of h_r = B/100 is selected in (4.20), then we obtain 99 orthogonal FH waveforms with frequency distance h_r, which, together, occupy the total bandwidth $B_g = 100 h_r$. Figure 4-20 shows the good spectral efficiency of FH. We see a virtually rectangular curve with a steep slope along the band limits. This means that the available bandwidth is well utilized by FH. In addition, FH lets us occupy noncontiguous portions of the spectrum, for example to avoid interference in a targeted way, or to exclude ranges with strong attenuation, or to skip certain frequency ranges in view of regulation and frequency allocation rules. This last issue will be discussed in detail in Chapters 5 and 6, where it will also become

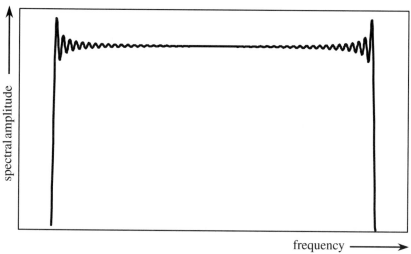

Figure 4-20 Example demonstrating the spectral efficiency of FH.

clear how closely FH is related with OFDM, the preferred method for powerline telecommunication.

To better understand later discussions, it is important to fully understand that, in order to fill a bandwidth B_g by use of FH, a hop rate of $h_r = B_g/2$ is by no means required, as one may assume in connection with PN-DS; in fact, $h_r = B_g/100$ or less is sufficient.

Finally, we will look at the difference between "fast" and "slow" frequency hopping (FH). Fast hopping distributes the information carried by a data bit over the available frequency band so that n FH waveforms from one orthogonal block are sent consecutively during the data bit duration T_b. "H" and "L" bits are generally distinguished by the sequence of these n frequencies. In fast hopping, the hop rate h_r is always an integer multiple n of the data rate $r_D = 1/T_b$.

Slow hopping transmits several data bits in one frequency validity interval with duration T_h. To achieve resistance to interference, the same data bits are repeated at one or several other frequencies. When using slow FH over problem channels like the mains network, there is a risk that phase fluctuations in the received signal can occur within one chip duration T_h (which is now fairly long in comparison with fast hopping). This leads to a degradation of the useful signal in the receiver, and possibly to complete disappearance of the desired autocorrelation maximum in the worst case [D3]. This is the reason why fast FH is the preferred method in mains networks, in particular for the low-voltage level, despite its higher requirements for synchronization accuracy. When robust data transmission over powerlines is the issue, then the generally lower requirements for synchronization precision are not sufficient to argue in favor of FH versus other methods, such as phase hopping (PH).[5] The transmission properties and the interference load of a channel supply additional arguments in favor of FH.

If we look at a powerline channel and its typical transmission properties, and if we assume that white Gaussian noise without mean value is the only interference that can occur, then we can achieve an equally good transmission quality with both PH and FH, provided that optimum receivers are used. If, in contrast, nonwhite noise and/or non-Gaussian interferers are present, then FH proves to be superior to PH. In such cases, which are naturally the rule in powerlines, there is a time-variant and frequency-dependent signal-to-noise ratio at the receiver input. The use of FH allows to beneficially utilize the statistical differences of the signal quality—expressed by the signal-to-noise ratio—at different frequencies. The following simple example demonstrates how this is possible:

---

[5] Chapter 6 explains that CDMA is closely related to PH.

A PH receiver converts each interferer approximately into white noise during the spectral compression of the desired signal [S4]. This means that PH normally does not allow the complete elimination of a received interferer's effect. The power of a strong sinusoidal interferer with frequency f_{dist}, for example, is distributed symmetrically to f_{dist} over a bell-shaped range in a PH receiver. Some amount of the interference always falls into the range of the spectrally compressed useful signal and contributes to the degradation of the signal-to-noise ratio at the receiver output. When transmitting digital information, the interferer increases the probability of bit errors. The interference effect is obviously higher, the closer f_{dist} is to the center frequency of the desired signal. In contrast, FH allows us to easily distribute a message bit over v FH waveforms at various frequencies within the transmission band. No bit error will occur, as long as at least $\lceil v/2 \rceil + 1$ undisturbed waveforms reach the receiver.

The effect of one sinusoidal interferer—or more generally one narrowband interferer—remains limited to a narrow frequency range in the FH receiver. This means that it can never disturb more than one out of v waveforms. Consequently, an FH system can be absolutely resistant to interference from up to $\lceil v/2 \rceil$ narrowband noise sources at different frequencies, because it can almost perfectly fade these interferers out. In contrast, when using PH, a single sinusoidal interferer with a power exceeding the (still spectrally spread) useful signal power at the receiver input could cause the message transmission to fail. The superiority of FH becomes obvious when looking at the typical interference scenario in mains networks at the medium- and low-voltage levels, as described in Chapter 2. However, it does not make sense for an FH transmitter to densely occupy the available transmission frequency band, because this would mean that numerous frequencies would be subject to the same negative impact, for example in positions with frequency-selective attenuation. On the other hand, it would be an advantage to distribute the information of a message bit over distant frequencies to avoid the concentrated impact of selective attenuation and/or selective interference power maxima. PH does not offer these or similar advantages.

In practice, an FH transmitter will always use only a fraction v of the maximum number N_{FH} (see (4.20)) of frequencies available within a transmission band with width B_g to represent one data bit. In mains networks with global synchronization based on the mains voltage, multiple access can be implemented collision-free, if $v \ll N_{\text{FH}}$ [ST4, ST5, ST8]. For binary data transmission in FH, each transmitter requires two of the N_{FH} frequencies at any given time. This means that, with a bandwidth B_g and equal hop rate h_r for all transmitters, and using (4.20), a maximum of

$$K_{\max} = \left\lceil \frac{N_{\text{FH}}}{2} \right\rceil = \left\lceil \frac{1}{2} \cdot \left(\left\lceil \frac{B_g}{h_r} \right\rceil - 1 \right) \right\rceil \qquad (4.22)$$

FH transmitters can be concurrently active without colliding. We use a "historical" example to emphasize the astonishing multiple access capacity—expressed by the FH channel number $K_{\max}$ according to (4.22)—in the A band (9–95 kHz):

Assuming a data rate $r_D = 60$ bits/s for each channel and mapping each data bit to five FH waveforms, we obtain the hop rate $h_r = 300$ s^{-1}, so that $\lceil B_g/h_r \rceil = 286$ and $N_{\text{FH}} = 285$; consequently, $K_{\max} = 142$. This means that the A band could be used to operate 142 FH channels concurrently without collision, each of them providing a data rate $r_D = 60$ bits/s. This is particularly impressive, because no tradeoff has to be made with regard to resistance to interference, as each FH transmitter can distribute its waveforms at large spectral distances within the transmission band. Functional samples of such systems were built and tested in the field at the end of the eighties [ST4, ST5, ST8]. Two European patents were issued for the corresponding transmission schemes and system implementation [B1]. Today, this type of low-data-rate transmission is no longer acceptable, not even in the A band. In addition, there is no mandatory need for multiple access, either in the implementation of the energy-related value-added services, or in building automation. Master-slave principles [D5] with relatively high data rates of over 1000 bits/s have been implemented and accepted instead. In such a master-slave environment, there is always only one active transmitter, which can use the entire channel capacity during its activity. The following section introduces these state-of-the-art principles by the example of implemented systems for both building automation and energy-related value-added services.

4.4 Examples of System Implementations

4.4.1 Matched Filtering and Synchronization—Introductory Remarks

The previous section described mainly the transmitted signals of different modulation schemes. It made very clear that "frequency-agile" methods offer several advantages. To utilize these advantages, however, additional requirements like optimum receiver technology and signal detection have to be met in addition to the transmission technology. Before we describe the technical implementation of the example systems, we will take a look at the principles of optimum reception of frequency-agile waveforms identified above as being beneficial. The theoretical basics of matched filtering are amply described in numerous standard data communication textbooks, e.g., [F10–F12]. This section provides a brief summary of the most important results needed in the next sections.

A matched filter in the classical sense exhibits its optimum properties provided only that interference concerns additive white Gaussian noise, which can be described by constant power density N_0 over all frequencies. In that case, we also speak of an AWGN (additive white Gaussian noise) channel. Although hardly any practical channel exactly suits this model, it can still be applied in most cases, at least in part, by splitting the real channel into suitable subchannels. This methodology is used mainly for powerline channels.

It is quite easy to mathematically describe a matched filter (MF) with impulse response $m(t)$ for a waveform $s(t)$ of duration T:

$$m(t) = K \cdot s(T-t) \quad (4.23)$$

Note that K is a constant. The important point is that $s(t)$ has to be time-inverted in (4.23). The filter becomes "causal" from a shift of duration T; i.e., no output signal occurs before the corresponding input signal. Although (4.23) appears to be simple, the practical construction of matched filters is generally a costly task. A simplification can be achieved for frequency-agile waveforms, which are of particular interest for powerline communication (see Figures 4-16 and 4-18), because the waveforms are symmetric in terms of time, so that no time inversion is needed.

When a signal $s(t)$ of duration T reaches the matched filter in (4.23), then we obtain

$$s_a(t) = s(t) * m(t) = K \int_{-\infty}^{\infty} s(t-\tau) \cdot s(T-\tau) d\tau = K \int_{-\infty}^{\infty} s(t+\tau) \cdot s(T+\tau) d\tau \quad (4.24)$$

by convolution at the output. The right-hand term in (4.24) describes a correlation, or more exactly, an autocorrelation.[6] The mathematical descriptions of convolution and correlation differ, as we know, only by a sign for the integration variable τ. This difference disappears in the matched filter in that its impulse response is a time-inverted version of the signal, to which it is matched. Equation (4.24) shows that the maximum output signal results for $t = T$. $s_a(T)$ is also called the maximum of the autocorrelation function (ACF) and represents physically the energy E_s of the waveform. For a sinusoidal signal with amplitude A, we obtain $s_a(T) = (A^2/2)T$, if K is assumed to be 1 for reasons of simplification. Following the general custom, we will call the time $t = T$ the sampling instant in our further discussion; this is the time when the MF output signal is

---

[6] The two waveforms underneath the integral are identical, except for a time shift.

evaluated and further processed. If additive interference in the form of white Gaussian noise with power density N_0 reaches the filter, in addition to the desired signal $s(t)$, we will always have the signal-to-noise ratio

$$\frac{S}{N} = \frac{E_s}{N_0} \qquad (4.25)$$

on the filter output at the sampling instant $t = T$. We can interpret (4.25) so that an MF is capable of evaluating the entire useful energy that the receiver gets from the corresponding transmitter. The power density N_0 is an invariable parameter determined solely by transmission-channel properties. The ratio E_s/N_0 turns out to be the essential figure determining the bit error rate in the transmission of digital information [F12]; this issue will be explained in detail later. This means that, with known N_0, we can design a transmission system so that certain requirements with regard to the bit error rate (BER), c.g., BER $< 10^{-5}$, can be achieved, by selecting signal amplitude and waveform duration appropriately.

Equation (4.24) shows that the MF function requires the integration over the product of two time-dependent functions. In digital signal processing, where signals must be discrete in both amplitude and time, a corresponding summation has to be done, so that a digital multiplication operation, followed by an accumulation, is required; this issue will also be explained in detail below. Note that a first view of (4.24) may not immediately disclose that there is a synchronization problem. This problem occurs because the desired ACF maximum occurs only at the MF output when the two multiplied signals are synchronous. This synchronism must always be produced before starting with the calculation of (4.24), regardless of whether the MF implementation is analog or digital. The synchronization cost involved should not be underestimated; it can account for two thirds of the entire receiver cost in some cases. On the other hand, it can always be dramatically reduced, if a global reference signal with sufficient precision is available to the transmitter and to the receiver. In general, we distinguish between initial synchronization—also called "acquisition"—and the constant maintenance of synchronism by appropriate corrective actions—called "tracking." Acquisition is a problem particularly in environments with strong interference, such as the mains network. The mains voltage as a global reference in transmission systems in the A–D bands has been used very successfully. The zero-crossings of the mains alternate voltage have proven to be relatively precise reference instants for acquisition purposes, with a few limitations. In many cases the entire tracking part can also rely on this reference. In addition, a number of measures can be taken to significantly improve the reference quality, which will be described later.

This simple synchronization method is, however, not suitable for telecommunication applications. First, the precision of the zero-crossing instants due to the very short waveforms required here, in comparison to the mains period (20 ms), is no longer sufficient. Second, the signal propagation delay plays a role now, so that one can no longer speak of a "quasi-contemporaneity" within the communication network, which means that global synchronization is impossible. In fact, all synchronization steps have to be performed exclusively on the basis of the more-or-less interfered receiver signal, which means in general a high additional cost in the receiver.

4.4.1.1 Frequency-Hopping Issues

For optimum detection of FH waveforms on the basis of (4.21), a matched filter with impulse response

$$h_{\text{FH}\nu}(t) = A \cdot \text{rect}\left(\frac{t}{T_h}\right) \cdot \sin\left(2\pi\left(f_0 + \left[\nu - \frac{N_{\text{FH}}+1}{2}\right] \cdot h_r\right)t\right),$$

$$\text{with } \nu \in \{1, ..., N_{\text{FH}}\} \tag{4.26}$$

can be used on the receiver side. This matched filter can be implemented in many different ways. The first functional samples of powerline modems were implemented as analog designs in the latter half of the eighties [ST5]. Initially we will introduce two examples: the "deattenuated resonance circuit" and the "lock-in amplifier." We will then move on to the digital technology, which will be described in more detail because it is the standard approach for current and future systems.

As briefly mentioned above, we obtain a signal

$$g_{i\nu} = \frac{A^2}{2} \cdot T_h \cdot \text{si}\left[\pi(i-\nu)\right], \text{ with } i, \nu \in (1, ..., N_{\text{FH}}) \tag{4.27}$$

at the output of a matched filter from (4.26) at the sampling instant, i.e., at the end of a frequency validity interval with duration T_h. This is an equivalent lowpass signal (envelope). We see in (4.27) that $g_{ii} = g_{\nu\nu} = (A^2/2)T_h$ only for $i = \nu$; i.e., the output signal corresponds to the signal energy E_s, while $g_{i\nu} = 0$ is true in all other cases with $i \neq \nu$. This statement follows immediately from the orthogonality of the waveforms. Table 4-3 is an example of a set of orthogonal waveforms with 30 frequencies. In this example, the frequency distance corresponds always exactly with the hop rate $h_r = 4800 \text{ s}^{-1}$. It could also be an integer multiple of this. Various optional variants will be described in detail later. The last column of the table lists the "number of samples." This issue will be described later.

For now, it is sufficient to know that the signals listed in the table have to be generated digitally. Analog approaches would not make sense, because we require a large number of different, highly accurate frequencies. This could be implemented only by use of a quartz oscillator for each single frequency. Of course, such a cost would make the implementation totally uneconomical. An elegant solution used in many powerline modem designs is the cyclic read-out of suitable samples from a wave table stored in a digital memory. Table 4-3 gives the required number of such samples to be read at a fixed rate of 600 kHz. This means that only one stable frequency, i.e., this *sampling frequency* from a quartz oscillator, has to be provided, e.g., by frequency division. The samples retrieved from the memory give the desired waveform with high frequency stability after digital/analog conversion and lowpass filtering. This way of signal generation will be discussed in detail below.

Table 4-3 Block with 30 orthogonal waveforms at different frequencies; start frequency = 9600 Hz; hop rate = 4800 Hz; sampling rate = 600 kHz.

No.	Frequency in Hz	Number of Samples	No.	Frequency in Hz	Number of Samples
0	9600	125	15	81600	125
1	14400	125	16	86400	125
2	19200	125	17	91200	125
3	24000	25	18	96000	25
4	28800	125	19	100800	125
5	33600	125	20	105600	125
6	38400	125	21	110400	125
7	43200	125	22	115200	125
8	48000	25	23	120000	5
9	52800	125	24	124800	125
10	57600	125	25	129600	125
11	62400	125	26	134400	125
12	67200	125	27	139200	125
13	72000	25	28	144000	25
14	76800	125	29	148800	125

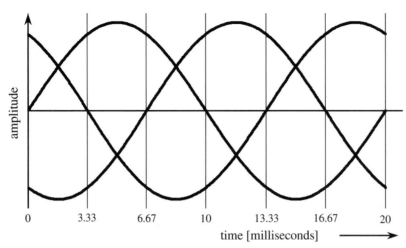

Figure 4-21 Temporal voltage courses in a three-phase system.

Today, an FH multiple access system built for applications such as measurement data transmission within buildings in 1986 can be considered historical [ST5]. The first modems used in such systems were presented at the Hanover Fair 1986 and offered the possibility to select among 30 channels for both the transmitting and the receiving direction. The modulation used was fast frequency hopping with three frequencies per bit. For this purpose, 60 orthogonal FH waveforms were available in the 30–146-kHz transmission frequency band. A global synchronization scheme based on the zero-crossings of the mains voltage was used. Because the system function had to also be guaranteed in three-phase supplies, where the zero-crossings occur each at a distance of 10/3 ms ≈ 3.33 at 50 Hz (see Figure 4-21), a "natural" data rate of $r_D = 300$ bits/s and hop rate $h_r = 900$ s^{-1} with $\xi = 3$ frequencies per data bit turned out as a result. We will first discuss a central—rather critical—problem, namely the transmitter/receiver synchronization. Subsequently, we will describe implemented systems.

4.4.1.2   Using the Mains Voltage as Synchronization Reference

A single modem "sees" zero-crossings always in a 10-ms raster (for 50 Hz), because it generally does not have a three-phase connection. The intermediate zero-crossings have to be generated locally, e.g., by use of appropriately programmed timers. This problem can be easily solved, because almost every modem contains a microcontroller with a sophisticated timer system. This means that the function of a global synchronization time raster is definitely based on the detection of 10-ms timing marks. We have already mentioned that this will work in practice as long as we have quasi-contemporaneity. The dimension of the network involved has to be limited, so that the propagation delay of electromagnetic processes is always much smaller than the duration T_h of a waveform.

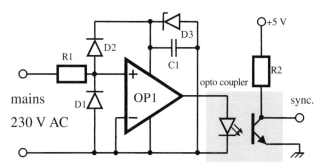

Figure 4-22 Extraction of the synchronization reference from the mains.

The simple circuit shown in Figure 4-22 has proven effective in practice for the preparation of the mains voltage as a synchronization reference with additional potential separation capability over an optocoupler. The operational amplifier OP1 is built in CMOS technology and works as a comparator. Due to its low current requirement, OP1 can be supplied directly from the 230-V mains over a large resistor R1 > 200 kΩ. Diode D2 serves for rectification, the Zener diode D3 serves for stabilization, and the capacitor C1 serves to filter the supply voltage for OP1. D1 protects the input of OP1 against excessive negative voltage. The optocoupler supplies a rectangular signal at 50-Hz frequency with a duty ratio of 1:1, which can be connected directly to the interrupt input of the microcontroller. The interrupt can be made edge-sensitive by programming. In this case, an interrupt is triggered in the 50-Hz mains every 20 ms. All required intermediate timing marks can be generated by means of timers integrated in the microcontroller.

As shown in Figure 4-21, most powerline networks are based on a three-phase system with three voltages, phase-shifted by 120 degrees. Ambiguities in global synchronization can occur due to the six zero-crossings, because one single modem sees generally only a 10-ms pattern, while there are three 3.33-ms shifted versions. On the other hand, if we divide the time between two zero-crossings (10 ms) into three identical time sections, then—at 50 Hz—we obtain a 3.33-ms time pattern, which has no ambiguity in a three-phase system. This approach was found useful in practice. If we allocate one data bit with duration T_b to each of the 3.33-ms intervals, then we obtain a good synchronization scheme that does not require any acquisition to find the beginning of a data bit. Figure 4-23 shows this scheme. We have a fixed data rate r_D, which is the six-fold of the mains frequency, i.e., r_D = 300 bits/s at 50 Hz. During the duration T_b of a data bit, ξ frequencies are sent, so that we obtain a chip duration of $T_h = T_b/\xi$.

Consequently, we obtain a chip duration of T_h = 1.1 ms in the 50-Hz mains for ξ = 3, which corresponds to a hop rate of h_r = 900 s^{-1}. The "historical" multiple-access system mentioned above was implemented exactly on this basis. The impact of syn-

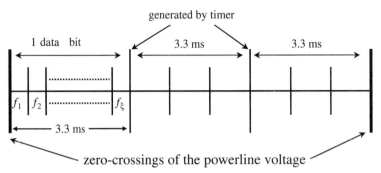

Figure 4-23 Time raster of an FH system for 300 bits/s.

chronization errors in the multiple-access environment emerged already at a small transmitter/receiver distance due to the relatively short chip duration. At that time, no sophisticated methods were available to improve the stability of the synchronization reference. An industrial realization of the system was out of the question due to considerable defects. In addition, an increase of ξ to improve the resistance to interference was impossible; because it would have further shortened the chip duration and reduced the stability of the reference (mains voltage), the global synchronization approach would tend to fail.

Larger distances of up to one kilometer, typical in low-voltage mains outside of buildings, require a different synchronization scheme than the one shown in Figure 4-23, because one has to work with longer chip duration due to unfavorable transmission and interference situations. Still, the division of a 10-ms interval into three equal parts is an advantage. Moreover, it would then be recommended to select a chip duration of $T_h = 3.33$ ms for the matched-filter receiver design. A detected zero-crossing of the mains voltage would then always coincide with the beginning of a chip. However, the beginning of a data bit is still ambiguous and cannot be found by use of the reference. To solve this problem, a fixed waveform sequence that does not depend on the message—a preamble—has to be sent in advance to mark the beginning of a data transmission for the receiver [ST12, ST15, D1, D2]. Once the beginning of a data bit has been determined correctly from the preamble, a data packet with a typical length of up to one hundred bits can be transmitted [D4, D5]. It is recommended to limit the data packet length to prevent loss of data if a synchronization loss is experienced, e.g., due to massive mains disturbances, such as a "brown out."[7]

Figure 4-24 shows an application example for the described synchronization scheme, in which one data bit has a duration of $T_b = 5 \cdot 3.33$ ms, resulting in a data rate

---

[7] Brief mains voltage decay of the kind that causes lamps to flicker.

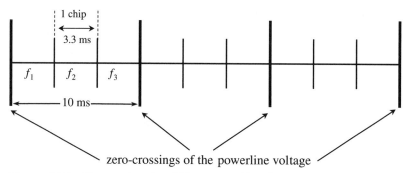

Figure 4-24 Time raster of an FH system with a fixed chip duration of 3.33 ms.

of $r_D = 60$ bits/s. This system concept was based on applications for remote power-meter reading using time-division multiplexing.

The prototype designs described in detail below were used in a 9-months field test over a 1-km transmission path, and they have proven effective in numerous trials over in-house mains with distances of over 100 m [D1].

Unfortunately, the mains voltage is not a reference signal with high precision. The instants of the zero-crossings of the mains voltage can deviate from the target values resulting from the 50-Hz or 60-Hz frequency due to interference, e.g., when large consumers switch on or off. Such a stochastic fluctuation, which is also called jitter, was measured in mains within buildings by way of example. Marginal cases showed jitter without mean value with standard deviations of up to 30 µs at the same phase. Depending on the measurement location, mean values of more than 100 µs were observed between different phases of the three-phase mains. The reasons are load-dependent neutral point shifts in the three-phase mains, so that the exact 120-degree phase relationship of the three-phase voltages is lost. When using the mains voltage as synchronization reference, one has to expect relatively large time errors of approximately ±130 µs over short transmission paths, i.e., <100 m. Over long distances, the delay due to electromagnetic wave propagation along the lines gains additional importance when energy and message flows are in opposite directions. This is naturally true in 50% of all duplex operation cases. With the type of ground cables normally used in low-voltage power distribution networks, the dielectric constant is $\varepsilon_r \approx 3$. In such cables, the propagation speed of electromagnetic waves is then the 0.577-fold of the speed of light. The result is a synchronization error of approximately 11.55 µs/km when energy and message flow in opposite directions. This means that synchronization errors due to propagation can be neglected over distances of a few kilometers. [N10] contains a theoretical study of the degradation in an FH system due to the impact of synchronization errors. This study was based on a global synchronization by means of mains voltage and matched-filter receivers. The salient figure for the signal quality is the respective signal-to-noise ratio at a matched filter receiver output. The most important findings are described briefly below:

If only one connection, i.e., one FH transmitter, was in operation at the network of interest, then the useful power at the matched filter output of the relevant receiver would decrease, as the synchronization error τ for a given waveform duration T_h would increase in proportion to the function $(1 - \tau/T_h)^2$. A synchronization error $\tau < T_h$ has no negative impact when there is no interference at the receiver input. In practice, however, there is always a degradation, because interference is always present at the matched filter output, leading to a noise power that is almost independent of τ. The useful power drops as τ increases, so that the signal-to-noise ratio defined as a quality measure also decreases. Still, for $T_h = 333$ μs and a relatively big synchronization error $\tau_1 = 130$ μs, for instance, composed of jitter and neutral point shift, we still obtain a degradation of only 0.345 dB. Assuming that $\tau_{max} = 0.1 T_h = 333$ μs is the upper limit for an acceptable synchronization error, the result is still a relatively low degradation of only 0.9 dB. Assuming that $\tau_1 \approx 130$ μs is an upper limit of the error due to jitter and neutral-point shift during zero-crossing detection, then an additional synchronization error of approximately 200 μs caused by propagation effects is admissible. This means that, with a delay time of 11.55 μs per kilometer, good global synchronization could be achieved over more than 17 km.

The impact of synchronization errors increases considerably in a multiple-access system with $Z_V \gg 1$ concurrently active transmitters. Normally, the worst case is considered when analyzing the impact of synchronization errors, where orthogonal waveforms are arranged at the distance of the hop rate h_r in an FH system. The worst-case conditions result for the transmission channel to be studied when the channel is roughly in the center of the transmission spectrum. For $\tau \neq 0$, the orthogonality of the waveforms can no longer perfect. The result is that the signals from undesirable neighboring transmitters no longer disappear at the receiver's matched filter output in the channel of interest. We obtain an interference power composed additively of two parts, N_1 and N_2, which can be calculated separately.

Under the simplified conditions that all Z_V studied transmitters are received with equal power, that no mains-generated interference is present, and that the same synchronization error occurs for all transmitters with regard to the receiver in the studied channel, then [N10] states an approach leading to the expressions

Examples of System Implementations

$$N_1 = \sum_{v=1}^{Z_V-1} |g_{T_h v}(\tau)|^2, \text{ with } g_{T_h v}(\tau) = \int_{\tau - \frac{T_h}{2}}^{\frac{T_h}{2}} e^{-j2\pi v \cdot \frac{t}{T_h}} dt \quad (4.28)$$

and

$$N_2 = \sum_{v=1}^{Z_V-1} |g'_{T_h v}(\tau)|^2, \text{ with } g'_{T_h v}(\tau) = \int_{\frac{T_h}{2}}^{\frac{T_h}{2}+\tau} e^{-j2\pi v \cdot \frac{t}{T_h}} dt \quad (4.29)$$

With a normalized useful power of

$$S_N = T_h^2 \cdot \left(1 - \frac{\tau}{T_h}\right)^2 \quad (4.30)$$

we finally obtain the (logarithmic) signal-to-noise ratio

$$a/\text{dB} = 10 \cdot \log_{10}\left(\frac{S_N}{N_1 + N_2}\right) \quad (4.31)$$

from (4.28), (4.29), and (4.30) at the receiver's matched filter output in the studied channel. Figure 4-25 shows the signal-to-noise ratio in dB as a function of the synchronization error τ/T_h normalized to the chip duration for four values of the number Z_V of simultaneously active transmitters.

It is obvious that the next neighbors of the studied channel contribute the largest interference. With a small synchronization error, e.g., $\tau \approx 0.001 T_h$, the signal-to-noise ratio deteriorates only by approximately 3 dB at the transition from 31 (curve b) to 65 (curve c) simultaneously active transmitters. With a very large number of transmitters $Z_V = 1000$, the signal-to-noise ratio drops by less than 10 dB. These results show clearly that the error impact decreases as the frequency distance increases. With higher values of τ, as they normally occur in practice, the b, c, and d curves approximate quickly and then merge at about $\tau \approx 0.03 T_h$.

Curve a shows that the signal-to-noise ratio drops to approximately 13 dB at $\tau = 0.1 T_h$ for only three channels. For example, detection problems can already arise if the desired transmitter signal arrives at the receiver only 10 dB weaker than the two undesired neighboring signals. The limit curve d for a high number of transmitters

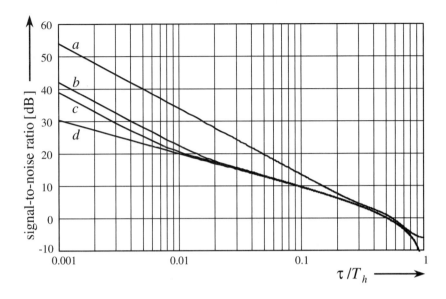

Figure 4-25 Signal-to-noise ratio as a function of the synchronization error with varying number Z_V of active transmitters: $a \Leftrightarrow Z_V = 3$, $b \Leftrightarrow Z_V = 31$, $c \Leftrightarrow Z_V = 65$, $d \Leftrightarrow Z_V = 1000$.

shows that, even with very small $\tau = 0.001\,T_h$, the signal-to-noise ratio can never be larger than 30 dB.

The following conclusions can be drawn from the above discussion. A close occupancy of the transmission band, e.g., with frequencies at the hop-rate distance, should be avoided in multiple-access systems with global synchronization by means of mains voltage, because the inaccurate reference will lead to considerable drop of the achievable signal-to-noise ratio on the receiver side, even when the hop rates are low and the distances are short. Note that this does not include interference generated by the mains operation, which is always present. If multiple access is required, then the frequencies for the transmitted waveforms should be selected in each chip interval so that they are as far apart from one another as possible. Thanks to global synchronization, each transmitter can distribute its information within the entire available frequency band. This simple type of synchronization entails the trade-off that the frequency band cannot be fully used at any given time. This means that it will be advantageous to select the frequency distance always larger than the hop rate, even with only one active transmitter. More precisely, even in an FSK system with only one frequency per bit, the synchronization error would lead to a loss of the perfect orthogonality (see Figure 4-17), and a signal will be present even at the nonaddressed receiver filter output, so that the decision which filter was actually addressed could be wrong. This is one reason why "Powernet-EIB," the building automation system described in a later section, introduces an FSK variant by the name of "Spread-FSK" (SFSK), where the frequency distance is a multiple of the hop rate.

The time-division multiplexing technology should ideally be used together with global synchronization by means of mains voltage, where only one transmitter may be active at a given time. In this case, even for $\tau \approx 0.1 T_h$, one will have to expect only a small degradation, i.e., less than 1 dB. In extensive supply networks between low-voltage transformer stations and the buildings connected to these networks, time-division multiplexing is virtually a must, e.g., for remote meter reading within energy-related value-added services. The individual buildings are arranged along a supply trunk with a maximum length of 1 km. With a receiver in the transformer station, level differences of up to 50 dB have been observed in the received signal, depending on whether the request served came from a building in the immediate neighborhood of the transformer station, or whether the consumer information comes from a building near the trunk end, i.e., the largest distance from the transformer.

4.4.1.3 A "Historical" Multiple-Access System with 30 Channels of 300-bits/s

This section introduces the historical 30-channel FH multiple-access system developed in 1986 [ST4, ST5] in connection with the synchronization scheme of Figure 4-23.

Figure 4-26 shows the front view of a device initially presented at the Hanover Fair 1986. The channel selector was designed for future expansions, so that more than 30 channels can be set. On the left side, there are a few flip switches for simple setting options. On the right side, there are a large number of BNC sockets used to observe the transmitter and receiver signals in various processing stages. Such control options have proven to be extremely helpful in this early phase of powerline modem development to analyze the channel properties and to check the function of the mostly analog system components.

A minimum frequency distance of 900 Hz was required due to the relatively short chip time of $T_h \approx 1.11$ ms for a set of orthogonal waveforms. Even relatively small synchronization errors would lead to considerable interference for neighboring channels. In addition, there are frequency errors, for example those originating from quartzes oscillating not exactly at their target frequencies in the transmitter and receiver frequency synthesizers. For this reason, the minimum frequency distance was defined as twice the hop rate, i.e., 1800 Hz; 32.4 kHz was selected as the lowest frequency, assigned to number 1. The frequencies numbered 2 through 60 follow in ascending order at a distance of 1800 Hz each. This means that frequency number 2 is 34.2 kHz, and the highest frequency, number 60, is 138.6 kHz. Table 4-4 shows the frequency allocation for the 30

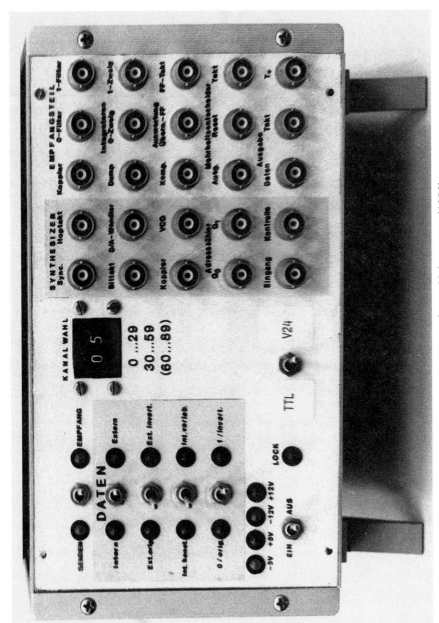

Figure 4-26 Front view of an early FH modem prototype for multiple access (1986).

channels in question; it is based on this numbering convention. We can see that no collision occurs in the frequency range, even when all 30 transmitters are active concurrently. This advantage stems from the global synchronization of all transmitters to a common reference. In general, and particularly in wireless transmission, global synchronization cannot be used, so that spectra will overlap, leading to a degradation of the system properties. However, the complete occupancy of the available transmission frequency band would not be a good idea, because synchronization based on the mains voltage is far from being perfect.

Table 4-4 Frequency allocation in a globally synchronized FH multiple-access system, including the "H" and "L" bit frequencies for the transmitter and the relevant "mixed frequencies" for the receiver for each of the 30 channels.

Channel Number	"H" Bit Frequencies			"L" Bit Frequencies			Receiver Frequencies		
Order →	1	2	3	1	2	3	1	2	3
1	11	31	51	1	21	41	60	40	20
2	12	32	52	2	22	42	59	39	19
3	13	33	53	3	23	43	58	38	18
4	14	34	54	4	24	44	57	37	17
5	15	35	55	5	25	45	56	36	16
6	16	36	56	6	26	46	55	35	15
7	17	37	57	7	27	47	54	34	14
8	18	38	58	8	28	48	53	33	13
9	19	39	59	9	29	49	52	32	12
10	20	40	60	10	30	50	51	31	11
11	31	51	11	21	41	1	40	20	60
12	32	52	12	22	42	2	39	19	59
13	33	53	13	23	43	3	38	18	58
14	34	54	14	24	44	4	37	17	57
15	35	55	15	25	45	5	36	16	56

Table 4-4 Frequency allocation in a globally synchronized FH multiple-access system, including the "H" and "L" bit frequencies for the transmitter and the relevant "mixed frequencies" for the receiver for each of the 30 channels (*Continued*).

16	36	56	16	26	46	6	35	15	55
17	37	57	17	27	47	7	34	14	54
18	38	58	18	28	48	8	33	13	53
19	39	59	19	29	49	9	32	12	52
20	40	60	20	30	50	10	31	11	51
21	51	11	31	41	1	21	20	60	40
22	52	12	32	42	2	22	19	59	39
23	53	13	33	43	3	23	18	58	38
24	54	14	34	44	4	24	17	57	37
25	55	15	35	45	5	25	16	56	36
26	56	16	36	46	6	26	15	55	35
27	57	17	37	47	7	27	14	54	34
28	58	18	38	48	8	28	13	53	33
29	59	19	39	49	9	29	12	52	32
30	60	20	40	50	10	30	11	51	31

The pertinent frequencies result from the frequency numbers listed in Table 4-4 by multiplying the frequency number by 1800 Hz and adding 30.6 kHz. For example, if we select channel number 26, an "H" bit will be sent in the frequency order 56, 16, 36 at 131.4 kHz, 59.4 kHz, and 95.4 kHz. For the "L" bit, the frequency order is 46, 6, 26, i.e., 113.4 kHz, 41.4 kHz, and 77.4 kHz. The receiver provides the frequencies number 15, 55, and 35 for channel 26, i.e., 57.6 kHz, 129.6 kHz, and 93.6 kHz for mixing with the receiver signal. If an "H" bit is sent, then we obtain a constant "sum frequency" of 189 kHz from the mixing procedure for the three pertaining waveforms in each of the three chip intervals of duration T_h = 1.11 ms. Accordingly, we obtain 171 kHz for an "L" bit. Table 4-4 shows that this fact is met for each of the 30 channels. The sum from

Examples of System Implementations

an "H"-bit frequency number and the pertinent receive frequency number is always 71, and 61 results always for an "L" bit. By multiplying 71 by 1800 Hz and adding 2·30.6 kHz, we obtain an "H" sum frequency of 189 kHz. Analogously, the calculation 61·1800 Hz + 2·30.6 kHz results in the "L" sum frequency of 171 kHz.

In addition to a mixer with the function described above, a receiver requires other components to be able to optimally detect waveforms with duration T_h = 1.11 ms, which have either frequency 171 kHz or 189 kHz. The mixer output signal is fed in parallel to two receiver branches. One receiver branch, the "H"-bit branch, has to be tuned to frequency f_H = 189 kHz, and the other one, the "L"-bit branch, has to be tuned to f_L = 171 kHz. For optimum reception, at the end of each chip interval we have a signal corresponding to the energy of the received waveform at the output of the addressed branch. The nonaddressed branch does not supply any signal due to the orthogonality of the waveforms, except for contributions from any interference that may be present. The output signals of the two receiver branches are compared at the end of each chip interval to determine whether the waveform just received belongs to an "H" bit or an "L" bit (chip decision). Upon expiration of three chip intervals, the decision is taken in favor of the ("H" or "L") bit, for which at least two chip decisions were taken. The correct instants for chip and bit decisions are supplied by a simple synchronization equipment. The acquisition is always achieved after a maximum delay corresponding to a bit duration of $T_b = 3T_h = 3.33$ ms.

Another interesting aspect of this implementation is the matched filter in both receive branches of this FH receiver. Two receive branches and one frequency synthesizer with only one output signal, used to mix the received signal in both branches, are sufficient. The mixer output reaches two clearable high-quality resonance circuits—one with its resonance at f_H = 189 kHz and the other at f_L = 171 kHz. The resonance circuits are the core components of the matched filters in both receive branches for "H" and "L" bits. "Clearable" resonance circuits means that they can be made "energyless" after each chip interval by use of a short dump impulse. "High quality" means that a resonance circuit loses almost none of the energy fed to it during one chip interval. In practice, this can be achieved only by electronic deattenuation by means of active elements, such as operational amplifiers [A1].

Figure 4-27 shows that matched filters of the type described above can be implemented very easily, and the function covers all theoretical expectations.

Figure 4-28 shows the technical implementation of two matched filters ("H"- and "L"-bit branches) as realized in a series of prototypes [ST4, ST5]. A study of detailed circuit diagrams would go beyond the scope of this book; instead, we will describe the basic function by means of Figures 4-27 and 4-29.

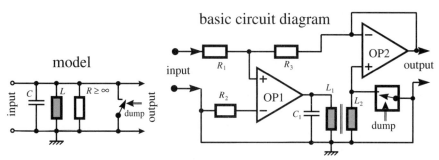

Figure 4-27 Model and basic circuit diagram of a "deattenuated" resonant circuit built as a matched filter.

Basically, this implementation requires only one parallel resonant circuit with an option to be dumped, consisting of an ideal coil with inductance L, an ideal capacitor with capacitance C, and a resistance $R \approx 10$ MΩ, where the losses are concentrated. An analog switch is connected in parallel to the resonant circuit for the dump operation. To build an "L"-bit receive branch, the circuit obtains resonance frequency $f_L = 171$ kHz, and for the "H"-bit branch it obtains $f_H = 189$ kHz. The resonant circuit's factor of quality $Q = R/\omega L$ has to be 10,000—a value that cannot be achieved with passive components. A proven circuit with operational amplifiers for deattenuating the resonant circuit and an analog switch for the dump operation is shown in a slightly simplified form on the right-hand side of Figure 4-27. The signal can be fed to the resonant circuit model from a current source. Assuming that a sinusoidal current with frequency $f_L = 171$ kHz—exactly at the resonance frequency—is fed to an "L"-bit receive branch, then we obtain a signal curve at the output over two chip intervals, as represented in the upper half of Figure 4-29.

We can see the linear rise of the voltage amplitude from zero to a maximum value at the end of the chip interval, where the dump impulse occurs; i.e., the analog switch at the resonant circuit is closed for a very short time. The "ON resistance" of the switch consumes the stored energy of the resonant circuit, so that it is always energyless when the analog switch opens at the beginning of a new chip interval; i.e., the amplitude starts again to rise from zero. Note that this absolutely requires an ON resistance that is clearly higher than zero, so that the circuit's energy can be "destroyed" quickly. Good results can be achieved with values of several hundred ohms.

Figure 4-29 also shows in a very impressive way how the resonant circuit, as a perfect matched filter, reacts to an FH waveform with the next neighboring frequency, namely $f_L + 900$ Hz $= 171{,}900$ Hz. The voltage amplitude passes the theoretically expected maximum in the center of the waveform interval, which is at $T_h/2$, and the desired zero points occur at the sampling instants T_h and $2T_h$.

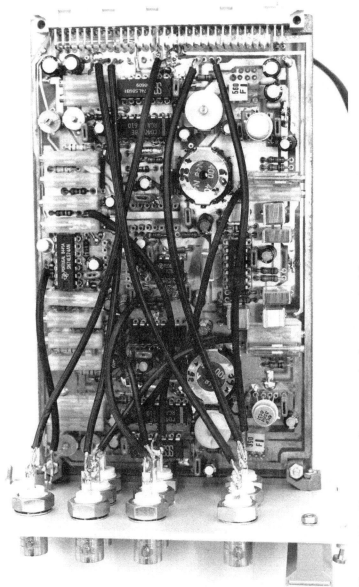

Figure 4-28 "Deattenuated" resonant circuits as matched filters.

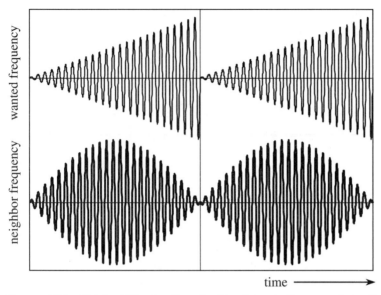

Figure 4-29 Matched filter function of a "deattenuated" resonant circuit.

Despite these apparently excellent results, this analog implementation does not represent an ideal solution for the general matched filter design of powerline modems—for example, modems suitable for use in large quantities—because the design of deattenuated resonant circuits is tied to a number of problems, the two major ones being as follows:

- First, setting the deattenuation by means of resistor R3 (top right in Figure 4-27) is critical in view of the fact that the factor of quality should reach 10,000, while avoiding a tendency to natural oscillation at the same time.
- Second, even small changes to the resonance frequency due to temperature, mechanical shock, and aging of components have dramatic consequences for a proper matched filter function. Imagine, for example, the impact of resonance-frequency errors in Figure 4-29: A frequency change of 900 Hz, which corresponds to only about 0.5% of the resonance frequency, is enough to cause the desired matched filter output signal to totally disappear.

Practice has shown that these problems are hard to control. For instance, it was not possible to ensure reliable receiver functionality over a lengthy period. By itself the transport of a modem or a big change in the environmental temperature would cause failures. Although bit error probability measurements also produced good results, the industrial realization of matched filters based on resonance circuits with dump capability was eventually given up. The problems described above appeared to be too difficult to be solved in a reliable and economical way for series production of powerline communication devices.

4.4.1.4   Second Generation: A Robust Remote Meter Reading System

Although the opening of the energy market, where automatic remote meter reading will be one of the most important applications for low-speed powerline communication, has become a highly explosive issue for German and other European power supply companies only recently, the author's research group had already built and tested the first transmission systems in the period from 1987 to 1990 [D1, ST12]. The data rate of 60 bits/s used then would hardly meet current demands, but the achieved resistance to interference was considerable, so that it would more than meet current expectations. At that time, both PSUs and the industry observed the development with interest, but mostly in a passive way. This was probably due mainly to the missing deregulation of the electricity market, which seemed to be far away, and to the then infant analog technology of modems. As mentioned in the previous section, the reproducibility required for series production could not be achieved with the receiver technology based on deattenuated resonant circuits.

The intensive search for better concepts led to one of the first implementations with matched filters for FH waveforms based on "lock-in amplifiers" at the beginning of 1987. Lock-in amplifiers are mainly known from measuring instrumentation where they develop their special advantages, when it is a matter of detecting weak sinusoidal signals of known frequency in an environment with heavy interference [A3], i.e., in situations generally occuring in the transmission of frequency-agile signals over powerlines. Lock-in amplifiers are often called synchronous demodulators or synchronous rectifiers. The core component of a synchronous rectifier is an amplifier that switches the sign of the amplification factor according to a digital periodic switching signal $s_{LI}(t)$ with frequency f_{LI} between the two states, +1 and −1 (see Figure 4-30).

Let's assume that a sinusoidal signal with amplitude A of frequency f_e and zero-phase angle φ_e reaches the input of a synchronous rectifier of the type shown in Figure 4-30. The analog switch controlled by the digital switching signal connects the noninverting input of the operational amplifier OP alternately with $s_e(t)$ and ground. OP has a gain of +1 at the "+1" level or −1 at the "−1" level of the switching signal. To better illustrate the interplay, Figure 4-31 shows a signal example, where $f_{LI} = f_e$, and where no phase shift is present between the input and the switching signal ($\varphi_e = 0$). In this case, we obtain a signal $s_a(t)$ at the output of the synchronous rectifier, similar to the one achieved from $s_e(t)$ by ideal two-way rectification. For evaluation purposes,

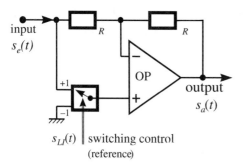

Figure 4-30 Synchronous rectifier to explain the lock-in principle.

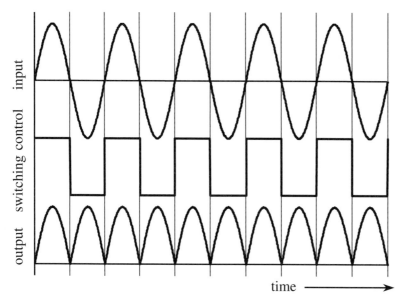

Figure 4-31 Synchronous rectifier signals, where input and switching signals coincide (frequency and phase equality).

$s_a(t)$ is subject to an arithmetic mean-value formation, preferably done by means of a clearable integrator.[8]

The arithmetic mean value of s_a is $A \cdot 2/\pi$. It is easy to understand that the mean value becomes smaller than $A \cdot 2/\pi$ at phase shifts different from zero between $s_e(t)$ and the switching signal; s_a becomes zero at $\varphi = 90$ degrees.

---

[8] The integration is done exactly over one waveform duration; subsequently, the integrator is cleared, i.e., set to zero.

Examples of System Implementations

Without limiting the generality, the zero phase is allocated to the switching signal $s_{LI}(t)$, i.e., $s_{LI}(t)$ jumps from -1 to $+1$ at the zero point of the time axis, as shown in Figure 4-31.

To study cases where the frequencies f_e and f_{LI} deviate, we have to look a little further. First, it is useful to develop the rectangular switching signal $s_{LI}(t)$ into a Fourier series. When $s_{LI}(t)$ assumes the levels 1 and -1, we obtain

$$s_e(t) = A \cdot \sin(2\pi f_e t + \varphi_e) \qquad (4.32)$$

$$s_{LI}(t) = \frac{4}{\pi} \cdot \sum_{n=0}^{\infty} \cdot \frac{1}{2n+1} \cdot \sin[(2n+1) \cdot 2\pi f_{LI} \cdot t], \text{ with } n \in N \qquad (4.33)$$

The output signal $s_a(t)$ of the synchronous rectifier is the product from $s_e(t)$ in (4.32) and $s_{LI}(t)$ in (4.33)—that is:

$$s_a(t) = A \cdot \sin(2\pi f_e t + \varphi_e) \cdot \frac{4}{\pi} \cdot \sum_{n=0}^{\infty} \frac{1}{2n+1} \cdot \sin[(2n+1) \cdot 2\pi f_{LI} \cdot t] \qquad (4.34)$$

If we form the mean value from $s_a(t)$, for example by means of a clearable integrator, we obtain

$$\overline{s_a} = \begin{cases} \dfrac{2}{(2n+1) \cdot \pi} \cdot A \cdot \cos\varphi_e & \text{for } f_e = (2n+1) \cdot f_{LI} \\ 0 & \text{for } f_e \neq (2n+1) \cdot f_{LI} \end{cases} \qquad (4.35)$$

We understand from (4.35) that, when using synchronous demodulators as matched filters for FH waveform detection, there may be problems when the received signal contains odd harmonics of the switching frequency f_{LI}. Such harmonics contribute to the mean value of the output signal, with only a very low attenuation, namely corresponding to the inverse value of their ordinal number $(2n+1)$, compared to the desired basic oscillation $(n=0)$. On the other hand, a suitable composition of the waveforms for an FH system helps avoid such problems [D1]. The synchronous rectifier behaves exactly like a matched filter when the above-mentioned harmonics are avoided.

We will now analyze the behavior with regard to the orthogonality, i.e., with relatively low frequency changes. We assume for this purpose that the frequency f_e of the received signal is near the switching frequency f_{LI}, and that it differs from it by v/T_h, where $v = 1, 2, 3, \ldots$; we study $s_a(t)$ from (4.34) and set $f_e = f_{LI} \pm v/T_h$. The zero-phase angle φ_e is assumed to be zero without limiting the generality, because its influence is eliminated by the quadrature receiver structure[9] that is required in any case [F10, F12]. Under these prerequisites, we obtain the product

$$s'_a(t) = \frac{4A}{\pi} \cdot \sin\left(2\pi\left(f_{LI} \pm \frac{v}{T_h}\right)t\right) \cdot \sum_{n=0}^{\infty} \frac{1}{2n+1} \cdot \sin[(2n+1) \cdot 2\pi f_{LI} \cdot t] \quad (4.36)$$

from (4.34). By trigonometric transformation and separation of the term for $n = 0$, we obtain

$$s'_a(t) = \frac{2A}{\pi} \cdot \left[\cos\left(2\pi \frac{v}{T_h} t\right) - \cos\left(\left(4\pi f_{LI} \pm \frac{2\pi v}{T_h}\right) \cdot t\right)\right. \quad (4.37)$$

$$\left. + \sum_{n=1}^{\infty} \frac{1}{2n+1} \cdot \left(\cos\left[4\pi\left(nf_{LI} \pm \frac{v}{T_h}\right) \cdot t\right] - \cos\left(\left[4\pi(n+1)f_{LI} \pm \frac{2\pi v}{T_h}\right] \cdot t\right)\right)\right]$$

from (4.36). Under the prerequisite that $f_{LI} \approx v/T_h$, the second cosine term in (4.37) above and the entire bottom part of (4.37) disappear when integrating $s'_a(t)$ over chip duration T_h. We obtain the mean signal value

$$\overline{s_a}(v) = K_I \cdot \frac{2A}{\pi} \cdot \int_{-\frac{T_h}{2}}^{\frac{T_h}{2}} \cos\left(2\pi \cdot \frac{v}{T_h} \cdot t\right) dt \quad (4.38)$$

on the integrator output at the end of the chip interval, where K_I is an integrator constant with dimension l/s. When calculating (4.38), we obtain

$$\overline{s_a}(v) = K_I \cdot \frac{2A}{\pi} \cdot T_h \cdot \text{si}(\pi v) \quad (4.39)$$

Equation (4.39) shows the perfect matched filter function with regard to the orthogonality of the FH waveforms very clearly. For example, using $K_I = 1/T_h$, we obtain the result

$$\overline{s_a} = A \cdot 2/\pi$$

for $v = 0$ for a desired FH waveform—see also the example in Figure 4-31 (two-way rectification); for all other values $v \neq 0$ we have

---

[9] The quadrature structure allows incoherent reception, i.e., independent of the phase angle of the received signal; it will be described in more detail in connection with digital implementations.

Examples of System Implementations

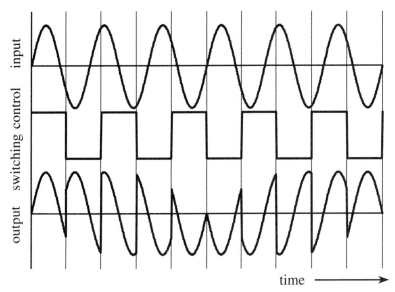

Figure 4-32 Synchronous rectifier signals at $f_e = f_{LI} + 1/T_h$, i.e., at the next orthogonal neighbor frequency.

$$\overline{s_a} = 0$$

The signal example in Figure 4-32 shows the above calculation result. In this example, the input frequency was increased versus Figure 4-31 by the inverse value of the waveform duration (= entire figure width). It leaps to the eye that the output signal now has positive and negative values.

A closer look discloses that the areas below and above the drawn center line are equal for $s_a(t)$, so that the mean value taken over one waveform duration becomes zero. This means that lock-in amplifiers or synchronous modulators in connection with clearable integrators are excellent to build matched filters for FH waveforms.

Lock-in amplifiers deliver a high dynamic range "by nature," because a good operational amplifier as the core element of a circuit like the one in Figure 4-30 has a dynamic range that reaches from its offset voltage to the maximum common mode input voltage, extending typically over more than 100 dB. In addition, lock-in amplifiers have a high long-term stability and well-reproducible properties. This represents considerable progress, compared to the clearable resonant circuits described above.

Finally we will look at the generation of the switching (reference) signal in the receiver. For the function of a lock-in amplifier as a matched filter in an FH receiver, the frequency f_{LI} of the switching signal $s_{LI}(t)$ in Figure 4-30 must always follow the frequency of the expected received signal. There are always different frequencies present for the "H" and "L" bits in each chip interval of duration T_h, so that a receiver requires at least two separate receive branches. We can understand from (4.35) that the zero-phase angle of the received signal influences the output signal of a lock-in amplifier. This is the reason why four separate receive branches are required for incoherent or phase-independent reception. Consequently, a frequency synthesizer on the receiver side has to supply four different output signals for switching. Switching signals at the corresponding frequency in sine and cosine positions are required for the "H"- and for the "L"-bit branch. The switching signals control four separate lock-in amplifiers, the output signals of which are integrated always over a chip interval of duration T_h for the purpose of forming mean values by means of clearable integrators. Further processing is purposefully digital; i.e., a suitable analog/digital conversion is required.

The fact that the lock-in amplifier output signals should be integrated means that an integrating analog-to-digital conversion method should ideally be applied. A clever approach is the conversion of the analog lock-in amplifier output voltages into corresponding frequencies by means of voltage-to-frequency converters (VFCs). All that is required for integration then are simple digital counters to count all impulses coming from the VFCs over the desired time interval of duration T_h. The counter contents are saved in latches upon expiration of one chip interval. The counters are reset and then ready for the acceptance of the VFC impulses from the next chip interval. The results are matched filter output signals in digital form. Processes like the respective geometric addition of in-phase and quadrature channels, the comparison of "H"-bit and "L"-bit branches, and everything entering into the bit decision, can then be digital, e.g., by use of a microcontroller. The enormous dynamic range of lock-in amplifiers cannot be handled with simple voltage-to-frequency converters. Linearity errors and drift effects of simple VFCs lead to significant degradation. Synchronous voltage-to-frequency converters (SVFCs), which do not have these drawbacks, have been in the market since 1986 [A5]. Such components have been successfully used in practice [D1, ST12] and allowed the use of the lock-in principle for powerline applications for the first time. Further details of the circuit structure, particularly further signal processing and data bit detection, are described in [D1]; they are not described here in detail, because even lock-in amplifiers were not successful for series production of powerline modems despite the benefits described above. Some of the reasons will be mentioned later. The photo in Figure 4-33 shows the prototype structure of a fourfold matched filter on lock-in basis for the incoherent reception of FH waveforms.

On the right-hand side of the figure are four operational amplifiers, which form the core components of lock-in amplifiers. To the left of these operational amplifiers are

Figure 4-33 Four-fold matched filter with lock-in amplifiers.

the analog switches, and the synchronous voltage-to-frequency converters are in the second column from the left.

One particular is the way the four switching signals for the analog switches can be generated precisely and at low cost. Without further consideration, it seems logical to yield the switching signals directly in the digital way by suitable frequency division from an appropriate "mother clock frequency." A short calculation example is enough, however, to prove that this idea is a one-way road. Let's assume that, with a hop rate of 300 s^{-1}, two neighboring waveforms have the frequencies 100 and 100.3 kHz, respectively; then the periods of these waveforms differ only by approximately 30 ns. Even if we were to allow a coarse period error of 10%, a mother clock frequency of over 333 MHz would be required.

A much more clever approach, invented by the author's research group at the beginning of 1987, which has proven to be excellent in practice, works as follows. First, the memory reading process[10] is used to generate analog FH waveforms and to subsequently convert them into the desired digital switching signals by means of a comparator. The required 90-degree phase shift can be achieved digitally, that is, independent of the frequency and free from errors, as follows: We generate the required FH waveforms at fourfold frequency on the receiver side. The comparator is followed by a flip-flop as a first frequency divider. An additional flip-flop is connected to the nonnegated and the negated output, respectively, of the first flip-flop for a second frequency division. The desired 90-degree phase-shifted switching signals at the correct frequency are available at the two nonnegated outputs of these flip-flops.

Figure 4-34 shows a complete FH receiver based on lock-in amplifiers. In 1988/89, this prototype was used to conduct extensive field trials over a supply cable at a length of approximately 1 km in a network structure like the one shown in Figure 2.11 [D1].

There was a turning away from full-duplex and multiple access operation. The first application concept was a remote power-meter reading system over low-voltage networks between consumers and the supplying transformer station. The lock-in principle combined with the synchronous voltage-to-frequency converters led to a high receiver dynamic range, which was a decisive prerequisite for the success of these field trials. Data transmission over the low-voltage mains section in Figure 2-11 was in permanent operation for about nine months. This mains section is part of the power supply grid for "Betzenberg," a residential area of Kaiserslautern, Germany; house number 31 near the end of the supply trunk was the transmission location for the test data. The

---

[10] This process will be explained later in connection with digital methods.

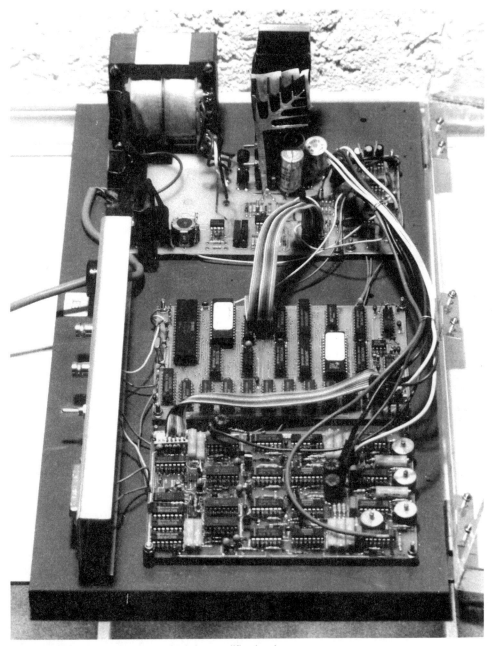

Figure 4-34 FH prototype on lock-in amplifier basis.

receiver was located in the transformer station. A 127-bit pseudorandom sequence was sent as the test data for bit error measurements. The received bits were compared against the "target bits" at the receiver side. The number of bit errors was saved, together with the time when the test data were transmitted for later evaluation. A data volume of 365,769 bits was transmitted daily to limit the storage and evaluation work. With a transmission voltage amplitude of 1 V, there was an average of only one error per week, which corresponds to a bit error rate of approximately $3.9 \cdot 10^{-7}$. This is certainly a satisfactory result, which proves the performance of the technology applied quite impressively. In fact, robust data transmission was achieved at very low transmission power at a very early time over a distance with the length corresponding to the maximum length in low-voltage networks. This means that it would have been possible to implement remote meter reading over the low-voltage mains already in 1989. While it is now seen as one of the key tasks in the open electricity market, PSUs did not consider its use then, although the technology was ready and available. In contrast, in today's deregulated electricity market there have been hectic activities in the search for quick solutions. We will not discuss this issue any further; instead, we will mention two main arguments that have always been brought up against automatic meter reading over powerlines:

- Reading is normally done only once a year, and it costs only about $0.50 (U.S.) per meter. No automatic reading system can be profitable on this basis.
- Most power meters are based on the so-called "Ferraris" measuring principle, coupled with a mechanical counter (based on rotating wheels), which cannot be read electronically. For cost reasons, neither complete replacement with electronic meters nor an upgrade, e.g., with optical sampling systems, is feasible.

For series production of modems, the lock-in principle does not yet appear to be an acceptable solution, despite satisfactory functionality and high mass production safety, for the following reasons. The operational amplifier (see Figure 4-30), which is the core element, has to be selected carefully and offset-compensated. Even more critical are the required compensation measures for the analog switches of the lock-in amplifiers. What has to be compensated here is the impact of "charge injections" caused by the switching signals. Charge injection means that an output signal occurs despite missing input signals and a perfect offset-compensated operational amplifier, for example in the circuit shown in Figure 4-30. Unfortunately, static compensation of the charge injection is not enough, so that a degradation of the receiver dynamics due to the charge injection has to be expected, as long as the compensation cost is moderate.

The search for system concepts that ensure constant quality in series production, and where no human specialist intervention is required during the entire production process, leads necessarily to all-digital solutions. Before we introduce the receiver concept used by most of the current solutions, we will introduce the digital signal synthesis for frequency-agile waveforms, considered to be optimal for powerline communication.

4.4.2 Digital Signal Synthesis by Wave Table

In a frequency-agile system for information transmission over powerlines, both the transmitter and the receiver generally have to generate numerous waveforms at closely adjacent frequencies, and frequency accuracy plays an important role for various reasons [D1, ST15]. The frequency change at the chip borders has to be quick, with continuous phase and without transients. A technically meaningful and economic solution requires a digital approach. One proven solution is reading appropriate samples from a wave table stored in a digital memory.

Figure 4-35 summarizes the example of a frequency-agile transmission system, as described above, for fast frequency hopping at five frequencies per data bit.

The core element of the transmitter is a frequency synthesizer, which will be the focus of the following discussion. The digital implementation of the receiver side will be discussed later.

The diagram of a digital frequency synthesizer based on memory reading is shown in the top part of Figure 4-36. The central element is a memory, e.g., in the form of a ROM (read-only memory), EPROM (erasable programmable read-only memory), or RAM (random-access memory). A programmable address counter, fed with a fixed and

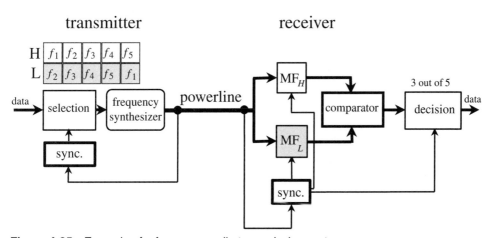

Figure 4-35 Example of a frequency-agile transmission system.

generally quartz-stable clock, addresses the memory and reads cyclically the contents stored within a specific address range. Observing the Nyquist criterion, the read-out clock frequency has to be selected to be higher than twice the highest frequency to be synthesized. Practical applications even work with at least fourfold oversampling. The reasons will become clear further below.

The samples for an FH waveform with duration T_h are stored in certain address ranges. The samples reach a D/A converter. A staircase signal $s_{Ai}(t)$ occurs at the D/A converter's output. Once the lowpass filtering is completed, $s_{Ai}(t)$ turns into the desired waveform $s_i(t)$. $s_{Ai}(t)$ and $s_i(t)$ will be analyzed mathematically further below to draw inferences about possible error sources and how to remove them. To generate FH waveforms, the addressed memory range has to be changed in the rhythm of the hop rate h_r, matching the desired frequency sequence. The single memory portions belonging to waveforms at different frequencies can differ in size.

There are many different solutions to control the programmable address counter shown in Figure 4-36. The use of microcontrollers is often a good idea, because the simple programming of such a component ensures high flexibility. For example, when an FH transmitter has to be implemented, one can deliver the binary data stream to be transmitted to the microcontroller over an integrated serial interface. Built-in timers enable the microcontroller to generate the time pattern for the transmission process [D1, ST15, ST17]. Also an FH receiver requires frequency synthesizers, as we will see later. Hereby the principle of the hardware structure shown in Figure 4-36 can be maintained. For example, the switching signals in the lock-in receiver described above require two frequency synthesizers, and each has to deliver its output signals in sine and cosine positions. An analog phase shifter cannot be used because of the accuracy required for a frequency range extending over approximately five octaves. On the other hand, the dig-

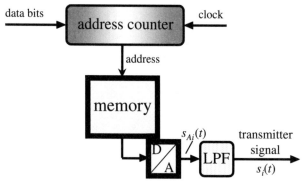

Figure 4-36 Principle of signal synthesis by memory read-out.

ital implementation does not require four separate synthesizers. The solution was described further above.

We will now look at some important aspects of digital frequency synthesis based on memory reading within a quantitative study to derive general guidelines for correct synthesizer design. More precisely, these are issues regarding the required quantization of the samples and optimization of the memory size for the samples. In a closer connection with the latter, there is the selection of a favorable read-out clock frequency, representing the sampling frequency f_A of the waveform to be generated. The ratio of f_A to the frequency f_i of a waveform $s_i(t)$ to be generated influences the amplitude of $s_i(t)$, which is an undesired effect, as we will see later. The problem to be solved here is to prevent the amplitude from decreasing significantly toward higher frequencies.

In practical applications, an 8-bit quantization of the samples has proven to be sufficient. More than 8 bits would have unpleasant consequences due to the usual memory organization in bytes ($\equiv$ 8 bits) with regard to circuit design costs. The theoretical argument that 8 bits are sufficient is substantiated by the general rule that each additional quantization bit used will improve the suppression of quantization errors by approximately 6 dB. This is plausible because the expansion of the quantization by one bit doubles the number of amplitude steps that can be represented. For example, the 8-bit quantization offers a distance (i.e., a signal-to-noise ratio) of approximately 50 dB between the powers of the desired waveform and the quantization noise. In addition, oversampling contributes to a further improvement, namely 3 dB for each factor of 2.

A problem in the design of digital frequency synthesizers like the one in Figure 4-36 is to minimize the number of samples to be stored. The following basic equation applies when a waveform at frequency f_i by use of sampling frequency f_A is to be synthesized:

$$\eta_i \cdot \frac{f_A}{f_i} = \left\lceil \eta_i \cdot \frac{f_A}{f_i} \right\rceil, \text{ with } \eta_i \in N \tag{4.40}$$

where $\lceil \cdot \rceil$ is the integer part of $(\cdot)$. The sampling frequency f_A must be bigger than $2f_{max}$ to fulfill the sampling theorem, where f_{max} is the highest frequency to be synthesized.

Note: We are talking here about synthesized frequencies in a simplified manner. Strictly speaking, this is not possible, because signals at these frequencies are generated, of course. As there is no risk of producing misunderstandings, we will use this simplification to keep the text easier.

Equation (4.40) says that every time the sampling frequency f_A is not an integer multiple of a frequency f_i, the fraction f_A/f_i has to be multiplied by constantly increasing positive integers η_i, until the product $\eta_i(f_A/f_i)$ is a positive integer. Otherwise, the generated waveform would exhibit undesired phase hops. Factor η_i has to be defined for each frequency f_i individually, and, of course, it should be as small as possible.

When a number N_{FH} of orthogonal waveforms has to be generated at a frequency distance equal to the hop rate h_r, then a total of

$$\Pi = \sum_{i=0}^{N_{FH}-1} \eta_i \cdot f_i \qquad (4.41)$$

samples have to be stored. It is intuitive to select the lowest frequency f_A as an integer multiple Θ of the hop rate h_r and the sampling frequency f_A as an integer multiple ζ of f_0, i.e., the lowest frequency to be synthesized. The result is then the sample sum

$$\Pi = \sum_{i=0}^{N_{FH}-1} \eta_i \cdot \frac{\zeta \cdot \Theta}{\Theta + i} \qquad (4.42)$$

for which a minimum has to be found. Currently, there is no analytical solution to this problem. Therefore, a computer program for memory optimization was developed for practical applications. Upon selection of the number N_{FH} of desired frequencies and hop rate h_r, it is a good idea to specify a variation range for the lowest frequency f_0. Subsequently, the program starts an optimization procedure. During the first run, $\zeta = 1$ is used and the sum Θ is searched for a minimum, while varying f_0 in increments of the hop rate h_r. Once the best value for f_0 is found, i.e., the value where Θ is as small as possible, then as small a value for ζ as possible is calculated:

$$\zeta_{min} \geq \frac{f_A}{f_{max}} \cdot \left[\frac{N_{FH}-1}{\Theta} + 1 \right], \text{ with } \zeta_{min} \in N \qquad (4.43)$$

Note that (4.43) does not necessarily supply a minimum for the sample sum Θ for each value of ζ_{min}. The reason is that common prime factors contained in η_i and ζ can be truncated (see (4.49) and (4.42)). For this reason, a second program run is started, where Θ is once again searched for a minimum, this time by varying ζ in the range of

$$\zeta_{min} \leq \zeta \leq \zeta_{min} + 5 \qquad (4.44)$$

In practice it was observed that increasing ζ beyond the upper limit given in (4.44) is not meaningful, because the prime-number distance grows drastically with larger numbers.

Once the desired frequencies f_i have been selected and the number of samples for each has been determined, the program calculates the samples as unsigned 8-bit hexadecimal numbers, or in two's-complement representation, if needed, and saves them in

Examples of System Implementations

a format that allows simple EPROM programming by use of commercial programming equipment.

The samples for the synthesis of the set with 30 orthogonal waveforms shown in Table 4-3 can be generated easily by means of the programs described above, using the following parameters: start frequency f_0 = 9600 Hz, hop rate h_r = 4800 Hz, sampling frequency f_A = 600 kHz.

Next, we want to study the influence of the f_A/f_i ratio on sinusoidal signals generated in the form

$$s_i(t) = \sin(2\pi f_i t) \qquad (4.45)$$

For easier understanding, we have set the amplitude of $s_i(t)$ to 1 and omitted the signal duration limit, because it is not relevant here. The signal $s_i(t)$ has to be generated from saved digital samples, where the samples should not require much memory space. We know from (4.41) that the memory-space requirements increase proportional to the sampling frequency f_A. For this reason, f_A should be as low as possible, taking the sampling theorem into account. f_A depends on the highest frequency f_{max} to be generated; i.e., the inequation $f_A > 2 f_{max}$ has to be met at all times. The sampling theorem says that it is sufficient to select $f_A > 2 f_{max}$ to be able to generate all desired waveforms at $f_i < f_{max}$ without any error; this is true as long as ideal sampling using "Dirac impulses",[11] is performed [F10]. This would require the output of samples with infinitely short duration from the D/A converter to the lowpass filter of the frequency synthesizer. Unfortunately, reality is far from that. The typical output waveforms of a real D/A converter consists of rectangular impulses with duration $T_A = 1/f_A$.

Figure 4-37 shows the relationships by a synthesis example that generates a waveform with frequency f_s = 72 kHz at a sampling frequency of f_A = 600 kHz. The smallest integer positive number $\{\eta_s \cdot (f_A/f_s)\}$ is 3(8.333) = 25. This means that a total of 25 samples have to be stored, corresponding to three periods of a signal according to (4.45) at frequency f_s. Figure 4-37 shows the typical staircase output signal delivered by a real D/A converter. Using the 8-bit quantization, the output signal can take on 256 different amplitude steps.

For a theoretical study of the influence of nonideal signal output to the D/A converter in the form of rectangular impulses instead of Dirac impulses, we first look at the sampling of a signal from (4.45), using a sampling rate $f_A = 1/T_A$ and an ideal sampler. First, we define an infinite Dirac impulse sequence.

---

[11] A Dirac impulse is a needle impulse with a width approaching zero, while the amplitude increases beyond all limits; it will be described in more detail later.

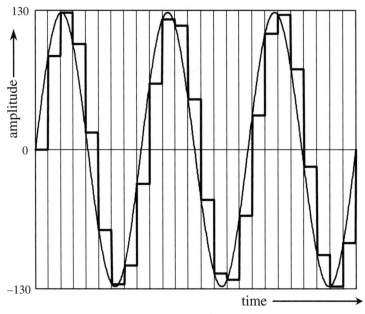

Figure 4-37 Synthesis example for f_s = 72 kHz at f_A = 600 kHz.

$$\text{Ш}(t) = \sum_{n=-\infty}^{+\infty} \delta(t-n) \qquad (4.46)$$

In (4.46), *n* is integer and real, and it progresses from $-\infty$ to $+\infty$ in increments of 1. A Dirac impulse $\delta(t)$ is a generalized mathematical function, also called distribution [F8]. Think of a Dirac impulse $\delta(t)$ as a needle impulse with an area of 1, infinite height, and disappearing width at $t = 0$. Using a Dirac impulse sequence from (4.46) for equidistant sampling of a signal at sampling frequency $f_A = 1/T_A$, we get:

$$\frac{1}{T_A}\text{Ш}\left(\frac{t}{T_A}\right) = T_A \cdot \sum_{n=-\infty}^{+\infty} \delta(t - nT_A) \qquad (4.47)$$

From (4.45) and (4.47), we obtain the discrete-time and continuous-value sampling sequence

$$s_{Ai}(t) = s_i(t) \cdot \frac{1}{T_A} \cdot \text{Ш}\left(\frac{t}{T_A}\right) \qquad (4.48)$$

Examples of System Implementations

for an ideally sampled signal $s_i(t)$. A sample of $s_i(t)$, taken from a real sampler at time instant nT_A, remains generally constant till time $(n+1)T_A$. So a sampled real signal corresponds to the staircase function represented in Figure 4-37 and is described by

$$s_{re}(t) = s_i(t) \cdot \frac{1}{T_A} \cdot \text{III}\left(\frac{t}{T_A}\right) * \text{rect}\left(\frac{t - \frac{T_A}{2}}{T_A}\right) \qquad (4.49)$$

The $*$ symbol stands for the convolution operation.

We will continue this study in the frequency domain for practical reasons. For this purpose, we apply the Fourier transform to (4.49). Using the following relationship between the time domain and the frequency domain (see for example [F8, F10])

$$\text{III}(t) \circ\!\!-\!\!\bullet \text{III}(f) \qquad (4.50)$$

and

$$s_i(t) \circ\!\!-\!\!\bullet s_i(f) \qquad (4.51)$$

we obtain the Fourier transform of (4.49), i.e.,

$$s_{re}(t) \circ\!\!-\!\!\bullet S_{re}(f) = \left[S_i(f) * \text{III}(fT_A)\right] \cdot T_A \cdot \text{si}(\pi f T_A) \cdot e^{-j\pi T_A} \qquad (4.52)$$

Then, analogous to (4.47),

$$\text{III}(fT_A) = \frac{1}{T_A} \cdot \sum_{n=-\infty}^{+\infty} \delta\left[f - \frac{n}{T_A}\right] \qquad (4.53)$$

With this result, we can rewrite (4.52) to

$$S_{re}(f) = \left[S_i(f) * \sum_{n=-\infty}^{+\infty} \delta\left[f - \frac{n}{T_A}\right]\right] \cdot \text{si}(\pi f T_A) \cdot e^{-j\pi f T_A} \qquad (4.54)$$

and

$$S_{re}(f) = \left[\sum_{n=-\infty}^{+\infty} S_i\left[f - \frac{n}{T_A}\right]\right] \cdot \text{si}(\pi f T_A) \cdot e^{-j\pi f T_A} \qquad (4.55)$$

follows with the "filtering property" of the Dirac impulse [F8].

If we apply the Fourier transform to (4.45), we obtain

$$s_i(t) \circ\!\!-\!\!\bullet \quad S_i(f) = \frac{1}{2}\left[\delta(f+f_i) - \delta(f-f_i)\right] \cdot e^{j\frac{\pi}{2}} \quad (4.56)$$

and thus

$$S_{re}(f) = \frac{1}{2}\left[\sum_{n=-\infty}^{+\infty} \delta\left(f+f_i-\frac{n}{T_A}\right) - \delta\left(f-f_i-\frac{n}{T_A}\right)\right] \cdot e^{j\pi/2} \cdot \text{si}(\pi f T_A) \cdot e^{-j\pi f T_A} \quad (4.57)$$

from (4.55).

We can see in (4.57) that a frequency-dependent weighting of the spectrum occurs from the function si($\pi f T_A$) due to the nonideal sampling. This effect should be eliminated, because we want to achieve equal amplitudes at all frequencies when synthesizing FH waveforms. The exponential functions $e^{j\pi/2}$ and $e^{-j\pi f T_A}$ in (4.57) represent phase shifts and, just like the factor 1/2, they can be neglected in the studies that follow, without limiting the generality.

We will now look at an example that uses the sampling frequency $f_A = 3f_i$, in order to better understand the relationships from (4.57). Figure 4-38 shows the influence of the amplitude weighting on the output signal of a frequency synthesizer according to Figure 4-36. The gray vertical lines show the position of the Dirac impulses from (4.57), which are within the bandwidth of an ideal lowpass filter with limiting frequency f_g, where $f_i = 1/3 f_A < f_g$. The weighting function of the amplitudes si($\pi f T_A$) from (4.57) is shown in black.

4.4.3 Digital Optimum Receiver Technology for Powerline Communication

The reception of heavily attenuated and interfered useful signals is typical in the transmission of information over powerlines. To recover the transmitted information without errors, we require methods known as "optimal reception techniques" or "matched filter reception." These methods not only evaluate the received signal briefly, but use the entire energy of the received waveform for data bit decision purposes. The following must be implemented to be able to use these methods:

- One correlator structure each for two receive branches for the "H" bit and the "L" bit.
- Synchronization equipment used by the receiver to exactly determine the beginning of the expected waveforms.

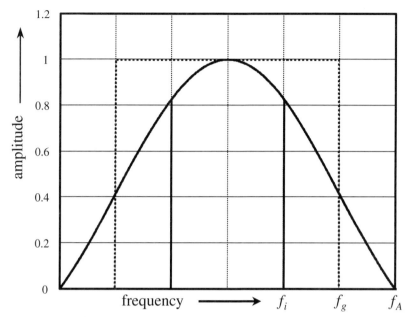

Figure 4-38 Synthesis example for $f_A = 3f_i$.

The synchronization can be implemented at low cost when a sufficiently reliable reference signal is available. If this is not the case, then the synchronization often accounts for over two-thirds of the total receiver cost. As described in detail above, powerlines use the mains voltage for reference.

The implementation of matched filters in analog technology was described in detail with several examples in the previous section. We also mentioned that incoherent reception has to be assumed in the applications studied; i.e., the phase angle of the received signal must not have any influence on the signal evaluation. So-called "quadrature receivers" are well known for incoherent reception from standard communications technology. The name derives from the fact that the reference signal on receiver side, which corresponds to the MF impulse response, as mentioned before, has to be supplied once in the "in-phase position," e.g., as a cosine function, and once in a 90-degree shifted "quadrature position" as a sinusoidal function. This means that the received signal is multiplied by four reference signals in a receiver that has two receive branches—one for the "H" bit and one for the "L" bit—and the results are integrated over one waveform duration. At the end of each waveform duration, the two components belonging to the "H" and "L" bits have to be added geometrically. Subsequently, the results are fed to the decision logic, which supplies the received data. Figure 4-39 illustrates these steps.

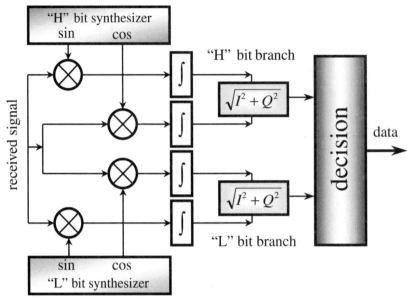

Figure 4-39 Principle of the quadrature receiver for incoherent reception.

The receiver in Figure 4-39 can basically be implemented all-analog or all-digital or as a combination of both. The lock-in principle introduced above is a good example of such a mixed variant. In the following, we are interested in purely digital variants, because they are the basis of current and future system generations for powerline communication.

Section 4.4.1 showed that the following (simplified) calculations have to be performed for a matched filter function:

$$K_{fi} = \frac{\text{si}(\pi f_{\max} T_A)}{\text{si}(\pi f_i T_A)} \qquad (4.58)$$

These are correlation operations characterized by the fact that an input signal is multiplied by a reference signal; the multiplication results are then integrated over one waveform duration.

A correlator can be thought of as a device used to numerically determine the similarity of two signals quantitatively. With maximum similarity, i.e., if the two signals are identical, we would obtain a very high output value (autocorrelation). With maximum dissimilarity, i.e., receipt of a waveform orthogonal to $s(t)$, we would obtain zero. In practical applications, there is an arbitrary number of intermediate values between the two extremes. This means that there is never autocorrelation, but always cross-correlation, in the reception of messages.

The term "similarity" as defined here should not be confused with optical similarity. For example, a sine and a cosine signal at the same frequency are very similar. But for a correlator they are "orthogonal" to one another, which means that one supplies the maximum correlation value, while the other one renders zero.

While the correlation process in an analog receiver can be described mathematically as a continuous integration of the product of input signal and reference over one waveform duration, equivalent operations with the appropriate discrete-time and discrete-value signals have to occur when we move to digital receiver technology. First, the received signal has to be sampled and quantized by means of an analog/digital converter. The reference can be generated digitally in the receiver. The correlation process can now be described by a series of consecutive multiplication and accumulation operations (multiply and accumulate $\equiv$ MAC). We will see later that such MAC operations play a key role in digital signal processing. This is why they are included in the standard instruction sets of digital signal processors (DSPs).

Figure 4-40 shows the principle of digital correlation. The received signal is sampled at a frequency f_A, which is at least twice as high as the highest expected frequency. After the A/D conversion the discrete-time and discrete-value received signal $s(k)$ according to the sampling pattern $T_A = 1/f_A$ is present at one input of the digital multiplier M, while the second input receives the reference $r(k)$ also in digital form. The products are added to a correlation value $z(u)$ in N steps in the accumulator, as described by the equation given in Figure 4-40. We can now easily imagine how the quadrature receiver in Figure 4-39 could be implemented. After the A/D conversion, the received signal would have to be fed to four digital correlators, replacing the four mixers and integrators. Of course, the output signals of the accumulators are also digital, so that the next processing steps (geometric addition, bit decision) can also be digital. We will describe design examples later. The use of four multipliers and accumulators seems exaggerated. It is, therefore, worth thinking about more elegant solutions that use less hardware. Results of these considerations will be presented further below.

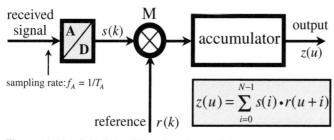

Figure 4-40 Principle of the digital correlator.

The abbreviated notation of variables used in Figure 4-40 is customary in digital signal processing (DSP), i.e., the time factor T_A is omitted in the variables k, i, and u. This convention does not cause any misunderstanding from the purely theoretical viewpoint. However, if you try to interpret the results physically, you have to take the time dimension into account. This means you should be aware that $k := k_A$, $i := iT_A$, and $u := uT_A$, and that the summation over N products requires a time of $NT_A = T$, where T is the waveform duration. The functionality of a digital correlator is described in more detail in the example below, because this is an important component for powerline communication modems.

In this example, an "H" bit, represented by a sinusoidal waveform at frequency 120 kHz and duration $T = 833$ µs, is received, sampled at $f_A = 600$ kHz, and quantized with an 8-bit resolution. With $T = 833$ µs, we obtain a chip rate of 1200 s^{-1}, which corresponds to the data rate $r_D = 1200$ bits/s in simple FSK. Assuming that the A/D converter is optimally exploited, so that its entire dynamic range is used, we get the following multiplier input signal (see Figure 4-40):

$$s(k) = 128 \cdot \sin(2\pi \cdot 120 \text{ kHz} \cdot kT_A) \qquad (4.59)$$

Given a length of $T_A \approx 1.66$ µs for the sampling interval, we obtain exactly $N = 500$ samples, when the waveform duration is $T = 833$ µs. This means that the correlator has to execute 500 MAC operations to completely calculate the output signal $z(u)$. The maximum value for $z(u)$ results when input signal $s(k)$ and reference $r(k)$ match and are synchronous, i.e., when there is no time shift between them. This is the case for $u = 0$; we then obtain

$$z(0)_{120} = \sum_{k=0}^{499} 128^2 \cdot \sin^2(2\pi \cdot 120 \text{ kHz} \cdot k \cdot 1.66 \text{ µs}) = \sum_{k=0}^{499} 128^2 \cdot \sin^2(1.256 \cdot k) \qquad (4.60)$$

The evaluation of (4.60) supplies the numerical value $z(0)_{120} = 4.096 \cdot 10^6$, representing the autocorrelation function (ACF) maximum.

We will now look at the reception of an "L" bit. Assume that the minimum frequency distance required for orthogonality, corresponding exactly to the chip rate—here 1200 Hz—is given. The frequency of the "L"-bit waveform is then 120 kHz + 1200 Hz = 121.2 kHz in this example, so that (4.60) takes the following form:

$$z(0)_{121.2} = \sum_{k=0}^{499} 128^2 \cdot \sin(2\pi \cdot 121.2 \text{ kHz} \cdot k \cdot T_A) \cdot \sin(2\pi \cdot 120 \text{ kHz} \cdot k \cdot T_A) \qquad (4.61)$$

The evaluation of (4.61) supplies now $z(0)_{121.2} \approx 0$, which means that the correlation signal disappears.

Examples of System Implementations

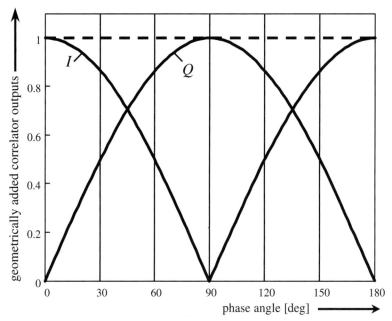

Figure 4-41 In-phase and quadrature signals and their geometric addition.

After these basic explanations, we will now study the behavior of the quadrature receiver at an arbitrary zero-phase angle of the input signal for the sake of completeness. We will use Figure 4-41 to show that the desired signal detection is achieved independently of the zero phase.

Figure 4-41 shows the output signals of two correlators, representing an in-phase branch (gray curve) and a quadrature branch (black curve). The dashed line drawn on top represents the geometric addition of the two branches; as expected, it is constant over the zero-phase angle α. The mathematical description of the in-phase component can look as follows:

$$z(\alpha)_I = \sum_{k=0}^{499} 128^2 \cdot \sin(2\pi \cdot f_H \cdot k \cdot T_A + \alpha) \cdot \sin(2\pi \cdot f_H \cdot k \cdot T_A) \tag{4.62}$$

In the left part underneath the sum, there is the digitized received signal with variable zero-phase angle α. The reference (on the right-hand side in (4.62)) is fed in sinusoidal position, so that we obtain the correlation maximum for $\alpha = 0$; see the gray curve in Figure 4-41. As α grows, the correlation value becomes smaller at the in-phase output until it eventually disappears for $\alpha = 90$ degrees. The inverse behavior is shown by the quadrature branch, the output signal of which is described by the equation

$$z(\alpha)_Q = \sum_{k=0}^{499} 128^2 \cdot \sin(2\pi \cdot f_H \cdot k \cdot T_A + \alpha) \cdot \cos(2\pi \cdot f_H \cdot k \cdot T_A) \quad (4.63)$$

The reference is fed in cosine position, in contrast to (4.62). Adding the two output signals geometrically, we use

$$\text{AKF}(\alpha) = \sqrt{z(\alpha)_I^2 + z(\alpha)_Q^2} \quad (4.64)$$

to eventually obtain the desired result, independent of the received signal's zero-phase angle. The quadrature receiver is always a good solution whenever the coherence of the received signal cannot be ensured. However, the most important drawbacks should also be mentioned:

- The cost for correlators doubles, regardless of whether the implementation is analog or digital.
- One channel is added to each bit branch of the receiver, so that the received interference power increases, which means that one has to expect twice the inference amount, so that there will always be a loss of about 3 dB, compared to coherent reception.

The examples presented in Equations (4.60) and (4.61) showed that the bit decision is always easy and free from errors when the received signal is undisturbed, because a numerical value of 4,096,000 can be easily distinguished from zero. Of course, this ideal example does not take any of the possible interference sources into account, and it also assumes that the channel attenuation is fully balanced, so that the receiver's dynamic range can be optimally used. The example in this form is certainly well suited to demonstrate how efficiently the correlation principle for signal detection works. You can see that the ACF maximum grows with the square of the signal amplitude and linearly with the waveform duration. The higher its value, the higher is also the "reserve" available with regard to the noise immunity. Note that an erroneous decision is taken, e.g., delivering an "L" bit when an "H" bit was sent, only if a correlation value can form in the receiver channel that was not addressed due to interference, and if this correlation value exceeds the ACF maximum of (4.60). Such a case cannot be generally excluded, because the transmitted signal never arrives at the receiver in nonattenuated condition, and because it is overlaid with many different kinds of noise, so that the two correlator branches for the "H" and "L" bits are always filled up to a certain amount, even when nothing at all is sent. This noise-specific "basic filling" reduces the "decision distance," so that faulty decisions can occur in cases where both high attenuation and high noise power are present. In this context, impulsive interference that can occur near the receiver where the desired useful signal is already heavily attenuated by the channel is particularly critical. In such cases, the correlation principle should generally be supported by suitable channel coding methods. Such methods allow data transmission in powerlines free from errors even under difficult conditions.

It was stated further above that the desired matched-filter function occurs only by synchronization of a correlator to the beginning of a waveform to be received. If the receiver had to rely exclusively on a received signal that generally arrived in very weak condition and overlaid by many different noise types, there would often be no way to produce reliable synchronization. On the other hand, when using synchronization schemes based on the mains voltage as a reference (see Section 4.4.1), synchronization will be successful even under heavy noise, so that even extremely weak received signals can be optimally evaluated. The systems presented in the following utilize this advantageous solution.

4.4.3.1 Special Digital Correlator Structures for Powerline Matched Filter Receivers

As mentioned before, the simple replacement of the four analog correlators in Figure 4-39 by the digital equivalents of Figure 4-40 is not really an elegant solution. Indeed, the digital technology offers much more. The starting point for our considerations is that a digital signal can basically be stored for an arbitrary time without loss and be used several times. Referred to the digitized received signal, this means that the immediate and parallel multiplication by the four reference waveforms is by no means a must. Instead, a sample of the received signal can be multiplied step by step with the four successive reference samples belonging to the "H" and "L" bits, each in sine and cosine positions, where the results are added in four separate accumulators. The four MAC operations required for this must be completed before the next received signal sample can be processed. The decisive advantage of this approach is that you require only one multiplier and one adder; also, the accumulators can be implemented by use of simple registers. However, the MAC operations will then run at a rate corresponding to at least four times the sampling rate for the received signal. On the other hand, this is hardly a drawback for the powerline applications studied here, because—as we will see later—this will not exhaust the working speed of appropriately designed microelectronic systems, even if certain applications require a much higher number of accumulators.

For the transmission system shown in Figure 4-35, two matched filters working in parallel are sufficient at the receiver side, but they each require one in-phase part and one quadrature part, i.e., a total of four correlators, due to the incoherence of the received signal. Figure 4-42 shows the basic setup of a fourfold correlator in the favor-

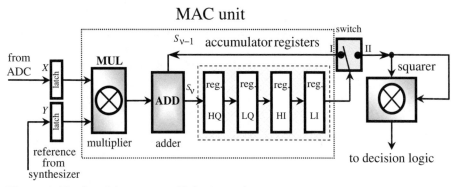

Figure 4-42 Special structure with four correlators.

able structure described above. A series of application-specific integrated circuits (ASICs), which will be described below, were implemented for powerline communication on this basis. The transmission of binary information—a stream of random "H" and "L" bits—with fixed data rate $r_D = 1/T$ and the use of frequency shift keying with two waveforms of duration T will be studied to explain how the structure in Figure 4-42 works.

In this example, the digitized received signal is called $X(i)$; i corresponds to the discrete time, with $i = 0, 1, 2, \ldots \equiv iT_A$, where $T_A = 1/f_A$ is the sampling interval. The Y samples for the reference are generally read from a memory at the fourfold sampling clock $f_r = 4f_A$ and written to another latch. Then the multiplier digitally multiplies each received signal sample $X(i)$ by four reference signal samples $Y(i + v/4)$ with $v = 0, \ldots, 3,$[12] and the four partial products are subsequently integrated over the waveform duration T (chip duration), i.e., added digitally to four separate accumulators. These accumulators are implemented in Figure 4-42 as follows. One adder is connected to four following registers HQ, LQ, HI, LI, and a switch in a ring structure. At the beginning of a waveform reception, the switch is in position II for the duration of four clocks at the frequency f_r. This means that the ring structure is broken up, so that zeros reach the upper input of the adder, while the other input receives consecutively four multiplication results $X(i) \cdot Y(i + v/4)$ with $v = 0, \ldots, 3$ from the multiplier, where the following allocation applies:

$Y(i + 0/4)$	½	"L"-bit sine sample	(in-phase component for "L")
$Y(i + 1/4)$	½	"H"-bit sine sample	(in-phase component for "H")

---

[12] Where v is an abbreviation for vT_A.

Examples of System Implementations

$Y(i + 2/4)$	½	"L"-bit cosine sample	(quadrature component for "L")
$Y(i + 3/4)$	½	"H"-bit cosine sample	(quadrature component for "H")

The registers contain the following results upon completion of four clocks of frequency f_r:

HQ	LQ	HI	LI
HQ_0	LQ_0	HI_0	LI_0

The following is true at the beginning of a waveform because $i = 0$:

$$HQ_0 = X(0) \cdot Y(0 + 3/4) + 0,$$
$$LQ_0 = X(0) \cdot Y(0 + 2/4) + 0,$$
$$HI_0 = X(0) \cdot Y(0 + 1/4) + 0,$$
$$LI_0 = X(0) \cdot Y(0 + 0/4) + 0.$$

The addition of 0 results because the ring structure is opened by the switch. The meaning of the above component sequence will become clear later when we take a closer look at the decision logic function. In this connection, we will also explain why the contents of the registers in position II of the switch are led onto a squarer.

Once the first four clocks at frequency f_r are completed, the switch is brought into position I, so that a ring structure forms, in which the contents of the register LI reach the adder. This activates calculation operations with the following effect on the contents of the registers HQ, LQ, HI, and LI:

	Newly calculated product		Currently accumulated value
Register HQ:	$HQ_i = X(i) \cdot Y\left(i + \dfrac{3}{4}\right)$	+	$\displaystyle\sum_{\xi=0}^{i-1} X(\xi) \cdot Y\left(\xi + \dfrac{3}{4}\right)$
Register LQ:	$LQ_i = X(i) \cdot Y\left(i + \dfrac{2}{4}\right)$	+	$\displaystyle\sum_{\xi=0}^{i-1} X(\xi) \cdot Y\left(\xi + \dfrac{2}{4}\right)$
Register HI:	$HI_i = X(i) \cdot Y\left(i + \dfrac{1}{4}\right)$	+	$\displaystyle\sum_{\xi=0}^{i-1} X(\xi) \cdot Y\left(\xi + \dfrac{1}{4}\right)$
Register LI:	$LI_i = X(i) \cdot Y\left(i + \dfrac{0}{4}\right)$	+	$\displaystyle\sum_{\xi=0}^{i-1} X(\xi) \cdot Y\left(\xi + \dfrac{0}{4}\right)$

Under the prerequisite that a waveform has chip duration $T = NT_A$, we obtain the four desired signal components in the registers HQ, LQ, HI, and LI after $i = N$ clocks at frequency f_A, i.e., after $4N$ clocks at frequency f_r:

HQ	LQ	HI	LI
H_Q	L_Q	H_I	L_I

Next we have to geometrically add the signal components according to the quadrature receiver principle to decide whether an "H" or an "L" bit was sent:

$$H = \sqrt{H_I^2 + H_Q^2}; \qquad L = \sqrt{L_I^2 + L_Q^2} \qquad (4.65)$$

Now we compare the results. If the numerical value H is larger than L, then we decide for an "H" bit; otherwise for an "L" bit.

The described ring structure can be thought of as a "pipeline," and it is by no means limited to four steps. Figure 4-43 describes the expansion for the parallel incoherent reception of ν waveforms. We know that we have to provide 2ν accumulators for this purpose. Also, the clock frequency f_r for the reference has to be increased to the 2ν-fold of the sampling clock for the received signal. The function of the pipeline can be represented in the form

$$S_n = X \cdot Y + S_{n-1} \quad \text{with} \quad S_{n-1} = \sum_{k=1}^{n-1} X(k) \cdot Y(k) \qquad (4.66)$$

where S_n is a newly calculated partial sum, resulting from the previous partial sum S_{n-1} and addition of the currently calculated $X \cdot Y$ product.

Each of the N samples from the received signal arising during one waveform duration must be multiplied by 2ν reference values, resulting in a total of 2νN MAC operations. The figure shows clearly that a circular memory structure is advantageous for the reference values. For each sample of the input signal, there is one complete circulation of this memory, and updating of the relevant registers in (4.66) occurs exactly in the sequence in which the reference values are output.

Basically, the structure of ν correlators for optimum reception of ν sinusoidal signals described in Figure 4-43 can be thought of mathematically as a "selective discrete Fourier transform" (DFT). This connection will be of great interest later in studies of OFDM for telecommunication applications and will, therefore, be described in more detail.

Applying the Fourier transform to a signal $s(t)$ given in the time domain allows the analysis of this signal in the frequency domain. This means that the Fourier transform

Examples of System Implementations

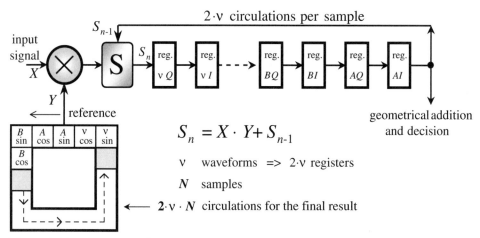

Figure 4-43 Expansion of the "pipeline" ring structure to an arbitrary number of correlations.

$S(f)$ of $s(t)$ represents the entire spectrum. In powerline communication by means of frequency-agile waveforms, it is important for the receiver to be able to properly detect the individual frequencies sent. So we are actually interested not in the analysis of a more or less broad continuous spectrum, but in individual spectral lines with known positions. This is exactly what the "selective discrete Fourier transform" does. The most important benefit versus a complete transform is the limitation to very few interesting spectral lines. The savings potential in hardware and/or computational effort will become clear in a few examples.

In general, you obtain the Fourier transform $S(f)$ of a continuous time signal $s(t)$ by calculating the following integral [F8, F10]:

$$S(f) = \int_{-\infty}^{+\infty} s(t) \cdot e^{-j2\pi ft} \, dt \quad (4.67)$$

With time-limited waveforms as they are of basic interest in powerline communication, the integration extends over the time domain, of course, in which $s(t)$ is present. At the transition to the discrete signal representation, which is mandatory for digital processing, the continuous Fourier transform becomes the discrete Fourier transform (DFT), which extends over N signal samples taken in the time domain; it is calculated by evaluating the following equation (see [F8]):

$$S\left(\frac{n}{NT_A}\right) = \sum_{k=0}^{N-1} s(kT_A) \cdot e^{-j2\pi \frac{n}{N} k} \quad \text{with } n = 0, 1, ..., N-1 \quad (4.68)$$

Here the variables n and k are written unabbreviated, so that we can see that $n: = n/NT_A$ has the dimension of a frequency and $k: = kT_A$ has the dimension of a time unit (in this case the sampling interval). $1/NT_A$ represents the frequency resolution. The example studied further above used $N = 500$ and $T_A = 1.66$ µs, resulting in a frequency resolution (distance of the spectral lines) of 1200 Hz. You can see that this resolution is sufficient to detect the transmitted waveforms, because the minimum frequency distance has to be 1200 Hz, or a multiple thereof, because the hop rate is 1200 s^{-1}.

The full calculation of (4.68) is expensive, because the sum over N complex products has to be calculated for each of the N spectral lines, so that a total of $2N^2$ multiplications and additions have to be executed. If this complete result is needed, then you can use the fast Fourier transform (FFT) [F8], as it can reduce the number of complex multiplications from N^2 to $1/2N \cdot \log_2(N)$. This would mean approximately 2242 complex multiplications and additions instead of about 250,000 for $N = 500$. To exploit these advantages, however, you would require rather expensive hardware in the form of a digital signal processor (DSP). We will see later that there is no workaround for powerline telecommunication to avoid such costs. The generation of OFDM signals at the transmitter side can be easily implemented by inverse FFT (IFFT), while FFT processing is necessary at the receiver side (see Chapter 6). If we were to achieve the same result on the basis of the pipelining principle in Figure 4-43, we would have to implement 1000 correlators for $N = 500$. From the hardware cost perspective, this may seem possible at first sight, because each additional correlator means merely that another register has to be added. Instead, the real problem is the working speed, because even with a moderate sampling rate of $f_A = 600$ kHz for the received signal, we would now need a 600-MHz MAC clock. In most cases, this low sampling rate is sufficient for building automation and energy-related value-added services, while it is absolutely insufficient for the high transmission rates required in telecommunication.

For the building automation and energy-related value-added services described here, the principle of Figure 4-43 has incontestable advantages, because only few of the N spectral lines have to be calculated. In the studied FSK example using $f_H = 120$ kHz and $f_L = 121.2$ kHz, only the positions $n = 100 \equiv 100 \cdot 1200$ Hz $= 120$ kHz and $n = 101 \equiv 101 \cdot 1200$ Hz $= 121.2$ kHz of (4.68) have to be evaluated.

Figure 4-44 shows the result from the evaluation of (4.68) for cosine input signals. To make the spectral lines visible separately, we increased the carrier distance to 12 kHz, which is the tenfold of our example, so that the calculation was done for $n = 100$ (shown in black) and $n = 110$ (shown in gray). We can see the theoretically expected result of the Fourier transform of a cosine in the form of two very sharp spectral lines (Dirac impulses) with "weight" 0.5. In the continuous transform, the lines are

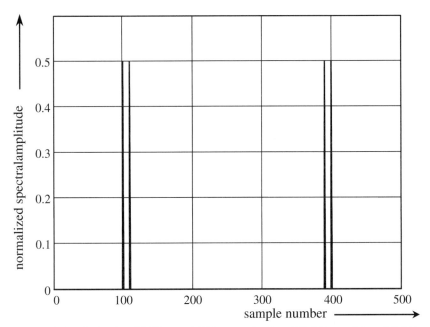

Figure 4-44 Selective DFT for an FSK example: f_H = 120 kHz and f_L = 132 kHz.

each at the positive and negative frequencies of the cosine, i.e., symmetrically to the zero point of the frequency axis. Due to the DFT algorithm from (4.68), where n is always positive, the spectrum appears here symmetric to $n = N/2$; i.e., one line results at $n = 100$ and the second at $n = N - 100 = 400$ for $f_H = 120$ kHz. Accordingly, we obtain a gray line at $n = 110$ and $n = 500 - 110 = 390$ for $f_L = 132$ kHz.

Before we study another important functional receiver block to evaluate the correlator output signals and the decision logic, it is useful to look at the numerical dimensions in the different positions which—as we will see in a moment—can be cut to achieve considerable hardware savings.

Under the assumption that a resolution of 8 bits is sufficient for the received signal in the applications studied here, a number range of −128–127 results at the multiplier inputs. This means that a 16-bit representation is required at the output. The word lengths for the adder and the associated registers (accumulators) should be dimensioned according to the maximum number N_{max} of the samples to be processed during one waveform duration. In practice, $N_{max} = 2000$ has been found useful, because this relatively high number of samples ensures good preamble detection even under difficult conditions; this issue will be discussed later.

We can now use $N_{max} = 2000$ to calculate the accumulator word width. The maximum numerical value for a sinusoidal 8-bit received signal is

$$Z_{max} = \pm \frac{1}{2} \cdot 128^2 \cdot 2000 = 1.6384 \cdot 10^7 = \text{FA0000}_{HEX} \qquad (4.69)$$

A word length of 25 bits would be sufficient to represent this value, including the signs. However, we would have to expand it to 26 bits to avoid overflow errors when the A/D converter is overdriven. Rounding is useful for the subsequent evaluation of the correlator contents. In this respect, we have to consider the shortest waveforms to be processed to ensure that no information is lost when the number range is truncated. N_{min} is about 200 in practical applications, so that $Z_{min} = 1.6384 \cdot 10^6 = 190000_{HEX}$. If the 8 or 9 least significant bits were truncated from this still large number, then we would surely not have to worry about losing information and thus confusing the bit decision, even if the received signal became so small that only the least significant bit of the A/D converter would be valid, because then $Z_{min} = 12{,}800$, and there would still be a value of 25 left after the 9-bit rounding. This means that, before the correlator output signals are passed on to the squarer in Figure 4-42, the 9 least significant bits could be omitted without any problem, so that a word width of 17 bits would be sufficient for the squarer. We will see below that this offers considerable hardware savings potential.

We will now describe the typical decision hardware shown in Figure 4-45, which has been used successfully in powerline communication.

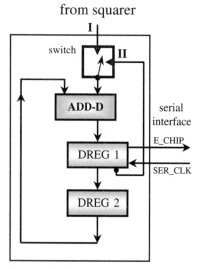

Figure 4-45 Typical decision logic.

To better understand, we assume that a waveform with duration $T = N \cdot T_A$ was just fully received. The switch in Figure 4-42 is now brought into position II for exactly four clocks at frequency f_r. This supplies the contents of the registers LI, HI, LQ, and HQ to the squarer in four clock steps. The squarer result reaches an adder ADD-D over the switch in Figure 4-45 which is in position I. This adder adds the content of register DREG 2 and writes the result to register DREG 1. The adder and the DREG 1 and DREG 2 registers are connected in a ring structure, allowing the calculation of the squared values H^2 and L^2 easily at low hardware cost. Another ring structure forms when the switch goes into position II. This ring structure is used to form the $L^2 - H^2$ difference in another clock step. This difference is then available with the most significant bit (MSB) in the DREG 1 register for serial output over a pin named E_CHIP. The MSB can be read directly. For serial read-out of the other bits of the $L^2 - H^2$ difference, a clock at an arbitrary frequency has to be applied to pin SER_CLK. It is obvious that the square root in (4.65) does not have to be calculated here. The decision can be taken without disadvantage, even with squared values, so that significant computation and hardware cost savings can be achieved. Note that we cannot always do without determining the square root. In such cases, we generally use an approximation solution, which fortunately can be reduced to a few simple mathematical operations; an example will be described below.

The meaning of the sequence of the register contents (see Figure 4-42) becomes clear in the following: The registers DREG 1 and DREG 2 are cleared at the end of the reception of a waveform with duration T. While the switch in Figure 4-42 is in position II, the switch in Figure 4-45 is in position I. With the first edge of clock (f_r), L_I^2 comes from register LI to the adder ADD-D. Zero is added from the DREG 2 register, and the result—L_I^2—is written to DREG 1. With the second clock edge, H_I^2 arrives at the adder, and zero is again added from DREG 2. The result—now H_I^2—is written to DREG 1, while the previous content, L_I^2, is transferred to DREG 2. With the third clock edge, L_Q^2, arrives at the adder from register LQ, together with L_I^2 from DREG 2, so that the sum $L_I^2 + L_Q^2$ reaches DREG 1, from where the old content, H_I^2, is shifted to DREG 2. Next, the fourth clock edge transports H_Q^2, together with H_I^2, to the adder. Now, $H^2 = H_I^2 + H_Q^2$ is in DREG 1 and $L^2 = L_I^2 + L_Q^2$ is in DREG 2. Subsequently, the switch in Figure 4-42 goes back to position I, and the accumulation of a new waveform starts as described above.

To determine the $L^2 - H^2$ difference, the switch in Figure 4-45 is brought to position II for the duration of a fifth clock step, so that the negated DREG 1 register outputs are connected to one adder input, while the content of DREG 2—namely L^2—is present at the second input. This means that the $L^2 - H^2$ difference is calculated in the adder and written to DREG 1 upon completion of the fifth clock step. The signal available at the E_CHIP output represents the received data bit directly due to the one's-complement representation, because the difference calculated for $H^2 > L^2$ is negative, so that the MSB is set; i.e., we have received an "H" bit. Normally, it is sufficient to read one single bit at the E_CHIP output to retrieve the received data. This means that the value of the difference is generally of no interest. Still, the adder ADD_D and the registers DREG 1 and DREG 2 were designed for a word width of 32 bits in the circuit of Figure 4-45; i.e., the two least significant bits are omitted after the squarer that supplies 34-bit results. The cost for the 32-bit processing in the decision logic is not particularly high; the implementation of this accuracy is worthwhile, as we will see below, for example in the detection of the beginning of a data transmission.

It is generally a good idea to send a preamble[13] to make sure that the receiver will recognize the beginning of a data transmission. A problem in the detection of this preamble is to decide only on the basis of a simple comparison of two numbers. The registers HQ, LQ, HI, and LI are always filled to a certain amount from the interference always present in the mains. Nevertheless, the receiver does not necessarily have to get irritated by this, because the interference is not correlated with the waveforms that carry messages, so that interference fills all registers to approximately the same amount. On the other hand, a simple numerical comparison of H and L for a preamble chip decision would be prone to errors, because statistical fluctuations in the register contents result from the stochastic nature of interference. This way, a random bit stream may be observed at the receiver output (E_CHIP pin) without any data being sent by a transmitter. This random bit stream can exactly match the preamble pattern in certain time intervals, which are relatively long for well-designed preambles. This produces a fake beginning of a data transmission from the receiver's viewpoint. This fact alone would not be critical, because the receiver can easily find out from checksums appended to the data whether it received useful data or a random pattern caused by interference. The critical point is that the receiver is blocked for the duration of an entire data frame. If a true preamble follows shortly after the receiver erroneously detects a supposed preamble, it will be "blind" and miss the entire useful data frame.

A receiver is always in search for a preamble during the ready-to-receive phase, because it is the preamble that marks the beginning of a transmission. Therefore, it is meaningful to provide a way to enhance the detection quality for each single preamble chip in this phase. To do this, we also have to evaluate the $L - H$ difference, in addition

---

[13] A signal pattern known to the receiver with particular aperiodic autocorrelation properties.

to the pure magnitude comparison. There is nothing against using the squared H^2 and L^2 values for this purpose.

For example, the $L^2 - H^2$ difference can be read serially—bit by bit—over output E_CHIP by use of a microcontroller, where the microcontroller supplies the read clock over SER_CLK. Subsequently, software is used to compare the amount of $L^2 - H^2$ with a certain threshold value. Ideally, this threshold is selected as a percentage of the maximum L^2 or H^2 value, which would result for undisturbed reception at optimum exploitation of the A/D converter's dynamic range. This percentage is between 2% and 10% in many practical applications. Of course, the best value depends on the real interference environment. For this reason, it is often a good idea to provide an adaptive threshold value, e.g., depending on the interference power received. This can be done by observing the $L^2 - H^2$ difference during transmission pauses. If relatively little noise is present, then the $L^2 - H^2$ difference is likely to be small and exhibit low variance. In this case, you would select a threshold value of about 2%. The threshold value is moved toward 10% as the $L^2 - H^2$ difference increases due to interference. Practical applications have shown that threshold values of more than 10% are not meaningful [D4, D5].

The beginning of a data transmission can be recognized with high safety by using the methods described above in different interference scenarios. Long-term trials confirmed that false alarms at the receiver are virtually excluded, even with high noise levels, so that there is no risk of undesirable blockages.

Based on the concept presented in Figures 4-42, 4-43, and 4-45, a series of application-specific integrated circuits (ASICs) were developed and used in extensive field tests at power supply networks. Meanwhile, the newest generation of these ASICs is the basis for industrial series production of powerline modems. These ASICs and their most important properties will be described shortly.

First, we will take a look at the signal generation for the transmitter and the receiver correlator references.

The functionality of the signal synthesizer shown in Figure 4-46 can be easily understood from the discussion in Section 4.4.2 and the explanations for Figure 4-36. A modem does not have to offer full-duplex capability for the current and—probably—future applications discussed here, because they do not require multiple access; a master-slave protocol [D5] can be used instead. Half-duplex operation will be sufficient, which means that transmitter and receiver do not work in parallel, but alternately. This means for the signal synthesis hardware that a single cyclical address counter (ring counter) and one memory for the samples will be sufficient. When switching from transmitter to receiver operation, an address offset should be provided by the microcontroller for the memory, so that the samples of transmitted and received reference signals

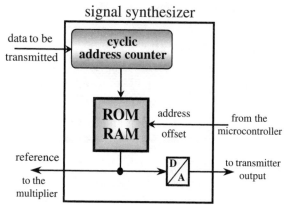

Figure 4-46 Block diagram of the signal synthesis equipment.

can be stored in separate memory locations. During the reception of data, the whole transmitter portion or at least the output stage is switched off. During transmission, the MAC structure of the receiver portion runs idle; i.e., its clock is switched off.

Figure 4-47 shows the entire structure of a half-duplex modem of the type normally used in applications for building automation and energy-related value-added services. The ASIC includes the functional units described in Figures 4-42, 4-45, and 4-46. One possible implementation for the sample memory can be in the form of an OTP (one-time programmable EPROM), which can be written only once. A microcontroller is responsible for further data transfer over its serial interface and, at the same time, controls the ASIC functions for both receive and transmit operations. In addition, the microcontroller is responsible for synchronization, preamble detection, and channel coding and decoding. To perform the synchronization, the microcontroller gets the precise time information, for example over the detection circuit shown in Figure 4-22, from the mains voltage zero-crossings. On this basis, it can then use its timer system to generate the required timing patterns.

In transmit operation, the ASIC works together with the sample memory to form a signal synthesizer, where the microcontroller selects the corresponding memory locations, depending on the data to be sent. After the D/A conversion and the lowpass filtering, the transmission signal reaches an output power stage and is fed over a coupler into the mains. The output power stage has to dispose of a mute[14] function that can be served by the microcontroller. In receive operation, the output stage can be totally switched off and does not influence the received signal.

---

[14] Muting deactivates the power transistors, so that we have a high-Z output.

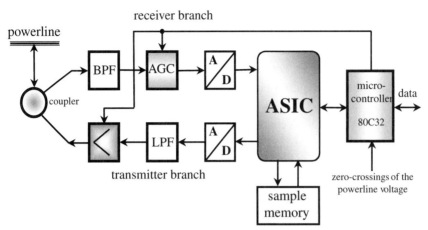

Figure 4-47 Complete setup of a powerline modem.

In receive operation, the received signal first reaches a bandpass filter over the coupler; this bandpass filter preselects the interesting frequency ranges. It is followed by an equipment for automatic gain control (AGC), which makes sure that the A/D converter is always exploited as far as possible to optimally use its dynamic range.

Based on the presented powerline data-transmission concept using frequency-agile waveforms, several industrial series devices for building automation applications have been developed during recent years. One representative example, the Powernet-EIB (European Installation BUS) system, which is based on a "spread" FSK technology, will be described below. The need for energy-related value-added services—currently a hot topic—has encouraged the development of corresponding systems, so that a mature result based on "modified frequency-hopping modulation" can be presented. The design of these two powerline communication systems will be discussed in the final section of this chapter.

4.4.3.2 Application-Specific Integration Issues

The immediate realization of suitable integrated circuits (ASICs) from the very beginning was an important prerequisite for the success of the concepts presented here. The requirements concerning the execution speed of the MAC operations rendered the use of simple microcontrollers impossible, even in FSK. Only digital signal processors (DSPs) would have managed the task. The first activities toward the digital implementation of powerline modems were started a decade ago. At that time, DSPs were characterized by high cost, complex system structure, and high power requirement. In addition, the industry suffered from a lack of trained developers in this complex field. Since then the landscape has changed, and many engineers are as familiar with DSPs today as with standard components like microprocessors. We will see later that this is an important prerequisite for the development and construction of future powerline telecommunication systems. For the applications studied here with their relatively low data rates, there is still a considerable gap between an adapted ASIC solution and the use of a standard DSP. The number of required equivalent gate functions[15] alone gives a good idea of how big this difference actually is: a typical powerline ASIC, for instance one that contains four to eight correlators, can be implemented with less than 20,000 equivalent gate functions, while standard DSPs require at least ten times this number. To explain this enormous difference, we will look at the hardware components that entail the highest gate cost—these are undoubtedly multipliers and squarers—in an example of an economic application-specific implementation.

The multiplier in Figure 4-42, for example, works with an input word width of 8 bits, and 16 bits are required at the output. In general, standard DSPs have word widths of 16–32 bits, so that there is already a high redundancy. Assuming that the gate cost increases exponentially to the word width, we already identified a high savings potential for an ASIC solution. Although the squarer in Figure 4-42 is required only for a very short time at the end of a received waveform, it has to work in parallel with the correlator's multiplier, because the next incoming waveform has to be processed without delay. Considering that a standard DSP uses only one multiplier, the tasks of a circuit from Figure 4-42 could generally be handled with only two DSPs. The first ideas about a favorable application-specific implementation of multipliers in powerline modems dates back more than ten years. The basic results, which will be presented below, have been applied in a series of implemented ASICs, and they are definitely still of interest.

Our starting point is the representation of the binary variables X and Y at the input of the multiplier in Figure 4-42 in the form of 8-bit two's-complement numbers. If we write X and Y in the form of row vectors, i.e.,

$$X = (x_7, x_6, x_5, x_4, x_3, x_2, x_1, x_0) \tag{4.70}$$

$$Y = (y_7, y_6, y_5, y_4, y_3, y_2, y_1, y_0) \tag{4.71}$$

then the binary positions $x_0, y_0, \ldots, x_6, y_6$ have the decimal values $2^0, \ldots, 2^6$. The decimal value $-2^7 = -128$ is allocated to the positions x_7 and y_7. Thus, the decimal values for X and Y can be calculated as follows:

$$X = -2^7 \cdot x_7 + \sum_{i=0}^{6} x_i \cdot 2^i \tag{4.72}$$

---

[15] Assuming here a NAND gate with two inputs in CMOS technology, containing four transistors.

Examples of System Implementations

and

$$Y = -2^7 \cdot y_7 + \sum_{j=0}^{6} y_j \cdot 2^j \tag{4.73}$$

The accumulated sum is called S_{AKK} in the following and represented as a binary variable with a width of 26 bits:

$$S_{AKK} = (a_{25}, a_{26}, ..., a_1, a_0) \tag{4.74}$$

The two's-complement representation is then

$$S_{AKK} = -2^{25} \cdot a_{27} + \sum_{i=0}^{24} a_i \cdot 2^i \tag{4.75}$$

Using the pipelining algorithm from (4.66), we obtain

$$S_{AKK,n} = X \cdot Y + S_{AKK,n-1} \tag{4.76}$$

Now the algorithm to be implemented can be stated in binary two's-complement notation

$$S_{AKK,n} = \left[-2^7 \cdot x_7 + \sum_{i=0}^{6} x_i \cdot 2^i \right] \cdot \left[-2^7 \cdot y_7 + \sum_{j=0}^{6} y_j \cdot 2^j \right] - 2^{25} a_{(n-1)25} + \sum_{i=0}^{24} a_{(n-1)i} \cdot 2^i \tag{4.77}$$

This means that $S_{AKK,n}$ in (4.77) is the sum accumulated in a register after n sampling clocks. The index $(n-1)$ at the components a_i indicates that these values are derived from the previous step $(n-1)$.

Before we can convert the calculation rule from (4.77) into digital hardware, we have to do a few elementary conversion steps. We obtain the result

$$S_{AKK,n} = \underbrace{-2^{25} \cdot a_{n25} + \sum_{i=0}^{24} a_{ni} \cdot 2^i}_{\text{new accumulated sum } S_{AKK,n}} = -2^{25} + \underbrace{\sum_{i=15}^{24} 2^i +}_{\substack{\text{sign extension} \\ \text{up to 26 bits}}}$$

$$\underbrace{x_7 y_7 \cdot 2^{14} + \sum_{i=0}^{6} \sum_{j=0}^{6} x_i y_j \cdot 2^{i+j} + 2^8 + \sum_{i=0}^{6} \overline{x_7 y_i} \cdot 2^{7+i} + \sum_{i=0}^{6} \overline{y_7 x_i} \cdot 2^{7+i}}_{\text{product } X \cdot Y}$$

continued →

$$\underbrace{-2^{25} a_{(n-1)25} + \sum_{i=0}^{24} a_{(n-1)i} \cdot 2^i}_{\text{old accumulated sum } S_{\text{AKK},n-1}} \quad (4.78)$$

From (4.78), we can directly derive rules for the construction of a fast digital circuit that executes the algorithm at high efficiency. Of particular interest is fast multiplication by means of parallel hardware, which we will describe below.

To understand how a fast parallel multiplier adapted exactly to the formulated task works, we take the section marked "Product $X \cdot Y$" from (4.78). We use AND gates to easily form the partial products $x_i y_j$ of two binary variables, which occur in (4.78). The bar, for example above $x_7 y_i$, means that the binary variables x_7 and y_i first have to be logically AND-ed, and that the result has to then be negated. $\overline{x_7 y_i}$ can thus be formed very easily by means of a NAND gate.

Table 4-5 shows a multiplication scheme in matrix form, which establishes directly the reference to digital hardware. For the two's-complement representation of the product $X \cdot Y$, the sign bit -2^{15} from the term "sign extension up to 26 bits" has to be added.

Table 4-5 Scheme for fast parallel multiplication.

$p_{15} \longleftarrow \qquad \qquad \qquad \qquad \qquad \longrightarrow p_0$

-2^{15}	2^{14}	2^{13}	2^{12}	2^{11}	2^{10}	2^9	2^8	2^7	2^6	2^5	2^4	2^3	2^2	2^1	2^0
									x_0y_6	x_0y_5	x_0y_4	x_0y_3	x_0y_2	x_0y_1	x_0y_0
							1	x_1y_6	x_1y_5	x_1y_4	x_1y_3	x_1y_2	x_1y_1	x_1y_0	
							x_2y_6	x_2y_5	x_2y_4	x_2y_3	x_2y_2	x_2y_1	x_2y_0		
						x_3y_6	x_3y_5	x_3y_4	x_3y_3	x_3y_2	x_3y_1	x_3y_0			
					x_4y_6	x_4y_5	x_4y_4	x_4y_3	x_4y_2	x_4y_1	x_4y_0				
				x_5y_6	x_5y_5	x_5y_4	x_5y_3	x_5y_2	x_5y_1	x_5y_0					
			x_6y_6	x_6y_5	x_6y_4	x_6y_3	x_6y_2	x_6y_1	x_6y_0						
		$\overline{x_7y_6}$	x_7y_5	x_7y_4	x_7y_3	x_7y_2	x_7y_1	x_7y_0							
1	$\overline{x_7y_7}$	$\overline{x_6y_7}$	$\overline{x_5y_7}$	$\overline{x_4y_7}$	$\overline{x_3y_7}$	$\overline{x_2y_7}$	$\overline{x_1y_7}$	$\overline{x_0y_7}$							

$P = XY$

$P = p_{15} - p_0$

$X = x_7 - x_0$

$Y = y_7 - y_0$

Table 4-5 has columns with a significance from 2^0 to 2^{15}. All entries in each column have to be added to calculate the multiplication result. The arising carry bits have to be taken into account in the columns with higher significance. The calculation speed achievable with conventional addition methods will be relatively low as the carry bits propagate. A basic concept for fast parallel multiplication that differs from conventional addition methods was published by Wallace and Dadda [F6, F9]; it uses "CARRY-SAVE" adders. CARRY means the carry bit that can result from binary addition of two variables when using a half adder, or from three variables when using a full adder. SAVE is self-explanatory; it means that carry bits are not immediately evaluated, e.g., by a more or less complex addition network, as is the case with a fast adder. Instead, an arrangement of CARRY-SAVE adders works for fast completion of an addition task, such as given in Table 4-5, in several steps. The carry bits arising in each step are not further processed immediately but are carried over to the following step.

Figure 4-48 shows the way an arrangement called "CARRY-SAVE ADDER ARRAY" works. The processing can be analyzed by use of Table 4-5, as described by Wallace and Dadda. In this connection, full adders are also called "(3, 2) counters" and half adders "(2, 2) counters." This means that a full adder reduces three inputs from one column in Table 4-5 to one output in the same column and supplies a carry to the next column with higher significance. Consequently, a half adder reduces two entries to one plus a carry. The carry bits from one step are transferred to the next step by arranging them with the remaining matrix elements that still have to be added and with the sum bits, so that a new matrix is built, which will be further reduced by the next step. Figure 4-48 shows how the scheme from Table 4-5 is processed with a large number of parallel half and full adders in four steps to a final addition of two 16-bit two's-complement numbers. At the beginning, a column contains up to 8 input bits waiting to be added. The large number of 39 full adders and 15 half adders can operate totally in parallel in a particularly efficient way, if "pipeline registers" are inserted between the steps [F9, D3, ST19, ST20]. The maximum delay time in the scheme of Figure 4-48 is the same for all four steps; it corresponds roughly to one full-adder delay.

The entire calculation of the product up to the final addition would require four full-adder delays. By inserting pipelining registers, the partial nets of the four steps are separated, so that they can process different operands independently of one another; i.e., after one full adder delay time, for example, the result from step 1 is latched into pipeline registers and available for processing as input of step 2, while new operands can enter step 1 at the same time. When the pipeline is filled, there is about the fourfold processing speed; i.e., after one full-adder delay plus the setup time of one pipelining register, the result appears at the final adder. The applications studied here do not require the

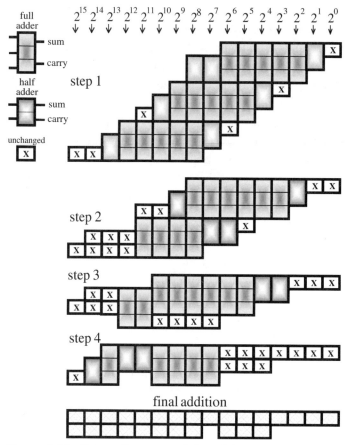

Figure 4-48 Schematic representation of the processing of a multiplication matrix by means of 15 half adders and 39 full adders in four steps.

use of pipelining. However, pipelining will be an important issue in telecommunication applications.

The fast parallel multiplication described above, which ended with a final addition of two binary vectors, requires another addition, namely that of the old accumulated sum, in order to complete the calculation of (4.78). As mentioned before, the speed requirements are not very high for the applications discussed here, so that you can use simple adders known as "RIPPLE-CARRY" adders [A1]. In this type of adder, the carry bits propagate ripplelike from the low-order to the higher-order positions. The hardware cost is low, which is at the cost of delay time, of course.

The result is a total gate number of only about 450 for the described parallel 8-bit multiplier without the final additions, if we assume approximately ten gate functions for a full adder and five for a half adder. This is a relatively small digital circuit. The complete structure, including input latches and final adders, can be implemented with less than 1000 equivalent gate functions, corresponding to about 4000 transistors.

The squarer in Figure 4-42 can also be implemented easily in an application-specific way. Although we now have an input word width of 17 bits, the circuit does not take on much more volume than the 8-bit multiplier described above. The reason is that, for the squarer, considerable simplification can be achieved in that its two input operands are always equal. First, an algorithm similar to the one in (4.78) can be formulated to square an N-bit two's-complement number

$$A = -a_n \cdot 2^{N-1} + \sum_{i=0}^{N-2} a_i \cdot 2^i \tag{4.79}$$

We obtain:

$$A^2 = -2^{2N-1} + a_{N-1} \cdot 2^{2N-2} + 2^N + \sum_{i=0}^{N-2} \overline{a_{N-1} \cdot a_i} \cdot 2^{i+N} + \sum_{i=0}^{N-2}\sum_{j=0}^{N-2} a_i \cdot a_j \cdot 2^{i+j} \tag{4.80}$$

For our application with $N = 17$, this results in:

$$A^2 = -2^{33} + a_{16} \cdot 2^{32} + 2^{17} + \sum_{i=0}^{15} \overline{a_{16} \cdot a_i} \cdot 2^{i+17} + \sum_{i=0}^{15}\sum_{j=0}^{15} a_i \cdot a_j \cdot 2^{i+j} \tag{4.81}$$

Based on (4.81), this gives a 34-bit result, which can again be entered in a calculation scheme like the one in Table 4-5.

The most important simplifications, compared to the multiplication, result from the fact that all partial products which have equal indices can immediately be written as $a_i \cdot a_i = a_i$, so that no gate is required for calculation, and from the fact that all partial products with interchanged indices are equal, i.e., $a_i \cdot a_j = a_j \cdot a_i$. This means that the latter have to be calculated only once in all cases and can then be inserted immediately into the column with the next higher binary significance. This executes the addition $a_i \cdot a_j + a_j \cdot a_i$ at no hardware cost. Therefore, all products with mixed indices appear in Table 4-6 only once—always in a column with the binary significance that is one higher than the sum of the indices. For example, column 2^{10} contains $a_0 \cdot a_9$, $a_1 \cdot a_8$, $a_2 \cdot a_7$, ..., $a_4 \cdot a_5$ as well as a_5, which represents $a_5 \cdot a_5$. The squarer can be implemented with about 1000 equivalent gates, so that it also represents a relatively small digital circuit.

The photos in Figure 4-49 show a series of eight ASICs for the construction of powerline modems, which were produced within the author's activities and tested in field trials during the past ten years. Instead of describing these devices in detail, they

Table 4-6 Calculation scheme for the squarer algorithm in (4.81).

	-2^{33}	2^{32}	2^{31}	2^{30}	2^{29}	2^{28}	2^{27}	2^{26}	2^{25}	2^{24}	2^{23}	2^{22}	2^{21}	2^{20}	2^{19}	2^{18}	2^{17}
				a_{15}	$a_{13}a_{15}$	a_{14}	$a_{11}a_{15}$	a_{13}	a_9a_{15}	a_{12}	a_7a_{15}	a_{11}	a_5a_{15}	a_{10}	a_3a_{15}	a_9	a_1a_{15}
				$a_{14}a_{15}$		$a_{12}a_{15}$	$a_{12}a_{14}$	$a_{10}a_{15}$	$a_{10}a_{14}$	a_8a_{15}	a_8a_{14}	a_6a_{15}	a_6a_{14}	a_4a_{15}	a_4a_{14}	a_2a_{15}	a_2a_{14}
							$a_{13}a_{14}$	$a_{11}a_{14}$	$a_{11}a_{13}$	a_9a_{14}	a_9a_{13}	a_7a_{14}	a_7a_{13}	a_5a_{14}	a_5a_{13}	a_3a_{14}	a_3a_{13}
								$a_{12}a_{13}$		$a_{10}a_{13}$	$a_{10}a_{12}$	a_8a_{13}	a_8a_{12}	a_6a_{13}	a_6a_{12}	a_4a_{13}	a_4a_{12}
										$a_{11}a_{12}$		a_9a_{12}	a_9a_{11}	a_7a_{12}	a_7a_{11}	a_5a_{12}	a_5a_{11}
												$a_{10}a_{11}$		a_8a_{11}	a_8a_{10}	a_6a_{11}	a_6a_{10}
														a_9a_{10}		a_7a_{10}	a_7a_9
																a_8a_9	
a_{16}																	1
1	$\overline{a_{16}a_{15}}$	$\overline{a_{16}a_{14}}$	$\overline{a_{16}a_{13}}$	$\overline{a_{16}a_{12}}$	$\overline{a_{16}a_{11}}$	$\overline{a_{16}a_{10}}$	$\overline{a_{16}a_9}$	$\overline{a_{16}a_8}$	$\overline{a_{16}a_7}$	$\overline{a_{16}a_6}$	$\overline{a_{16}a_5}$	$\overline{a_{16}a_4}$	$\overline{a_{16}a_3}$	$\overline{a_{16}a_2}$	$\overline{a_{16}a_1}$	$\overline{a_{16}a_0}$

2^{16}	2^{15}	2^{14}	2^{13}	2^{12}	2^{11}	2^{10}	2^9	2^8	2^7	2^6	2^5	2^4	2^3	2^2	2^1	2^0
a_8	a_0a_{14}	a_7	a_0a_{12}	a_6	a_0a_{10}	a_5	a_0a_8	a_4	a_0a_6	a_3	a_0a_4	a_2	a_0a_2	a_1		a_0
a_0a_{15}	a_1a_{13}	a_0a_{13}	a_1a_{11}	a_0a_{11}	a_1a_9	a_0a_9	a_1a_7	a_0a_7	a_1a_5	a_0a_5	a_1a_3	a_0a_3		a_0a_1		
a_1a_{14}	a_2a_{12}	a_1a_{12}	a_2a_{10}	a_1a_{10}	a_2a_8	a_1a_8	a_2a_6	a_1a_6	a_2a_4	a_1a_4						
a_2a_{13}	a_3a_{11}	a_2a_{11}	a_3a_9	a_2a_9	a_3a_7	a_2a_7	a_3a_5		a_3... a_2a_5							
a_3a_{12}	a_4a_{10}	a_3a_{10}	a_4a_8	a_3a_8	a_4a_6	a_3a_6		a_3a_4								
a_4a_{11}	a_5a_9	a_4a_9	a_5a_7	a_4a_7		a_4a_5										
a_5a_{10}	a_6a_8	a_5a_8	a_6a_7	a_5a_6												
a_6a_9		a_6a_7														
a_7a_8																

Examples of System Implementations 171

Figure 4-49 Steps in ASIC development for powerline communication.

will be presented in several pictures. A more detailed presentation will follow for those prototypes, which became the immediate forerunners of devices currently produced in series in the industry.

ASIC 1 was manufactured by Lasarray of Bienne, Switzerland, on a 2-μm CMOS gate array in November 1989 [A7]. It includes the receiver functions of Figure 4-42 without squarer. It was produced by means of a "laser direct writing process" on a wafer called XLS 3600[16] [A7]. Laser direct writing is a specialty used by only few semiconductor manufacturers worldwide for rapid prototyping. One of the major benefits is that the production of an integrated circuit does not require previous expensive mask production. Instead, a laser is used to write the customer-specific wiring directly on a special wafer, fully metallized and coated with photoresist. The remaining production steps

---

[16] 3600 means the maximum available number of equivalent gates ≡ approximately 14,400 transistors.

consist of relatively simple photoresist development and etching processes. Customers obtain their integrated circuits quickly and at relatively low cost [A7].

The drawbacks of this method are that production of large quantities is uneconomic, because each individual component has to be written, and the CMOS technology used cannot keep up with the development in semiconductor industry toward smaller and smaller structures due to the limited resolution of the laser exposure. There was no multilayer metallization for the XLS 3600 wafer. One tried to offset this disadvantage at least in part by use of fixed polysilicon paths underneath the metallization level, which are accessible at certain points from the metallization level. Note, however, that an extensive use of this polysilicon has a degrading effect on speed due to the high specific resistance. The component named PRP 8/28 (short for Pipelining Ring Processor with 8-bit inputs and 28-bit output) occupies a silicon area of about 83 mm^2.

The PRP 8/28 was used to build the first fully digital FH receivers for powerline data transmission. A few additional digital components for reference signal synthesis at the receiver side and evaluation of the correlator output signals for bit decision as well as a separate transmitter were required to complete a modem. Figure 4-50 shows the important functional groups of the first modems for a data rate of 60 bits/s and the use of fast frequency hopping with 5 frequencies per bit.

The upper half of the picture shows the transmitter, containing the core elements—a microcontroller and an EPROM to store the samples. One particularity of this setup is that the microcontroller program is also stored in this sampling memory. The bottom half of the picture shows the receiver with its components grouped around the ASIC. We can see a microcontroller (in DIL-40 package) and two EPROMs—one for the reference signal samples and the other for the microcontroller program. We will not describe this circuit board any further, because this setup is long obsolete in view of the state-of-the-art technology.

In autumn 1990, FH receivers like those in Figure 4-50 were successfully put into operation in powerlines [D2, ST12]. From the very outset the digital concept allowed achieving a quality that compares well to similar devices based on lock-in amplifiers, which, however, have to be built with careful selection of the components and time-consuming compensation work.

A look at Figure 4-50 makes it clear that we are far from dealing with a satisfactory hardware solution. Therefore, it was desirable to encourage further integration, as all important components were already digital. This was done in a first step by implementing the ASICs 2, 3, and 4 (see Figure 4-49) during the period from 1990 to 1994. These three chips have basically the same contents, but they use different production technologies. Comparing the contents to those of ASIC 1, an output register with a stan-

transmitter

first digital receiver with ASIC

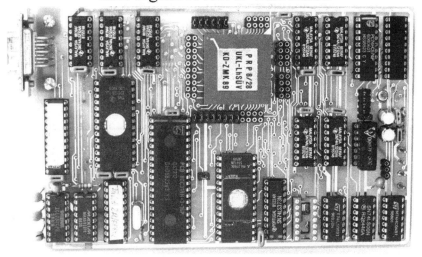

Figure 4-50 The first purely digital FH modem prototypes [D2].

dard 8-bit microprocessor interface was added, so that the correlators can be read over an external microcontroller independently of the current correlation running inside. In addition, the hardware of Figure 4-46 for signal synthesis was also integrated both for the transmitter signals and for the correlation reference. This improvement allowed building complete modems in fully digital construction for the first time. In 1990, ASIC 2 was still manufactured by the laser direct writing method, like ASIC 1, but on a slightly larger wafer with 5000 equivalent gate functions.

ASIC 3 contains exactly the same functional units as ASIC 2, but it was manufactured in 0.7-μm BiCMOS[17] technology within a sample production run. Upon successful testing of ASIC 3, approximately 500 pieces were manufactured in a different package (ASIC 4), but with the same contents and also in 0.7-μm BiCMOS technology. This was the first time that a sufficient number of low-cost chips were available for extensive experiments. In 1991, roughly at the same time, the European standard EN 50065 came into force and replaced the former very restrictive regulations of Deutsche Bundespost [NO1, NO2]. Eventually, the new side conditions made it possible to increase the data rate to 1200 bits/s.

Figure 4-51 shows the prototype of a complete half-duplex modem for a data rate of 1200 bits/s—the first prototype implemented on the basis of ASIC 2. The progress in signal processing enabled by the digital technology now permitted achieving high noise immunity, even with simple FSK modulation. This approach had been closely followed during the period from 1992 to 1994 in the development of low-cost and relatively fast building automation systems. One important goal was to achieve a small modem design, so that the modem would fit into a standard concealed wall socket usually found in most building installations. A closer look at Figure 4-51 shows that this modem still required two eurocards,[18] which means that there was extensive development work ahead. Consequently, the next goal was to achieve a more compact packing of components onto one chip. This was not realized in building automation systems for cost reasons, but was acheived later when systems for energy-related value-added services in the form of ASIC 8 (see Figure 4-49) were developed.

The ASIC series 5, 6, and 7 in Figure 4-49 characterizes the development path of building automation systems. The three chip variants include basically the same functional units. The entire decision logic from Figure 4-45 and the squarer from Figure 4-42 were integrated in a further development of chip 4. The hardware description language VHDL (Very High Speed Hardware Description Language) was used for the entire design, because it offers easily readable circuit designs and supports easy and safe changes and expansions. ASICs 5 and 6 were produced within university programs called EUROCHIP or EUROPRACTICE in 1994 and 1995. These chips were realized in standard cell arrays based on 1-μm CMOS technology, and they comprise approxi-

---

[17] Combination of bipolar and CMOS technologies.
[18] 10×16 cm in size.

Examples of System Implementations

Figure 4-51 Progress in the development of digital powerline modems.

mately 5200 equivalent gate functions on a silicon area of approximately 14 mm^2. A small series of approximately 1000 pieces was produced in the form of ASIC 7 upon successful testing of samples.

Eventually, the modems for the Powernet-EIB building automation system were developed on the basis of these ASICs, and by 1996 the modems had reached a development stage allowing their installation in standard concealed wall sockets. Since 1997, corresponding series devices have been on the market [ST25]. Due to the underlying significance of the technology implemented here, we will summarize the most important ideas that led to this success in the following section. Next, we will discuss recent further developments for energy-related value-added services, for which a considerable modification of the transmission scheme was required. The resistance to interference achievable by use of FSK has proven to be insufficient for a reliable connection from the transformer station to the customer's appliances. This means that extensive revision and expansion of then-current solutions were required for the design of ASIC 8.

4.4.4 Powernet-EIB in Building Automation

Studies in numerous power supply networks of various lengths and properties inside and outside of buildings have shown that frequency hopping (FH) is generally a good modulation scheme. This holds true even if only two different waveforms at relatively distant frequencies per transmission channel are used. The frequency distance should always be a multiple of the data rate, for example between 5 and 10 kHz at 1200 bits/s.

Until 1990, the frequency band available for powerline communication was specified for 30 to 146 kHz by the regulations of Deutsche Bundespost. This resulted in a relatively large available bandwidth of $B = 116$ kHz. Since December 1, 1991, the European standard EN 50065, described further above, has been in force [NO3]. This standard allows private users a frequency range of 95–148.5 kHz at a transmission level of 122 dBµV for in-house applications not subject to approval. The relatively low admissible transmission level is accompanied by a much smaller bandwidth, which is now 53 kHz instead of 116 kHz. These new side conditions impose high requirements upon devices for sufficiently undisturbed communication over indoor powerlines.

Signal generation in the transmitter and signal processing in the receiver normally require some form of synchronization. At the receiver side, the received signal is synchronized to a reference signal generated locally in the receiver. This reference signal is supplied by a frequency synthesizer. At the other end, the transmitter also requires a synchronized frequency synthesizer; its information-carrying output signals are fed into the mains. The synchronization problem can be solved easily and at low cost by means of the mains voltage.

The modem concept presented in the following combines transmitter and receiver components, which are synchronized by globally using the mains voltage as the reference, allowing the transmission or reception of message-carrying waveforms (half-duplex operation). This allows high flexibility with regard to the selected data rate, the noise immunity (preamble code, number of waveforms per data bit), and the selected frequency (position of the frequencies used in the transmission band). The prerequisite is a mostly digital modem construction.

The modem's transmitter equipment is responsible for the generation of waveforms at the correct frequency and time, and for sufficient filtering and amplification, so that the maximum admissible transmission level is fed into the mains. It should be possible to change the frequency in a phase-continuous way and without transients.

The modem's receiver equipment has to be capable of perfectly separating weak signals which are relatively close in frequency, even under heavy interference, to recover the transmitted information. This requires matched filtering. The principle of suitable quadrature receiver structures was described in further detail above.

The following description is based on the modem concept introduced in Figure 4-47. This digital concept generates signals at the transmitter side and handles correlative signal processing at the receiver side. At the receiver input, after analog amplification and filtering, the received signal is sampled at the sampling frequency and digitized with an 8-bit resolution. For correlative processing, the received signal is digitally multiplied, using samples of a reference waveform. Subsequently, the product is integrated over the waveform duration T; i.e., it is digitally summed up in an accumulator. The required reference signal samples are stored in a memory, e.g., an EPROM. They are read out by means of an address counter. There are several reference signal samples for one received signal sample; i.e., each received signal sample is multiplied by four reference values consecutively. The multiplication results are transferred to four digital accumulators, where they are summed, so that there are one in-phase and one quadrature signal each (HI, HQ; LI, LQ) in digital form for the "H" and the "L" data bit channels at the end of a chip duration. The processes of evaluation of these correlation results to the bit decision were described earlier.

As mentioned above, the reference signals are read from a memory, which is addressed by a cyclic counter. The address spaces for the reference values of each waveform should ideally have the same length. In the modem design described here, this was achieved by setting the counter cycle to 500. The desired reference values are selected by the microcontroller (see Figure 4-47) over the ASIC by use of address lines leading to the memory's address bus. This implements a digital frequency synthesizer for the reference waveforms required at the receiver side. After each waveform duration, the microcontroller reads one received result from the ASIC and eventually delivers the received data stream over its serial interface to an arbitrary message sink.

Considering that we want to send and receive, it is a good idea to also use the microcontroller, the sample memory, and the cyclic counter integrated in the ASIC to implement the transmitter function, allowing bidirectional data traffic in half-duplex mode. The implemented solution is strikingly simple and inexpensive, and it supports modification of the most important system parameters, such as data rate, waveform, and frequency selection, without the need to change hardware. All that is required is to change the content of the sample memory and the microcontroller program. Also, note that a method for intelligent preamble detection was implemented in the existing microcontroller in hardware and software.

For the following example, we assume a transmission of binary information with a fixed data rate of $r_D = 1/T = 1200$ bits/s by use of FSK with two waveforms of duration T, where the "H" bit is mapped to frequency $f_H = 120$ kHz and the "L" bit is mapped to $f_L = 132$ kHz (see also Figure 4-44). As the signals are received, they pass from the mains over the coupler to the input bandpass filter, which suppresses interference from undesired frequency ranges. As described in further detail above, the coupler consists of a highly permeable ring core made of ferromagnetic powder carrying two windings. It provides for galvanic separation between the mains and the modem and, at the same time, represents a highpass filter that prevents all signals with frequencies of up to several kilohertz from reaching the modem. The coupler is used for both incoming and outgoing data traffic. Next, the filtered received signal passes on to an AGC amplifier, which ensures that the dynamic range of the following 8-bit analog/digital converter is optimally used. This analog/digital converter samples and quantizes the received signal and supplies the digitized values to the ASIC, where the correlative processing follows. The microcontroller obtains the result, which is generally the received data bit, after each waveform duration. There are some noteworthy particulars during the reception of a preamble, which will be described in detail below.

The modem's transmit operation is started when the microcontroller receives an appropriate command and data to be transmitted over its bidirectional serial interface. Next, the memory location where the samples for the waveforms to be transmitted are stored is selected. The receiver components of the ASIC switch into a wait state, while the integrated counter continues working. The transmission requires two waveforms at different frequencies (e.g., f_H = 120 kHz and f_L = 132 kHz). Their samples are read from the memory during the data bit duration, depending on the current data bit ("H" or "L"). The samples reach an 8-bit digital/analog converter, followed by a reconstruction (lowpass) filter. An amplifier supplies the required transmission voltage, and the transmission signal reaches the coupler, which feeds the signal into the mains.

Synchronization is done over a zero-crossing detector that synchronizes the program for transmission in the microcontroller over an interrupt input with the mains voltage. Figures 4-21, 4-23, and 4-24 show that the beginning of a data transmission should be coupled to a mains zero-crossing to prevent ambiguities in three-phase systems. Operating within a three-phase system results in the known time pattern of 3.3 ms, which has to be generated by a timer integrated in the microcontroller. This pattern would correspond to a data rate of 300 bits/s. Of course, this does not mean that no other data rates could be realized. For example, the basic pattern of 3.3 ms can be divided into four equal parts of 833.3 µs each for a data rate of r_D = 1200 bits/s. This task can also be performed by a timer integrated in the microcontroller. A typical transmission frame consists of a preamble and a data field, both inserted into the time pattern derived from the mains voltage. As mentioned before, it is very important for the preamble detection to provide for high safety against both detection losses and false alarms. Either case would lead to the loss of a large quantity of data, because the typical data field lengths for technical applications are 100 bits and more. The task of achieving high detection safety for the preamble and, at the same time, low false alarm probability is very complex. One solution now commonly used in industrial series production will be described below.

The assurance of correctly detecting a transmitted waveform at the receiver side depends mainly on the energy at which this waveform arrives. As the admissible transmission voltage is limited and should purposefully be utilized to its maximum, a waveform's energy can be increased only by extending the time. In a modem for 1200 bits/s, this is done by building the preamble in several sections with a total duration of 3.3 ms; these sections are called "preamble chips." During these time periods, waveforms at the frequencies defined for the "H" and "L" bits are transmitted quasi-randomly and consecutively based on a specific preamble code. Detection becomes safer because one preamble chip has now the fourfold duration of a data bit. In addition, it is important to select a "code" with good aperiodic autocorrelation properties for the preamble. This means that a high correlation peak forms, provided that the preamble chips are widely free from errors and appear in the correct time position in the receiver. In all other cases, there must be no significant peak, i.e., the so-called "sidelobes" have to be as small as possible, to achieve good aperiodic autocorrelation properties.

When it detects this peak, the receiver knows that it has to start detecting data. This can be done after the preamble reception much more advantageously than otherwise, e.g., at the beginning of a transmission. The reason is that the automatic gain control (AGC) has adapted the receiver input amplifier to the existing receive level, and the microcontroller disposes of safe preliminary information to know when to expect data to arrive. Basically, all it has to do after expiration of each bit interval (with duration $T = 833.3$ μs) is to read the bit information waiting at the output pin E_CHIP of the ASIC (see also Figure 4-45). Given that no comparison with a threshold value is required during the bit decision, no bit errors are to be expected as long as the receiver's dynamic range is maintained, even during heavily fluctuating receive levels. A suitable design of the automatic gain control (AGC) will ensure that a dynamic range of far more than 80 dB can be covered by a low-cost 8-bit analog/digital converter.

In contrast, conditions are relatively complex at the beginning of the preamble detection. First, the problem of detecting the beginning of a transmission has to be solved. The interference level permanently present at the mains causes the accumulators in the ASIC to always fill to a certain degree. When the interference level is low, then AGC will raise it considerably. Note that this is not desirable, but it cannot be avoided. Nevertheless, this does not necessarily have to irritate the receiver, because the interference is not correlated with message-carrying waveforms in general, so that all accumulators get filled to about the same amount. On the other hand, an exact magnitude comparison of the accumulator contents for a preamble chip decision would be problematic and error prone, because there are always slight statistical fluctuations. It is therefore meaningful to do an intelligent evaluation of the detection quality for each individual preamble chip. Otherwise, there may be a risk that the random bit stream from the ASIC gets very close to a preamble in certain time intervals, causing a false alarm.

The intelligent evaluation of the detection quality for each preamble chip is shown in Figure 4-52 and described in the following. We assume that in-phase and quadrature components for "H" and "L" bit branches were added geometrically and are available in an "H" and "L" register, respectively. The correlation results are delivered from these registers to a comparator. This comparator has two different outputs, **(a)** and **(b)**. Output **(a)** supplies logical "1" levels, when the "H" register has the higher value (magnitude); otherwise, output **(a)** will supply a logical "0" level. The output values of the compara-

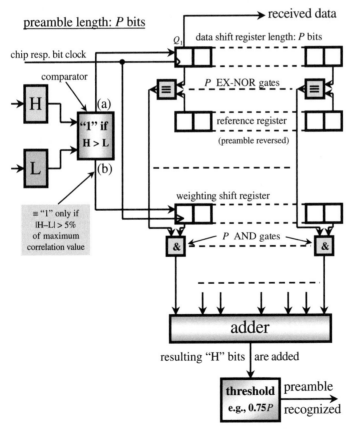

Figure 4-52 The principle of intelligent preamble detection.

tor are fed to a clocked data shift register over output **(a)**. The clock corresponds to the chip or bit clock. While data are received, output Q_1 of the shift register directly supplies the received data stream.

The other functional units represented in the figure are important for safe preamble detection. The reference register contains the reversed preamble used as reference. An equivalence operation (using logical EX-NOR gates) is executed bit by bit after each chip interval. The results of these operations have to be added for preamble decision. Before discussing this important point, we will show the underlying functionality of the blocks represented in Figure 4-52, using an example for better understanding. In this example, we use a so-called "Thue-Morse Code" [D4, B1] with a length of 16 as a preamble; this code is stored in the reference register.

Ref. Reg.   0 1 1 0 1 0 0 1 1 0 0 1 0 1 1 0

We further assume that the preamble has been received up to the last chip. The data shift register then contains the represented values ("x" marks the missing chip):

Data Reg.   1 1 0 1 0 0 1 1 0 0 1 0 1 1 0 x

In the case of bit-by-bit match, the equivalence operation supplies a logical "1"; otherwise "0". The character "–" is entered for the missing preamble chip. The bit position marked in this way is ignored in the summation. We obtain the following result:

Equivalence   0 1 0 0 0 1 0 1 0 1 0 0 0 1 0 –

The summation can now be done simply by weighting the "1" bits in the above equivalence register with +1 and adding. The resulting sum is 5.

If the last preamble chip is correctly detected in the next interval, the content of the data shift register is shifted one position to the right, and a new "0" is written to the register. The summation after the equivalence operation now supplies the value 16, according to the preamble length. At this point, one preamble was detected, and all subsequent information supplied by the comparator over output **(a)** will be interpreted as received data. The selected Thue-Morse code has relatively good aperiodic autocorrelation properties, but it is not optimal, because it supplies a side maximum of 9, which occurs at 13 received chips. The following example shows this situation.

Ref. Reg.   0 1 1 0 1 0 0 1 1 0 0 1 0 1 1 0
Data Reg.   0 1 0 0 1 1 0 0 1 0 1 1 0 x x x
Equivalence   1 1 0 1 1 0 1 0 1 1 0 1 1 – – –

The above explanations assume implicitly that the receiver knows how many preamble chips were already sent, and that it has detected them correctly. Neither of these two assumptions is met in practice. This has critical consequences for the receiver concept, leading to either high cost or bad receive quality. The next section introduces a better solution to this problem.

The comparator in Figure 4-52 has a second output (**b**), which supplies the result from a weighted comparison of the contents in the "H" and "L" registers. On output (**b**), the logical level "1" occurs at the end of a chip interval only if the contents of the registers (H and L) differ by a certain percentage of the maximum value that would occur if the reception of a chip were optimal. This percentage is roughly between 5% and 10% in practical applications. The right choice depends on the respective noise environment. For this reason, it should be set adaptively, e.g., depending on the received interference power. This can be done by observing the register contents during transmission pauses. Both registers fill up to about the same amounts due to interference. If these amounts are very small relative to the maximum value in an optimum reception of a preamble chip, then the above-mentioned threshold should be selected to be around 5%. As the interfered register content increases, the threshold is moved toward 10%. Threshold values much beyond 10% have not proven to be useful in practice.

Output (**b**) supplies the evaluated comparison results to the input of an evaluation shift register, to which they are written at the chip clock. This means that the relevant weighting of each chip reaching the data shift register is mapped to the weighting shift register. If the chip was received safely (big difference between the H, L register contents), then the weighting shift register contains a "1"; otherwise it contains "0". Now, the content of the weighting shift register serves to allow safely distinguishing between the reception of a useful signal and the reception of interference. This is achieved by AND-ing the results of the equivalence operations with the corresponding positions from the weighting shift register, before the summation is executed in the adder. The sum is then transferred to a threshold decider, which indicates the complete reception of a preamble. When this message occurs, the receiver is synchronized and accepts the received data over output Q_1 from the data shift register.

When determining the threshold for the decision logic, it is not recommended to select the entire preamble length (here 16). First, if the noise level is high and/or the attenuation of the useful signal is high, interference could be interpreted as preamble chips. Second, preamble chips arriving at the receiver at a very low level could be erroneously classified as invalid chips. It has proven to be useful to set the threshold value between 12 and 14 with a preamble length of 16 chips. Though an adaptive setting is

Examples of System Implementations

possible, it appears worthwhile only for larger preamble lengths. The functions of Figure 4-52 could be easily implemented in an 80C32 microcontroller.

Table 4-7 shows an example of a typical telegram structure, e.g., for building automation applications. The telegram starts with a preamble with length 16 (Thue-Morse sequence); the chips of the preamble are transmitted in the known 3.3-ms pattern. As the microcontroller is always synchronized to this pattern over the mains zero-crossings, no preamble chip can be lost due to poor synchronization. At this point, the enormous value of the global reference (mains voltage) as a synchronization basis becomes very obvious. Without this reference, the correct synchronization instants would have to be determined from a generally much more interfered signal arriving at low level. This problem can be solved, if at all, only via a much more expensive and error-prone procedure versus the exact detection of mains zero-crossings. Unfortunately, the conditions are not as ideal in practice as we describe them here, so that one has problems with jitter, neutral-point shifts, and signal-delay effects when the waveform duration becomes very short. For this reason, telecommunication applications unfortunately cannot make use of this simple and reliable synchronization method.

Table 4-7 Typical telegram structure for building automation applications (total duration of one frame: 153.3 ms).

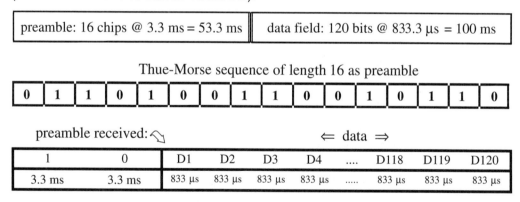

Figure 4-53 shows the digital part of the final prototype built within the scope of university research work on building automation applications, which formed the basis for the development of Powernet-EIB. The ASIC (see No. 5 in Figure 4-49) was produced within the EUROPRACTICE university program in the form of a standard cell; it contains only digital functions. The A/D converter and its reference source are shown as external components above the ASICs. On the right-hand side, there is the microcontroller with an RS.232 interface chip arranged above it to connect to data source and data sink.

184   Chapter 4 • New Usage Possibilities of the Low-Voltage Level Based on European Standards

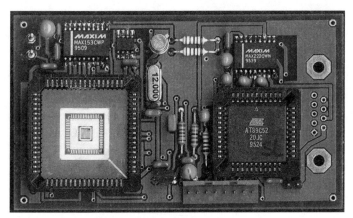

Figure 4-53   Prototype as immediate predecessor of Powernet-EIB.

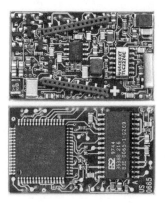

Figure 4-54   Powernet-EIB main printed circuit board; analog part (top) and digital part (bottom).

The industrial development led to a significantly smaller size of the components, as the next figures demonstrate impressively.

Figure 4-54 shows a standard modem printed circuit board used in different Powernet-EIB system components. With its dimensions of 4 cm × 2.3 cm, the modem can be installed easily in any concealed wall socket. In addition, it can be easily integrated in any device that connects to the mains. The entire electric installation of a building becomes "intelligent"; i.e., it is capable of communicating. This means that the "smart home" is no longer a vision, but can be easily turned into reality today [ST25].

The upper side of the board contains the analog portions of the modem, i.e., filters for transmit and receive equipment, AGC, and transmitter output stage. The bottom side contains the digital components, i.e., the ASIC (right) and the microcontroller (left). The A/D and D/A converters are integrated on the ASIC. "Continuous Phase FSK" (CPFSK) is the applied modulation for a data rate of r_D = 1200 bits/s, where each frequency change is basically free of phase hops. Another particularity is the large frequency distance of the two carriers, which can be selected in the range from $4r_D = 4.8$ kHz to $16r_D = 19.2$ kHz. For this reason, the method is also called "Spread FSK" (SFSK). The samples for the following three FSK frequency pairs are stored in a ROM area of the ASIC: 100.8/120 kHz, 105.6/115.2 kHz, and 129.6/134.4 kHz.

What advantages are offered by SFSK? In the "classical" form of FSK, the two carriers have a frequency distance corresponding exactly to the data rate r_D. This type of dense frequency packing is always meaningful when the available spectrum is to be utilized optimally and when the transmission channel allows this. The latter is definitely not the case for powerline, because selective attenuation effects and interference can impair individual portions of the transmission band to such an extent that they become unusable for transmission. For this reason, we can basically not use the entire spectrum in the same way, because the prevailing transmission properties will not remain uniform. The required noise immunity and system reliability can be achieved only by high spectral redundancy, i.e., large unused gaps in the transmission band. A big gap between the two FSK frequencies can thus offer the following advantages:

- In general, only one of the two carrier frequencies is affected by a strong selective attenuation or narrowband interference; the other carrier is almost undisturbed. An intelligent receiver equipped with a microcontroller can detect this condition and ignore the reception from the interfered carrier. This means that we have to look only at the undisturbed carrier. If it is present, the receiver decides about the pertaining data bit, e.g., "H". Otherwise, it decides in favor of the complementary data bit, "L". Now the transmission is no longer as noise-immune as before, because the "L" bit is passively transmitted, i.e., whenever the receiver gets "nothing," it interprets this as an "L" bit. The important point is, however, that the transmission does not fail even under bad conditions.
- It is known that the receiver synchronization relies on the mains voltage as a reference. This means that the zero-crossings are detected to generate the appropriate time pattern by means of a microcontroller. Due to interference, the zero-crossings of the mains voltage can never be fixed exactly. Instead, they are subject to a certain jitter. Deviations of 10–100 µs have to be expected in practical applications. Synchronization errors lead to crosstalk between the two FSK carriers during signal detection, because the orthogonality is partially lost. It is obvious that the negative effects of crosstalk become stronger the closer the carriers are to each other. For example, one bit takes approximately 830 µs at 1200 bits/s. A synchronization error of 100 µs would result in almost 1/8 of the bit duration and lead to noticeable degradation. If the two carriers are far apart, then there is a good separation in the

frequency domain by nature, so that synchronization errors may still decrease the usable power, but there would be almost no crosstalk. This can be a big advantage in critical situations.

In practice some cases were observed where one of the two frequencies was totally removed by attenuation or hit exactly by strong narrowband interference. On the other hand, these cases are infrequent, so that the higher cost for a more resistant broadband modulation with more than two frequencies, e.g., frequency hopping, would not be justified in the design of a building automation system. The modems described on SFSK basis have meanwhile gone through extensive practical testing, and their performance is excellent for building automation applications. Considerable synergies were released by integrating powerline communication into the EIB concept. The basic idea behind EIB is to make the entire electric installation of a building capable of communicating. The most important problem in Standard EIB is that it requires separate wiring with a two-wire bus, e.g., twisted pair. In general, the installation required in existing buildings would not be justified. Even in new buildings for residential purposes, an extensive communication network is still rather the exception than the rule.

These boundary conditions and the high development status achieved are decisive prerequisites for the global success of Powernet-EIB. All options of the EIB system technology are available at no additional installation cost [ST25].

In closing this section, we will describe another important problem in detail, because it has impaired the comprehensive use of powerline communication systems in the past: the effect of impulsive interference from many different types of phase-control devices. Of course, the problem is particularly critical in the interior of buildings due to devices like light dimmers, which are found in almost every household. But even a dimmer equipped with standard EMI[19] suppression components still cause impulsive noise. Although the impulses are now smaller in comparison to the mains voltage, they cannot be neglected in respect to the signal amplitudes that a powerline modem has to process. Of course, the situation becomes particularly critical when a dimmer is to be remote-controlled via powerline. Figure 4-55 shows a Powernet-EIB modem variant for dimmer applications. We know from predecessor systems, e.g., TIMAC-X10 [A2, A6], that a dimmer remote-controlled in this way can merely be started; no further manipulation is possible after that. The described phenomenon becomes clear when you consider the signal conditions at the input of a modem that is supposed to control a dimmer during active dimmer operation. While signal levels down to a few hundred microvolts have to

---

[19] Electromagnetic interference.

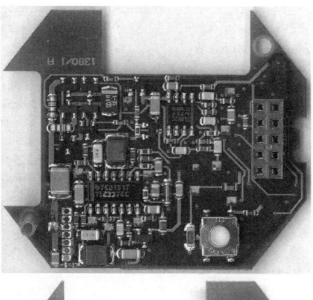

Figure 4-55 Powernet-EIB modem board for dimmer applications; analog part (top) and digital part (bottom).

be processed in the modem's receiving branch, the interference impulses coming from the dimmer are so big that the transient protection equipment at the receiver's input will clip the peaks. This means that a level of several volts is present during such an interference impulse. This simple example is enough to show that "normal" signal processing as described so far cannot solve the problem. Even if an interference impulse is short, compared to the typical waveform duration, it can cause bit errors, e.g., whenever the received signal has such a low level that the AGC amplification is no longer sufficient to match the level fully to the dynamic range of the A/D converter. This constellation has been observed frequently in practice, where an interference impulse leads generally to a bit error with a 50% probability. There is no way to prevent this on the signal processing level (e.g., filtering, amplification, modulation, correlation), so that intervention at the digital side by means of suitable channel coding is required.

4.4.5 Error-Correction Coding Against Impulsive Noise

A coding process adds redundancy bits intentionally to the "net data" for the receiver to use them to both detect and correct errors; this is called forward error correction (FEC) [F4, F7]. Simple forms of block codes, so-called "modified Hamming codes," are well suited to low-speed indoor powerline communication. Such a coding method decomposes the data stream into blocks of suitable length and then appends the required redundancy bits to these blocks. Block codes can be easily implemented in software, e.g., on microcontrollers, and they can generally correct one error and detect two or more errors within one block. It is important for FEC that the net data rate should not be excessively reduced. The three block codes below (each marked by a number pair, which will be explained later) were successfully tested in connection with Powernet-EIB:

- (7, 4) cyclic block code; efficiency: 4/7; net data rate at 1200 bits/s: 686 bits/s.
- (12, 8) linear block code; efficiency: 2/3; net data rate at 1200 bits/s: 800 bits/s.
- (15, 11) cyclic block code; efficiency: 11/15; net data rate at 1200 bits/s: 880 bits/s.

The first digit of the code number designates the block length and the second the number of data bits in one block. In the (12, 8) case, for example, this means that 8-bit blocks have to be formed from the data to be transmitted, where 4 appropriately calculated redundancy bits are added, so that 12 bits are actually sent instead of 8. The net data rate is generally lower when coding is used. We can see from the above list how much of the gross data rate of 1200 bits/s remains, depending on the code used. This means that transmission speed is traded off against zero errors, even under extreme conditions. For example, the (7, 4) code makes the transmission immune against periodic dimmer interference. Note that needle impulses far beyond 1000 V with a repetition rate of 100 Hz can occur at the receiver location, without causing bit errors, even when the received signal is weak. This astonishing noise immunity can be explained by the fact that the (7, 4) block code can always correct one error per block, and that a block of

7 bits with 7·833 μs = 5.83 ms is always shorter than the interference impulse distance of 10 ms.

In less extreme cases, the other two codes perform well and allow a higher remaining net data rate. In particular, the (12, 8) code is used in practical applications, because it can be easily implemented on simple 8-bit microcontrollers. The block length of 12 bits offers a duration of exactly 10 ms, so that two errors caused by dimmer interference could get into one block under unfavorable conditions, and there would be no way to correct one of them. On the other hand, practical applications have proven that such cases are rather unlikely to occur. In addition, there are other effective protection mechanisms, because errors that cannot be corrected can normally be detected by the code, so that retransmission of a faulty telegram can be requested.

Suitable channel coding is obviously an effective approach to combat impulsive interference with high repetition rate and high energy. Figure 4-56 explains the typical problem arising from periodic noise impulses. A data bit stream with 1200 bits/s ($T = 833$ μs) is hit by impulses every 10 ms, corresponding to half a mains voltage period. In such a scenario, every twelfth data bit can normally be corrupted, while no interference occurs in between. This situation can be modeled by a symmetric binary channel with the two states "good" and "bad." The channel remains for one bit duration in the bad state every 10 ms and stays in the good state the rest of the time. When SFSK waveforms are used for transmission, one can assume that two narrowband subchannels are present, so that the normally colored noise in the mains can be assumed as approximately white in each of these narrow channels.

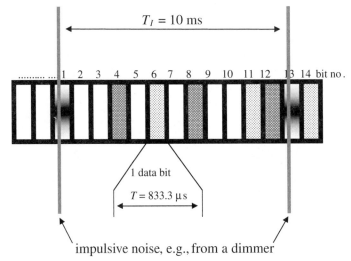

Figure 4-56 Effect of typical impulsive noise.

Similar factors apply to frequency-selective attenuation. If N_0 is the noise-power density and E_b the energy per information bit, then you have an AWGN scenario with the following bit error probability, assuming incoherent reception:

$$P_e = \frac{1}{2} \cdot e^{-\frac{E_b}{4N_0}} \qquad (4.82)$$

When an interference impulse occurs, then N_0 rises drastically and normally causes a bit error, because E_b cannot be increased at the same time to overcome this interference. In such cases a suitable channel coding method can offer elegant and low-cost solutions.

Extensive studies of the distribution of bit errors caused by impulsive noise have been undertaken in order to set up suitable coding schemes. Figure 4-57 shows several important results. The accumulation of the distances between the bit errors (stated in multiples of the bit time T) was registered over a range of 1–37. Obviously, there are peaks at multiples of 12, which can be attributed to the periodicity represented in Figure 4-56. Note that single-bit errors dominate in general. There are virtually no error bursts. These observations led to the following ideas with regard to suitable channel coding, taking into account the limited possibilities offered by a low-cost 8-bit microcontroller.

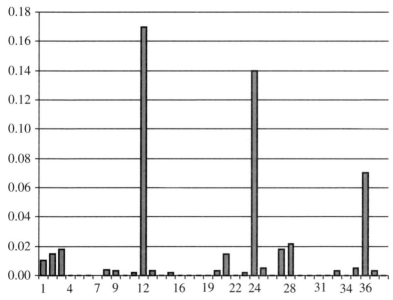

Figure 4-57 Accumulation of bit error distances.

Block coding based on Hamming codes is normally the first choice when coding and decoding have to be executed as real-time routines in software on a low-cost microcontroller. Hamming codes are systematic and linear. They are capable of removing a single bit error within an n-bit frame. Table 4-8 lists the most important properties of Hamming codes.

Table 4-8 Overview of the most important properties of Hamming codes.

Basic properties of (n, d, h) Hamming codes	
Codeword length:	$n = 2^m - 1, \quad m \in \{1, 2, 3, ...\}$
Number of data bits:	d
Hamming distance:	$h_d = 3$ (typical)
Number of redundant bits:	$r = n - d$
Number of correctable error words:	$f_{\text{corr}} = \sum_{i=0}^{\lfloor \frac{h_d-1}{2} \rfloor} \binom{n}{i}$; e.g., 32 for a (31, 25, 4) code
Number of detectable error words:	$f_{\text{det}} = \binom{n}{\frac{h_d}{2}}$; e.g., 465 for a (31, 25, 4) code
Calculation of r according to the Hamming limit:	$2^r \geq \sum_{i=0}^{\lfloor \frac{h_d-1}{2} \rfloor} \binom{n}{i}$

examples of Hamming codes: (7, 4, 3) code, (31, 26, 3) code. For the (31, 26, 3) codes we have $r = 5$

codeword generation: $\quad \vec{c}_n = \vec{d}_d \cdot [\mathbf{I}_{d,d}, \mathbf{P}_{d,r}] = \vec{d}_d \cdot [\mathbf{G}_{n,d}]$

$\vec{c}_n \equiv$ codeword of length n
$\vec{d}_d \equiv$ data bit vector of length d
$r \equiv$ number of redundant bits

Steps for the adaptation of Hamming codes

1. Code extension:
A parity bit is added to each row of the parity matrix, i.e., $\mathbf{P}_{d,r} \Rightarrow \mathbf{P}_{d,r+1}$. In this way a (32, 26, 4) code is generated from a (31, 26, 3) standard Hamming code.

2. Code shortening:
Deleting a row from the parity matrix $\mathbf{P}_{d,r} \Rightarrow \mathbf{P}_{d-1,r}$ will, e.g., generate a (31, 25, 4) code from the extended (32, 26, 4) code.

We know from Figure 4-56 that obviously this frame must be limited to 12 bits or less to take account of all conceivable worst-case conditions. The code words for Hamming codes have always 2^{m-1} bits ($m = 1, 2, 3, ...$). This limits our case to the selection of $n = 7$. A (7, 4, 3) Hamming code is composed of $d = 4$ data bits and 3 redundancy bits. This gives a code rate of $d/n = 0.571$, which leaves a net data rate of only 686 bits/s at 1200 bits/s. The implementation of such a Hamming code on a standard microcontroller is easy. As there are $r = 3$ redundancy bits, the parity matrix $\mathbf{P}_{d,r}$ for this code consists of 4 lines and 3 columns. One code word $\vec{c}_n$ of length n is generated as follows:

$$\vec{c}_n = \vec{d}_d \cdot \left[\mathbf{I}_{d,d}, \mathbf{P}_{d,r}\right] = \vec{d}_d \cdot \left[\mathbf{G}_{d,n}\right] \qquad (4.83)$$

where $\vec{d}_d$ is the d-bit data bit vector, $\mathbf{I}_{d,d}$ is the unit matrix with dimension $d \cdot d$, $\mathbf{P}_{d,r}$ is the parity matrix with dimension $d \cdot n$, and $\mathbf{G}_{d,n}$ is the generator matrix.

A data vector $\vec{d}_d$ is encoded by modulo-2 adding all lines of $\mathbf{P}_{d,r}$, for which a "1" bit is contained in $\vec{d}_d$. Table 4-9 provides an overview of the coding steps.

Table 4-9 Calculation steps during the coding process.

$$\{c_1, c_2, ..., c_n\} = \{d_1, d_2, ..., d_d\} \cdot \begin{bmatrix} 1_{11} & 0 & .. & 0 & 0_{1d} \\ 0_{21} & 1 & 0 & .. & 0 \\ .. & 0 & .. & .. & .. \\ 0 & .. & .. & 1 & 0 \\ 0_{d1} & 0 & .. & 0 & 1_{dd} \end{bmatrix} \cdot \begin{bmatrix} r_{11} & .. & .. & r_{1r} \\ r_{21} & .. & .. & r_{2r} \\ .. & .. & .. & .. \\ .. & .. & .. & .. \\ r_{d1} & .. & .. & r_{dr} \end{bmatrix}$$

codeword $\vec{c}_n$ — data bit vector $\vec{d}_d$ — unity matrix $\mathbf{I}_{d,d}$ — parity matrix $\mathbf{P}_{d,r}$

↑ generating matrix $\mathbf{G}_{d,n}$ ↑

Necessary mathematical operations for coding:

Wherever the data bit vector contains a "1" bit:

Calculate the modulo-2 sum of the corresponding row of the parity matrix $\mathbf{P}_{d,r}$.

If $r \leq 8$, this can be easily performed by an EXOR operation within an 8-bit microcontroller.

For the (31, 25, 4) code we have, e.g., $d = 25$ and $r = 6$

The parity matrix $\mathbf{P}_{d,r} \equiv \mathbf{P}_{25,6}$ thus can be placed into a 25-byte portion of a microcontroller's memory; these locations can be used unchanged also for decoding.

if $\vec{d}_d \equiv \{1,1,...,1\}$, 25 EXOR operations must be executed (worst case).

The first step in decoding a received code word is to calculate the syndrome $\vec{s}_r$ by means of a parity-check matrix $\mathbf{H}_{n,r}^T$, consisting of the parity matrix $\mathbf{P}_{d,r}$, extended by a unit matrix $\mathbf{I}_{r,r}$:

Examples of System Implementations

$$\vec{s}_r = \vec{c}_{nR} \cdot \begin{bmatrix} \mathbf{P}_{d,r} \\ \mathbf{I}_{r,r} \end{bmatrix} = \vec{c}_{nR} \cdot \begin{bmatrix} \mathbf{H}_{n,r}^T \end{bmatrix} \qquad (4.84)$$

Table 4-10 lists the parity-check matrix for the (31, 25, 4) code. The right column states the error word numbers.

Table 4-10   Parity-check matrix H^T for the (31,25,4) code.

x: 1...6							Error Word
P_{1x}	0	0	0	1	1	1	7
P_{2x}	0	0	1	0	1	1	11
P_{3x}	0	0	1	1	0	1	13
P_{4x}	0	0	1	1	1	0	14
P_{5x}	0	1	0	0	1	1	19
P_{6x}	0	1	0	1	0	1	21
P_{7x}	0	1	0	1	1	0	22
P_{8x}	0	1	1	0	0	1	25
P_{9x}	0	1	1	0	1	0	26
P_{10x}	0	1	1	1	0	0	28
P_{11x}	0	1	1	1	1	1	31
P_{12x}	1	0	0	0	1	1	35
P_{13x}	1	0	0	1	0	1	37
P_{14x}	1	0	0	1	1	0	38
P_{15x}	1	0	1	0	0	1	41
P_{16x}	1	0	1	0	1	0	42
P_{17x}	1	0	1	1	0	0	44
P_{18x}	1	0	1	1	1	1	47
P_{19x}	1	1	0	0	0	1	49
P_{20x}	1	1	0	0	1	0	50
P_{21x}	1	1	0	1	0	0	52
P_{22x}	1	1	0	1	1	1	55
P_{23x}	1	1	1	0	0	0	56
P_{24x}	1	1	1	0	1	1	59
P_{25x}	1	1	1	1	0	1	61
I_{26x}	1	0	0	0	0	0	32
I_{27x}	0	1	0	0	0	0	16
I_{28x}	0	0	1	0	0	0	8
I_{29x}	0	0	0	1	0	0	4
I_{30x}	0	0	0	0	1	0	2
I_{31x}	0	0	0	0	0	1	1

As in the encoding process, the lines corresponding to "1" bits in the received code word $\vec{c}_{nR}$ of $\mathbf{H}_{n,r}^T$ have to be EXOR-ed. If an error occurs, then it is corrected by complementing the bit specified by the calculated syndrome. By storing a pointer indicating the matching code word bit for each of the syndromes, it is easy to implement the error correction in the microcontroller program. Table 4-11 summarizes the steps in detail.

Table 4-11 Decoding and error-correction steps.

Decoding and error correction:

$\vec{c}_{nR} \equiv$ received codeword of length n

Calculation of the syndrome: $\quad \vec{s}_r = \{s_1, s_2, ..., s_r\} = \vec{c}_{nR} \cdot \begin{bmatrix} \mathbf{P}_{d,r} \\ \mathbf{I}_{r,r} \end{bmatrix} = \vec{c}_{nR} \cdot \begin{bmatrix} \mathbf{H}_{n,r}^T \end{bmatrix}$

↑ parity check matrix

In case of a (31, 25, 4) code we have: ↓ parity matrix

$$\{s_1, s_2, ..., s_6\} = \{c_{1R}, ..., c_{25R}, c_{26R}, ..., c_{31R}\} \cdot \begin{bmatrix} r_{1,1} & .. & .. & .. & .. & r_{1,6} \\ .. & .. & .. & .. & .. & .. \\ .. & .. & .. & .. & .. & .. \\ .. & .. & .. & .. & .. & .. \\ r_{25,1} & .. & .. & .. & .. & r_{25,6} \\ 1 & 0 & 0 & 0 & 0 & 0 \\ 0 & 1 & 0 & 0 & 0 & 0 \\ 0 & 0 & 1 & 0 & 0 & 0 \\ 0 & 0 & 0 & 1 & 0 & 0 \\ 0 & 0 & 0 & 0 & 1 & 0 \\ 0 & 0 & 0 & 0 & 0 & 1 \end{bmatrix}$$

Operations for the calculation of the syndrome:

For each "1" bit within the first 25 bits of the received codeword:
• modulo-2 addition of the corresponding rows of the parity matrix $\mathbf{P}_{d,r}$
• then: modulo-2 addition of the 6 redundant bits $\{c_{26R}, ..., c_{31R}\}$

If $\vec{c}_{31R} \equiv \{1, 1, ..., 1\}$, 26 EXOR operations are necessary (worst case).

Final steps:

Fetching the error word belonging to the syndrome which has been calculated from a table containing 64 words of 4 bytes (31 bits are necessary). For 32 syndromes error correction is possible by EXOR-ing the data portion $\{c_{1R}, c_{2R}, ..., c_{25R}\}$ of the received codeword with the error word.

The correctable bit errors are bold gray and underlined in the two error word tables, Tables 4-12 and 4-13, and the corresponding syndromes are also highlighted in gray. All other syndromes differing from zero report an error that can no longer be corrected.

To demonstrate the capabilities of a (7, 4, 3) code, a light dimmer, from which the EMI suppression components had been previously removed, for a 500-W lamp was operated in the immediate neighborhood of the receiver, while the strongly attenuated useful signal was transmitted 800 m over a ground cable. At 1200 bits/s and with the coding disabled, the resulting bit error rate was $4 \cdot 10^{-2}$, which means that approximately every 24th bit was corrupted. This shows very clearly that in reality not every impulse causes an error, but that the error probability is ½. Considering that a noise impulse occurs every 12 bits, it is easy to understand the above result. As soon as the coding was enabled, the bit error rate dropped below 10^{-7} [D4, ST21].

A low code rate like the one offered by a (7, 4, 3) code undoubtedly means an impairment for practical applications. In addition, undetected errors cannot be tolerated, for example in commercial applications of PSUs. For this reason, special block codes with a higher rate and better error-detection capabilities are generally desirable. The modified (31, 25, 4) Hamming code described here has proved to be well suited for applications outside of buildings, where strong periodic impulsive interference cannot be totally excluded but is by no means dominant. Starting from the standard (31, 25, 3) Hamming code, one has a parity matrix $\mathbf{P}_{d,r}$ consisting of 25 lines and 6 columns. First, the code is extended to (32, 26, 4) by adding one parity-bit line to $\mathbf{P}_{d,r}$. This step increases the Hamming distance to four, allowing the detection of numerous uncorrectable errors, namely

$$\begin{bmatrix} 32 \\ 2 \end{bmatrix} = 496$$

in addition to the 33 correctable errors.

Note that the code concatenation is also an effective method for reliable detection of uncorrectable errors, even when relatively simple concatenation schemes like the following are used. In this concatenation scheme, N code words $\vec{c}_1$, ..., $\vec{c}_N$ are transmitted in the following frame, and an additional check word $\vec{c}_{\text{CHK}}$ is appended to this frame.

| $\vec{c}_1$ | $\vec{c}_2$ | $\vec{c}_3$ | ... | $\vec{c}_N$ | $\vec{c}_{\text{CHK}}$ |

Table 4-12 Error words belonging to the syndromes 0–31.

$\vec{s}$								
0	0000	0000	0000	0000	0000	0000	0000	0000
1	1000	0000	0000	0000	0000	0000	0000	0000
2	0100	0000	0000	0000	0000	0000	0000	0000
3	0000	0000	0000	0000	0000	0000	1111	1111
4	0010	0000	0000	0000	0000	0000	0000	0000
5	0000	0000	0000	0000	0000	0000	1111	1111
6	0000	0000	0000	0000	0000	0000	1111	1111
7	0000	0000	0000	0000	0000	0000	0000	0010
8	0001	0000	0000	0000	0000	0000	0000	0000
9	0000	0000	0000	0000	0000	0000	1111	1111
10	0000	0000	0000	0000	0000	0000	1111	1111
11	0000	0000	0000	0000	0000	0000	0000	0100
12	0000	0000	0000	0000	0000	0000	1111	1111
13	0000	0000	0000	0000	0000	0000	0000	1000
14	0000	0000	0000	0000	0000	0000	0001	0000
15	0000	0000	0000	0000	0000	0000	1111	1111
16	0000	1000	0000	0000	0000	0000	0000	0000
17	0000	0000	0000	0000	0000	0000	1111	1111
18	0000	0000	0000	0000	0000	0000	1111	1111
19	0000	0000	0000	0000	0000	0000	0010	0000
20	0000	0000	0000	0000	0000	0000	1111	1111
21	0000	0000	0000	0000	0000	0000	0100	0000
22	0000	0000	0000	0000	0000	0000	1000	0000
23	0000	0000	0000	0000	0000	0000	1111	1111
24	0000	0000	0000	0000	0000	0000	1111	1111
25	0000	0000	0000	0000	0000	0001	0000	0000
26	0000	0000	0000	0000	0000	0010	0000	0000
27	0000	0000	0000	0000	0000	0000	1111	1111
28	0000	0000	0000	0000	0000	0100	0000	0000
29	0000	0000	0000	0000	0000	0000	1111	1111
30	0000	0000	0000	0000	0000	0000	1111	1111
31	0000	0000	0000	0000	0000	1000	0000	0000

Table 4-13 Error words belonging to the syndromes 32–64.

$\vec{s}$								
32	0 0 0 0	0 1 0 0	0 0 0 0	0 0 0 0	0 0 0 0	0 0 0 0	0 0 0 0	0 0 0 0
33	0 0 0 0	0 0 0 0	0 0 0 0	0 0 0 0	0 0 0 0	0 0 0 0	1 1 1 1	1 1 1 1
34	0 0 0 0	0 0 0 0	0 0 0 0	0 0 0 0	0 0 0 0	0 0 0 0	1 1 1 1	1 1 1 1
35	0 0 0 0	0 0 0 0	0 0 0 0	0 0 0 0	0 0 0 1	0 0 0 0	0 0 0 0	0 0 0 0
36	0 0 0 0	0 0 0 0	0 0 0 0	0 0 0 0	0 0 0 0	0 0 0 0	1 1 1 1	1 1 1 1
37	0 0 0 0	0 0 0 0	0 0 0 0	0 0 0 0	0 0 1 0	0 0 0 0	0 0 0 0	0 0 0 0
38	0 0 0 0	0 0 0 0	0 0 0 0	0 0 0 0	0 1 0 0	0 0 0 0	0 0 0 0	0 0 0 0
39	0 0 0 0	0 0 0 0	0 0 0 0	0 0 0 0	0 0 0 0	0 0 0 0	1 1 1 1	1 1 1 1
40	0 0 0 0	0 0 0 0	0 0 0 0	0 0 0 0	0 0 0 0	0 0 0 0	1 1 1 1	1 1 1 1
41	0 0 0 0	0 0 0 0	0 0 0 0	0 0 0 0	1 0 0 0	0 0 0 0	0 0 0 0	0 0 0 0
42	0 0 0 0	0 0 0 0	0 0 0 0	0 0 0 1	0 0 0 0	0 0 0 0	0 0 0 0	0 0 0 0
43	0 0 0 0	0 0 0 0	0 0 0 0	0 0 0 0	0 0 0 0	0 0 0 0	1 1 1 1	1 1 1 1
44	0 0 0 0	0 0 0 0	0 0 0 0	0 0 1 0	0 0 0 0	0 0 0 0	0 0 0 0	0 0 0 0
45	0 0 0 0	0 0 0 0	0 0 0 0	0 0 0 0	0 0 0 0	0 0 0 0	1 1 1 1	1 1 1 1
46	0 0 0 0	0 0 0 0	0 0 0 0	0 0 0 0	0 0 0 0	0 0 0 0	1 1 1 1	1 1 1 1
47	0 0 0 0	0 0 0 0	0 0 0 0	0 1 0 0	0 0 0 0	0 0 0 0	0 0 0 0	0 0 0 0
48	0 0 0 0	0 0 0 0	0 0 0 0	0 0 0 0	0 0 0 0	0 0 0 0	1 1 1 1	1 1 1 1
49	0 0 0 0	0 0 0 0	0 0 0 0	1 0 0 0	0 0 0 0	0 0 0 0	0 0 0 0	0 0 0 0
50	0 0 0 0	0 0 0 0	0 0 0 1	0 0 0 0	0 0 0 0	0 0 0 0	0 0 0 0	0 0 0 0
51	0 0 0 0	0 0 0 0	0 0 0 0	0 0 0 0	0 0 0 0	0 0 0 0	1 1 1 1	1 1 1 1
52	0 0 0 0	0 0 0 0	0 0 1 0	0 0 0 0	0 0 0 0	0 0 0 0	0 0 0 0	0 0 0 0
53	0 0 0 0	0 0 0 0	0 0 0 0	0 0 0 0	0 0 0 0	0 0 0 0	1 1 1 1	1 1 1 1
54	0 0 0 0	0 0 0 0	0 0 0 0	0 0 0 0	0 0 0 0	0 0 0 0	1 1 1 1	1 1 1 1
55	0 0 0 0	0 0 0 0	0 1 0 0	0 0 0 0	0 0 0 0	0 0 0 0	0 0 0 0	0 0 0 0
56	0 0 0 0	0 0 0 0	1 0 0 0	0 0 0 0	0 0 0 0	0 0 0 0	0 0 0 0	0 0 0 0
57	0 0 0 0	0 0 0 0	0 0 0 0	0 0 0 0	0 0 0 0	0 0 0 0	1 1 1 1	1 1 1 1
58	0 0 0 0	0 0 0 0	0 0 0 0	0 0 0 0	0 0 0 0	0 0 0 0	1 1 1 1	1 1 1 1
59	0 0 0 0	0 0 0 1	0 0 0 0	0 0 0 0	0 0 0 0	0 0 0 0	0 0 0 0	0 0 0 0
60	0 0 0 0	0 0 0 0	0 0 0 0	0 0 0 0	0 0 0 0	0 0 0 0	1 1 1 1	1 1 1 1
61	0 0 0 0	0 0 1 0	0 0 0 0	0 0 0 0	0 0 0 0	0 0 0 0	0 0 0 0	0 0 0 0
62	0 0 0 0	0 0 0 0	0 0 0 0	0 0 0 0	0 0 0 0	0 0 0 0	1 1 1 1	1 1 1 1
63	0 0 0 0	0 0 0 0	0 0 0 0	0 0 0 0	0 0 0 0	0 0 0 0	1 1 1 1	1 1 1 1

The check word $\vec{c}_{CHK}$ is calculated as a modulo-2 sum from the N code words as follows:

$$\vec{c}_{CHK} = \vec{c}_1 \oplus \vec{c}_2 \oplus \vec{c}_3 \oplus ... \oplus \vec{c}_N = \left(\vec{d}_1 \oplus \vec{d}_2 \oplus \vec{d}_3 \oplus ... \oplus \vec{d}_N\right) \cdot \left[\mathbf{G}_{d,n}\right] \quad (4.85)$$

where $\vec{d}_1, ..., \vec{d}_N$ are the N data vectors with length d and $\left[\mathbf{G}_{d,n}\right] = \left[\mathbf{I}_{d,d}, \mathbf{P}_{d,r}\right]$ is the generator matrix. The receiver decodes the $N + 1$ code words $\vec{c}_{1r}, ..., \vec{c}_{Nr}$ and $\vec{c}_{CHK}$, as described above, and corrects errors, if applicable. Subsequently, the data portions of all code words, including the received check word $\vec{c}_{CHK}$, are modulo-2 added. The result has to be zero if there was no error:

$$\vec{d}_{1c} \oplus \vec{d}_{2c} \oplus \vec{d}_{3c} \oplus ... \oplus \vec{d}_{Nc} \oplus \vec{d}_{CHKc} = \vec{0} \quad (4.86)$$

A result other than zero means that an uncorrectable error has occurred. Normally, the telegram has to be retransmitted. In this case, the receiver has to output an automatic retransmission request (ARQ).

The concatenation scheme described above worked in an excellent way with $N = 8$ in field trials as long as periodic impulse interferers were not dominant. Unsatisfactory results were observed as soon as such interferers were present. The reason is the length of the code words, because with a 32-bit code word and a 12-bit periodic impulsive interference, every third code word is impaired in exactly the same bit positions, so that many errors delete themselves during summation and slip through detection.

A minor change in the code word length, namely to $n = 31$ (a prime number), can lead to a considerable improvement, because now every twelfth code word will be impaired in the same bit positions. Table 4-14 demonstrates this situation, where all cells containing "E" are error positions, and 0 means that there is no error.

The required (31, 25, 4) Hamming code can be yielded easily by removing one line from the parity matrix $\mathbf{P}_{d,r}$ to shorten the (32, 26, 4) code. This gives us a code rate of $25/31 = 0.806$, and with the described concatenation of $N = 8$ code words and a net data rate of approximately 850 bits/s, excellent immunity against undetected errors was achieved. For example, in field trials in a problematic environment, no error slipped through detection during an entire week, although strong periodic impulsive interference was introduced artificially. The described coding scheme was accepted for industrial series production [D5].

4.4.6 Behavior of Coded Transmission under White Noise (AWGN Environment)

Fortunately, the described coding schemes not only are effective against impulsive interference but also offer the benefits analyzed below in an AWGN environment.

Table 4-14 Influence of the code word length when periodic impulsive interference is dominant.

	Error Pattern for a 32-bit Code Word in a Periodic Impulsive Interference Scenario	Error Pattern for a 31-bit Code Word in a Periodic Impulsive Interference Scenario
1	E0000000000E0000000000E0000000	E0000000000E0000000000E000000
2	0000E0000000000E0000000000E000	00000E0000000000E0000000000E0
3	00000000E0000000000E0000000000	0000000000E0000000000E0000000
4	E0000000000E0000000000E0000000	000E0000000000E0000000000E000
5	0000E0000000000E0000000000E000	00000000E0000000000E0000000000
6	00000000E0000000000E0000000000	0E0000000000E0000000000E00000
7	E0000000000E0000000000E0000000	000000E0000000000E0000000000E
8	0000E0000000000E0000000000E000	00000000000E0000000000E0000000
9	00000000E0000000000E0000000000	0000E0000000000E0000000000E00
10	E0000000000E0000000000E0000000	000000000E0000000000E000000000
11	0000E0000000000E0000000000E000	00E0000000000E0000000000E0000
12	00000000E0000000000E0000000000	0000000E0000000000E00000000000
13	E0000000000E0000000000E0000000	E0000000000E0000000000E000000
14	0000E0000000000E0000000000E000	00000E0000000000E0000000000E0

From (4.82) the probability P_{ed} for a single bit error in a block of d data bits would result in

$$P_{ed} = 1 - (1 - P_e)^d \tag{4.87}$$

Curves 1 and 2 in Figure 4-58 show these error probabilities for $n = 25$ and $n = 4$ as a function of the ratio E_b/N_0 of the bit energy to the noise power density in the range of 6–20 dB. Compared to this, curves 3 and 4 show the probability P_{ec} for the occurrence of two bit errors in blocks with a length of $n = 7$ and $n = 31$ bits, respectively. This means that one uncorrectable error each occurs when using a (7, 4, 3) or (31, 25, 4) code, respectively. P_{ec} can be calculated as follows:

$$P_{ec} = \sum_{i=2}^{n} \binom{n}{i} \cdot P_{er}^{\,i} \cdot (1 - P_{er})^{n-i} \tag{4.88}$$

where P_{er} is calculated by means of (4.82), taking into account that the effective energy per bit during encoding has to be reduced by factor d/n (number of data bits / block length). This results in

$$P_{er} = \frac{1}{2} \cdot e^{-\frac{E_b}{4N_0} \cdot \frac{d}{n}} \tag{4.89}$$

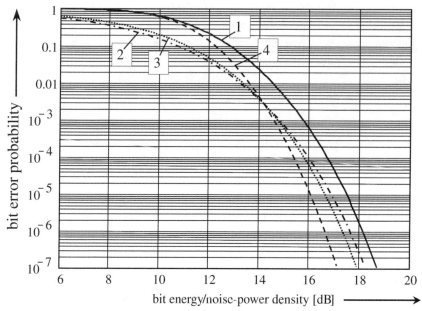

Figure 4-58 Bit error probability with and without coding.

Despite this drawback, a general superiority remains when using coding at bit error probabilities below 10^{-3}, or even 10^{-1} in case of the (31, 25, 4) code. Coding yields about 2 dB at very low error probabilities. Channel coding can be used in power-line communication to solve a series of serious transmission problems, which could not be solved by other methods, such as modulation or adaptation. Even when hardware resources are limited, block codes can realize very efficient solutions that can be implemented easily in software on low-cost 8-bit microcontrollers. Field tests confirmed the effectiveness of properly used coding [D4, D5, ST21].

4.4.7 Communication System for Energy-Related Value-Added Services of PSUs

The European standard EN 50065, which also has the status of a German norm, came into force in December 1991 [NO3]. This standard regulates the use of the frequency range from 3 to 148.5 kHz for signal transmission purposes in electric energy distribution networks. As described in Section 4.1, the available band is roughly divided into two areas: the frequency range from 3 to 95 kHz is reserved for power supply utilities (PSUs); the maximum admissible transmission amplitude is 134 dBµV. The remaining frequency range from 95 to 148.5 kHz is available for private users, where a transmission amplitude of 122 dBµV must not be exceeded. Fulfilling EN 50065 specifications puts high requirements upon methods and equipment for robust transmission of information over powerlines. In particular, the area between a transformer station and the customer's appliances, which has to be bridged in a reliable way within the scope of energy-related value-added services, often causes problems, so that only a careful and channel-adapted design of transmission systems can achieve the required reliability and availability of the connections. Basically the transmission methods that can always use the full amplitude have the best chance to be successful, owing to the amplitude limitation. This includes the modulation schemes with frequency-agile waveforms described earlier.

Probably the simplest method in this connection from the technical view is FSK (frequency-shift keying). The simple implementation of FSK is surely one important reason why the majority of the systems for information transmission over powerlines currently offered in the market use FSK. In the practical environment, owing to increased requirements with regard to the data rate and reliability even on critical paths, serious drawbacks of FSK have been discovered. Today it is found that a sufficient reliability within the scope of the services to be handled by PSUs over their networks in the future will most likely not be achieved with FSK. The main problem of FSK found in numerous field trials is that the transmission can fail due to interference hitting only a single carrier frequency. It is widely known that powerlines, particularly within the EN 50065 range of application, are time-variant transmission channels where a selective attenuation can occur in any point at any time, or where a narrowband interferer, e.g., a television set or a switching power supply, can disturb. Band-spreading, frequency-agile methods like frequency hopping (FH) allow many different frequency variations, which is very helpful in such situations. Although the cost is higher, compared to an FSK system, it is worthwhile to consider FH in view of the progress achieved in modern microelectronics. Nevertheless, FH schemes have not made their way from the trial stage to practical implementations, so that no usable systems were available until recently.

Although FH is basically capable of overcoming all problems of FSK observed in practice, important drawbacks were found in state-of-the-art FH schemes and their technical implementations. These drawbacks delayed the construction of systems with flexible use of several consecutively transmitted carrier frequencies, though this situation has changed.

The main drawback in the transition from FSK to standard FH is the necessary increase of the chip rate at equal net data rate. With a chip rate of 1200 s^{-1}, an FSK system supplies exactly this speed for the data rate, while a classical FH system with four carriers per bit can offer only 300 bits/s. This means that a chip rate of 4800 s^{-1} is required to achieve the net data rate of FSK in FH under the same conditions. This leads to higher cost for signal generation and signal processing in general, but in particular for the receiver synchronization, which would then have to be more precise by a factor of four.

The author's research group worked on better solutions in close cooperation with the industry and PSUs during the period from 1995 to 1998. At the end of 1996, a so-called "symbol-processing multicarrier scheme" was developed, which dramatically reduced the drawbacks described above, offering an excellent outlook for practical utilization of the benefits of frequency-agile transmission technology.

When N carrier frequencies are used in an FH system, then $N!$[20] different combinations can be formed, referred to here as "symbols." In this way, six symbols can be represented with $N = 3$ frequencies, and $N = 5$ offers 120. The 6 symbols from $N = 3$ would be capable of transmitting $\lceil \log_2(6) \rceil = 2$ bits,[21] while $\lceil \log_2(120) \rceil = 6$ bits could be achieved with $N = 5$. As explained further above, however, these options are not fully utilized in conventional fast FH schemes. Instead, only two symbols (combinations) are evaluated, corresponding to an information content of one bit.

This is the way in which regular FH technology supports redundancy, for example to make a transmission more robust against the failure of single frequency ranges. The resulting drawback is obvious: the transmission speed drops in line with the number of carrier frequencies used. The innovative idea introduced here as the "symbol-processing multicarrier scheme" avoids this speed drawback and ensures at the same time unlimited maintenance of the noise immunity inherent in the FH principle. Table 4-15 compares several examples to better illustrate this important issue.

Table 4-15 FH technology and symbol processing for $N = 3$ carrier frequencies.

Symbol Number	Frequency Sequence	FH Technology	Symbol Processing
S_1	$f_1 f_2 f_3$	L	LL
S_2	$f_1 f_3 f_2$	–	LH
S_3	$f_2 f_1 f_3$	–	HL
S_4	$f_2 f_3 f_1$	–	HH
S_5	$f_3 f_1 f_2$	H	–
S_6	$f_3 f_2 f_1$	–	–

---

[20] N factorial = $N(N-1)(N-2)(N-3)\ldots 1$
[21] $\lceil \cdot \rceil$ means the maximum integer part of $(\cdot)$.

If we denote the three frequencies used for $N = 3$ as $f_1, f_2,$ and f_3, we obtain six symbols S_1 through S_6, each of which can be used in a different way. The FH Technology column shows that the symbols are not heavily used. This high redundancy accounts eventually for the robustness of the classical FH. However, note that the sheer number of unused symbols is no indication of the general amount of the achievable noise immunity, which is determined rather by the number of chips at different frequencies for the symbols used to *represent information*. We can see in the FH Technology column that the symbols S_1 and S_5 marked *L* or *H*, respectively, differ in all three chips, so that the receiver can separate them very easily.

A look at the Symbol Processing column discloses that only the symbols S_1 and S_4 marked with *LL* and *HH* as well as S_2 and S_3 marked with *LH* and *HL*, differ in all three chips, while the *LL* and *LH* or *LL* and *HL* combinations are different in only two chips. The noise immunity suffers from this fact. If we used all six symbols for data transmission, we would have only one different chip each as the decision criterion. This would make the transmission no more reliable than FSK, because the failure of a single frequency could lead to a system failure.

Table 4-15 shows that the use of three carrier frequencies obviously does not lead to a satisfactory solution. In contrast, Table 4-16 represents a sequence of four frequencies allowing 24 symbols. However, only the four that differ in all four chips are used for data transmission. This would result in the same noise immunity normally obtained in conventional FH, but twice the transmission speed would result, because now each symbol represents two data bits.

Table 4-16 Optimum symbol processing with $N = 4$.

Frequency Sequence	Data Symbols
$f_1 f_2 f_3 f_4$	LL
$f_2 f_3 f_4 f_1$	LH
$f_3 f_4 f_1 f_2$	HL
$f_4 f_1 f_2 f_3$	HH

The principle represented in Table 4-16 is applicable not only to $N = 4$ but to all cases where exactly *N* out of *N*! possible symbols are used for data representation. This principle works best if *N* is a power of two, because $\log_2(N)$ will then result in an integer bit number. If $\log_2(N)$ is not an integer, it is possible to round only to the next smaller integer number, not up to the next higher number; i.e., the usable net data rate is reduced. This means that, after $N = 4$, $N = 8$ would be another suitable value, where the use of eight out of $8! = 40{,}320$ possible symbols would transmit four bits with each symbol used. For practical use in powerline communication, a system relying on four carrier frequencies will definitely offer the safety required to ensure the new services planned by the PSUs with sufficient reliability. We can see from Table 4-16 that up to three of the four carrier frequencies could fail in the worst case without causing the occurrence of a bit error. This is an enormous progress versus FSK, where the suppression or disturbance of a single frequency generally leads to a system failure. When using conventional FH, a comparable result would require $N = 7$ carrier frequencies, where the net data rate would be only 1/7 of the chip rate. The new symbol-processing multicarrier scheme offers the same noise immunity at a data-rate/chip-rate ratio of 1/2.

Practical field trials and extensive channel analyses have shown that, if the carrier frequencies are properly selected, the probability that two of them will fail simultaneously due to attenuation or interference is negligible. The case that this would occur for three carriers can virtually be excluded. However, the failure of single carriers has been observed relatively often. Therefore, the selection of $N = 4$ carrier frequencies is a good choice for the practical implementation of this new scheme [ST22, D8].

A digital concept that allows high integration is of utmost significance for a low-cost and reliable reproducible system. In addition, it is desirable that various parameters can be modified without hardware change, and that most of the analog circuitry required for the complete construction of a modem can be integrated monolithically together with the digital portions. The next section describes a solution that considers all these factors.

The starting point is the transmission of binary information ("H" and "L" bits), which are mapped onto frequency sequences (symbols) according to Table 4-16. During time T_s of one symbol, the transmission frequency changes here rapidly four times. The higher the N value that was selected, the more immune the data transmission will be against interference, although at the drawback of increasing system costs. For the reasons previously mentioned, $N = 4$ different and appropriately selected frequencies are sufficient for most applications. Appropriate selection means that you distribute the frequencies used within the available transmission band so that a strong selective attenuation or a strong narrowband interferer can never disturb two or more frequencies simultaneously. Proper selection requires sufficient experience in network measurements, field trials, and theoretical network modeling based on the measurement results.

The distribution network of a PSU has the frequency band from 3 to 95 kHz for data transmission according to EN 50065, which means that a bandwidth of $B \approx 92$ kHz is available. Now, if binary information with a net data rate (bit rate) of $r_D = 1200$ bits/s has to be transmitted by use of the symbol-processing multicarrier scheme, where $N = 4$ carrier-frequency hops occur during a symbol interval T_s, then the frequency-hop rate is

$h = 2400$ s^{-1}. The inverse value $T = 1/h$ of the frequency-hop rate specifies the duration of the time interval in which one of the N frequencies is transmitted. As described in Section 4.3, a maximum of $\lceil B/h \rceil = \lceil 92000/2400 \rceil = 38$ waveforms at a frequency shifted by 2400 Hz each can be transmitted concurrently and detected free from errors without mutual interference by receivers, which execute correlative signal processing.

This means that it would be possible to operate up to $\lceil 38/4 \rceil = 9$ modems with the symbol-processing multicarrier scheme as described here concurrently and without mutual interference in a power distribution network, e.g., between a transformer station and the connected households. Note that the decisive prerequisite is not only precise signal generation, but also the exact insertion of all transmission signals into a global time pattern and perfect correlative signal processing in the receiver of each modem. As we already know, the latter depends essentially on precise synchronization.

Precise transmission-signal generation not only is required when the channel loading is dense, but also is an advantage in a single point-to-point connection. Moreover, with regard to the implementation cost, it is important that all signals are derived from a fixed and very constant basic frequency and that during frequency change no phase hops occur, but there is a continuous phase transition. In contrast, the frequency has to change rapidly without generation of transients. Continuous phase transition is required to be able to meet the strict limits for out-of-band noise specified in the EN 50065 standard with acceptable filter cost.

Signal generation and signal processing basically require synchronization. The synchronization of the received signal to a reference signal generated locally in the receiver is required in particular during a correlative reception. In the case of the symbol-processing multicarrier scheme, this reference signal comes from a frequency synthesizer. A synchronized frequency synthesizer is also required during transmission to feed its information-carrying output signals into the mains. As described in detail above, the synchronization problem can be solved easily and at low cost by means of the mains voltage. A later part of this section will describe a correlative synchronization independent of the mains voltage, which results as an advantage inherently provided by the properties of the symbol-processing multicarrier scheme.

The standard hardware available for data transmission over electric distribution networks does not allow the implementation of a symbol-processing multicarrier scheme. Therefore, realization of the ideas in industrial series products requires the production of an ASIC. This is why we will introduce a manufactured ASIC that includes all important parts for transmission-signal synthesis and correlative receiver signal processing for the proposed symbol-processing multicarrier scheme. This ASIC includes monolithically integrated analog and digital circuit parts, so that only few external components—primarily capacitors—are required to complete a modem. Maximum flexibility is offered with regard to the frequency selection (position of the frequencies used within the allowed transmission band).

The transmission equipment has to ensure precise generation of a series of waveforms with relatively closely neighboring frequencies. In this context, depending on the information to be transmitted, a quick phase-continuous frequency change should be possible, but without transients.

The receiver equipment has to be capable of perfectly separating signals at neighboring frequencies of unknown phase position. This requires *matched filtering* in several parallel receiver branches. As described in an earlier section, a matched filter for waveforms of a certain chip duration T is formed by an active correlator, consisting of a multiplication and an integration unit, if the synchronization is such that the integration unit is reset to zero after each chip interval, once the value for further processing integrated during the chip duration was sampled and stored. The implementation of such functions in analog circuits is out of the question today for the reasons described earlier.

A receiver for a symbol-processing multicarrier scheme with $N = 4$ frequencies for use in powerline communication requires eight separate correlators due to incoherent reception. The necessary hardware was not available in the market. This is why it was necessary to first build a suitable hardware basis in the form of an ASIC. This seemed to offer the best chance of ensuring simple and reproducible production of modems based on the symbol-processing multicarrier scheme, where growing production numbers would achieve continually decreasing costs. These are the prerequisites to encourage a breakthrough of new energy-related value-added services of PSUs based on EN 50065 on a broad level. PSUs still complain that there are no reliable modems in the market, either with conventional modulation schemes or multicarrier technology, for their imminent tasks of bidirectional communication with their customers. This situation is expected to change soon.

4.4.8 Integrated Microsystem Concept as a Mixed-Signal ASIC

Before we describe the technical details of the modem structure for realization of the symbol-processing multicarrier scheme, we will first introduce the overall modem concept represented in Figure 4-59. This presentation shows that it is meaningful and also technically feasible to combine analog and digital signal processing on a single piece of silicon in CMOS technology in such a way that virtually the entire modem integration is monolithic. It is also useful to include the microcontroller, which has proven itself extensively as an interface and timing component, in this integration. This controller (Intel 80C51 core) uses so-called memory mapping for integrated peripheral compo-

Examples of System Implementations

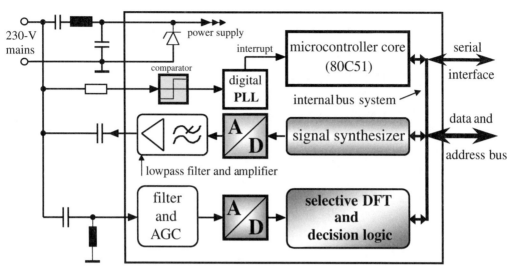

Figure 4-59 ASIC overall concept for symbol-processing multicarrier technology.

nents; i.e., the control and operation of all functions of these components is done by write and/or read operations in certain reserved memory locations. This eventually allows the control of all functions of the modem over the microcontroller's standard instruction set. Note that this offers an enormous simplification for the user. All new functions contained in the ASIC for modem operation can be fixed in a so-called "include" file; i.e., they are mapped into specific memory locations. Now the programmer can include this file among the standard assembler tools of the microcontroller and use them to program all conceivable modem applications.

The most important digital signal processing components included in Figure 4-59, i.e., the signal synthesizer and the selective DFT (here eight correlators), and the decision logic, were described briefly in Sections 4.3 and 4.4.1 based on Figures 4-36, 4-42, 4-43, and 4-45. Compared to the FH system with five frequencies per bit in Figure 4-35, which still requires only two matched filters in the receiver, we are dealing here with a real multicarrier scheme. While only one of two frequencies can occur within one chip interval in Figure 4-35, the technology discussed here uses one of four frequencies; i.e., four locations in the spectrum have to be observed. This is why the selective DFT has to be expanded accordingly.

It is also intuitive to include A/D and D/A converters in the integration. With regard to the speed and resolution requirements (maximum 10^6 8-bit conversions per second), there are no problems for the current standard CMOS technology (0.5–0.8-μm processes). In contrast, the inclusion of amplifier and filter functions is critical. A discussion of the details would go beyond the scope of this book. The most important point is the limiting frequency of the amplifiers for applications in active filters. Even if the signals to be processed do not exceed 150 kHz, the amplifiers in filter applications need transition frequencies[22] of more than 10 MHz to ensure the required factors of quality. One of the main problems is to guarantee sufficiently low parameter variations, because the analog components have to be produced in a CMOS process, which was designed for digital circuits. The tolerance of digital functions versus parameter variations in semiconductor production is generally much higher than that of analog functions. This is why considerable safety reserves have to be taken into account when selecting manufacturers and processes.

To synchronize by use of the mains voltage, the ASIC also has a feature that has proven effective against the jitter of mains zero-crossings, namely a digital phase-locked loop (PLL). This PLL activates a kind of flywheel, which can perfectly eliminate high-frequency jitter to yield a much better quality of synchronization reference [D8].

Figures 4-60 and 4-61 form the basis of the following description of the most important functional units to implement the symbol-processing multicarrier technology. We assume a transmission of binary information at a bit rate of $r_D = 1/T_b = 1200$ bits/s and use four orthogonal waveforms at different frequencies with chip duration $T = T_b/2$. This means that we are looking at a symbol-processing multicarrier system with $N = 4$. We need eight parallel correlators in the receiver for $N = 4$, because incoherent reception is the rule in powerline communication within the frequency range specified in EN 50065. $T = T_b/2$ results in a chip rate of $h_r = 2400$ s^{-1}. In the frequency range from 9.6 to 148.8 kHz we can find 60 frequencies forming a set of orthogonal waveforms with a distance of 2400 Hz each. With a sampling rate of 600 kHz, we require a maximum of 125 samples for an error-free representation (see also Table 4-3). Due to the incoherent reception, eight reference values are required for each sample of the received signal, so that we need a clock frequency of $f_r = 4.8$ MHz to output the reference signal samples.

With the symbol-processing multicarrier scheme, an optimum with regard to the noise immunity and at the same time manipulation safety can be achieved if the frequencies, onto which the user information is distributed, are as far apart as possible. It is then improbable that several frequencies will be subject to the same attenuation and/or interference at the same time. For example, the frequency selection in Table 4-17 is useful for an $N = 4$ system:

---

[22] The frequency at which the amplifier gain drops to 1.

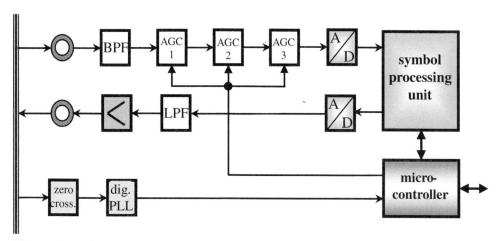

Figure 4-60 Complete structure of a modem for symbol-processing multicarrier technology.

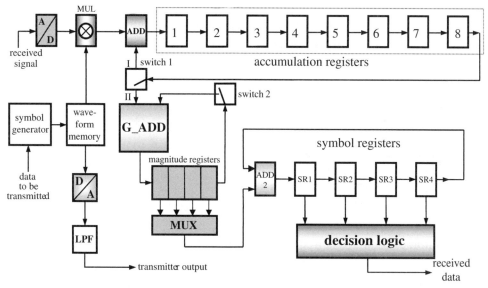

Figure 4-61 Integrated microcomputer system as ASIC component.

Table 4-17 Example of a frequency definition for $N = 4$.

f_1	f_2	f_3	f_4
52,800 Hz	62,400 Hz	72,000 Hz	86,400 Hz

Figure 4-60 shows the entire structure of a modem. We will now describe the function of this modem, starting with the receiver branch. The incoming signal reaches a bandpass filter (BPF) from the mains over the coupler. This BPF is permeable over a range of four desired frequencies f_1–f_4, but ideally suppresses the other frequency range well. Separation of the mains and lower frequencies is handled mainly by the coupler, which functions as a highpass. Next follows an equipment for automatic gain control (AGC), consisting of three operational amplifier stages. Their gain can be set digitally over the integrated microcontroller. The amplified received signal reaches the analog/digital converter, which supplies the digitized samples to the symbol-processing system; its results are accepted and further processed by the integrated microcontroller. Finally, the microcontroller supplies the received data over a serial interface and calculates the gain to be set for the three AGC stages. Note here that two of the three amplifiers are "able to rapidly and considerably change" their gains, while the gain of the third changes from symbol to symbol by a maximum factor of 2. The rapidly changing amplifiers are responsible for quick response to sudden considerable channel changes, while the slowly changing amplifier should compensate low fluctuations. The control algorithm implemented in software in the microcontroller attempts always to set the fast amplifiers to the highest possible gain to ensure that a large response flexibility is available when channels change quickly. A level-estimation unit, which supplies the calculation basis for the amplifier settings to the microcontroller, is executed as an averaging processor in digital hardware and is an integral part of the symbol-processing system. The level-estimation unit obtains Z digitized values $x(k)$ of the received signal and uses them to determine the estimated value

$$X_{\text{sch}} = \frac{\pi}{2} \cdot \sum_{k=1}^{Z} |x(k)| \qquad (4.90)$$

after completion of each symbol duration. The ratio from the estimated value X_{sch} to a target value X_{soll} supplies the basis for the generation of the switching commands for the gain control by the microcontroller. The maximum total gain is variable from 1 through 4096, where each of the three amplifier stages can be set to the gain values 1, 2, 4, 8, or 16.

Another functional unit within the symbol-processing system is represented in Figure 4-61 and will be described in detail. It handles the transmit signal preparation, where the digital data stream to be sent, which reached the microcontroller from the data source, is prepared such that the D/A converter is fed directly with the samples of the waveforms to be transmitted. After the digital/analog conversion, the signal to be transmitted is filtered in the reconstruction lowpass (LPF), then amplified in the transmitter power stage, and fed over the coupler into the mains. The two functional blocks, "zero-cross." and "dig. PLL", in Figure 4-60 serve for synchronization with the mains voltage. Hereby, "zero-cross." does a precise mains zero-crossing detection, and at the same time the zero-crossing information is galvanically separated from the mains by means of an optocoupler. The block labeled "dig. PLL" is a digital phase-lock loop used to remove high-frequency jitter from the zero-crossing information, so that a steep-edged and stable synchronization signal is eventually fed to an interrupt input of the microcontroller. As mentioned before, synchronization to the mains voltage is generally not an optimal solution. We will, therefore, describe an option to implement a correlative synchronization based on the received signal to also achieve perfect synchronism independent of the mains voltage in a later section.

Figure 4-61 shows an integrated microcomputer system including the functional blocks A/D converter, symbol-processing system, and D/A converter from Figure 4-60.

In the technical execution of a complete symbol-processing multicarrier system, it is a good idea to integrate all functional units from Figure 4-60, except for the couplers and the transmission power amplifier, monolithically in the form of a mixed-signal ASIC. At this point, we discuss only the section shown in Figure 4-61, because it is better to explain the important functions than an extensive and confusing complete circuit diagram.

In Figure 4-61, the received signal, now separated from the mains and filtered and amplified, first reaches the analog/digital converter. Incoherent reception is the rule in powerline communication, so that the receiver requires eight parallel correlators for $N = 4$ to execute a selective DFT for four spectral lines. Due to the incoherent reception, eight reference values are required for each received signal sample. This means that a maximum of 125 samples are required per frequency with a 600-kHz sampling rate (corresponding to a sampling time raster of $T_A = 1.66$ µs), so that a memory for 2000 samples at a clock frequency of $f_r = 4.8$ MHz is necessary to output the reference signal samples.

We call the digitized received signal $E(i)$ in the following discussion, where i is the discrete time with $i = 0, 1, 2, \ldots$ in increments of the sampling time raster T_A. Each digitized sample $E(i)$ of the received signal is multiplied by eight reference signal samples $R(i + \nu/8)$, with $\nu = 0\text{--}7$, from the waveform memory. The eight partial products are subsequently integrated over a waveform duration T (chip time); i.e., they are added to separate accumulators. These accumulators were implemented by the principle shown in Figure 4-61 and described in Section 4.4.1: an adder (ADD) is connected to eight following registers and a switch, all together in a ring structure. At the beginning of a waveform reception, switch 1 is in position II for the duration of eight clocks at frequency f_r; i.e., the ring structure is opened, so that zeros reach one input of the adder, while the other input receives eight multiplication results consecutively, $E(i) \cdot R(i + v/8)$ with $E(i) \cdot R(i + v/8) = 0-7$ from the multiplier, where the allocation is as follows:

$R(i + 0/8) \equiv$ sine sample f_1 chip (in-phase component for f_1 chip),
$R(i + 1/8) \equiv$ sine sample f_2 chip (in-phase component for f_2 chip),
$R(i + 2/8) \equiv$ sine sample f_3 chip (in-phase component for f_3 chip),
$R(i + 3/8) \equiv$ sine sample f_4 chip (in-phase component for f_4 chip),
$R(i + 4/8) \equiv$ cosine sample f_1 chip (quadrature component for f_1 chip),
$R(i + 5/8) \equiv$ cosine sample f_2 chip (quadrature component for f_2 chip),
$R(i + 6/8) \equiv$ cosine sample f_3 chip (quadrature component for f_3 chip),
$R(i + 7/8) \equiv$ cosine sample f_4 chip (quadrature component for f_4 chip).

The registers 1–8 contain the following results after eight clocks at frequency f_r:

Register →							
1	2	3	4	5	6	7	8
Q_4	Q_3	Q_2	Q_1	I_4	I_3	I_2	I_1

I_j is the in-phase component and Q_j the quadrature component of the chip at frequency f_j, with $j = 1-4$. Because $i = 0$, the following applies at the beginning of a waveform:

$Q_4(0) = E(0) \cdot R(0 + 7/8) + 0$
$Q_3(0) = E(0) \cdot R(0 + 6/8) + 0$
$Q_2(0) = E(0) \cdot R(0 + 5/8) + 0$
$Q_1(0) = E(0) \cdot R(0 + 4/8) + 0$
$I_4(0) = E(0) \cdot R(0 + 3/8) + 0$
$I_3(0) = E(0) \cdot R(0 + 2/8) + 0$
$I_2(0) = E(0) \cdot R(0 + 1/8) + 0$
$I_1(0) = E(0) \cdot R(0 + 0/8) + 0$

The addition of 0 results, because the ring structure is opened by switch 1. After eight clocks at frequency f_r, switch 1 goes to position I, so that the described ring structure forms, where the contents of register 8 now reach the adder. Subsequently, the calculation operations are executed, and the contents of registers 1–8 are accumulated as follows:

Examples of System Implementations

Register 1: $Q_4(i) = E(i) \cdot R\left(i + \frac{7}{8}\right) + \sum_{\xi=0}^{i-1} E(\xi) \cdot R\left(\xi + \frac{7}{8}\right)$

Register 2: $Q_3(i) = E(i) \cdot R\left(i + \frac{6}{8}\right) + \sum_{\xi=0}^{i-1} E(\xi) \cdot R\left(\xi + \frac{6}{8}\right)$

Register 3: $Q_2(i) = E(i) \cdot R\left(i + \frac{5}{8}\right) + \sum_{\xi=0}^{i-1} E(\xi) \cdot R\left(\xi + \frac{5}{8}\right)$

Register 4: $Q_1(i) = E(i) \cdot R\left(i + \frac{4}{8}\right) + \sum_{\xi=0}^{i-1} E(\xi) \cdot R\left(\xi + \frac{4}{8}\right)$

Register 5: $I_4(i) = E(i) \cdot R\left(i + \frac{3}{8}\right) + \sum_{\xi=0}^{i-1} E(\xi) \cdot R\left(\xi + \frac{3}{8}\right)$

Register 6: $I_3(i) = E(i) \cdot R\left(i + \frac{2}{8}\right) + \sum_{\xi=0}^{i-1} E(\xi) \cdot R\left(\xi + \frac{2}{8}\right)$

Register 7: $I_2(i) = E(i) \cdot R\left(i + \frac{1}{8}\right) + \sum_{\xi=0}^{i-1} E(\xi) \cdot R\left(\xi + \frac{1}{8}\right)$

Register 8: $I_1(i) = E(i) \cdot R\left(i + \frac{0}{8}\right) + \sum_{\xi=0}^{i-1} E(\xi) \cdot R\left(\xi + \frac{0}{8}\right)$

Under the prerequisite that a waveform has a chip duration $T = NT_A$, we obtain the desired eight signal components in the registers 1 through 8 in Figure 4-61 after $i = N$ clock periods at frequency f_A, i.e., after $8N$ clock periods at frequency f_r:

Register →							
1	2	3	4	5	6	7	8
Q_4	Q_3	Q_2	Q_1	I_4	I_3	I_2	I_1

For the symbol decision that now follows, we first have to do the geometric addition of the signal components I_j, Q_j according to the quadrature receiver principle:

$$B_1 = \sqrt{I_1^2 + Q_1^2}$$
$$B_2 = \sqrt{I_2^2 + Q_2^2}$$
$$B_3 = \sqrt{I_3^2 + Q_3^2} \qquad (4.91)$$
$$B_4 = \sqrt{I_4^2 + Q_4^2}$$

resulting in the amounts B_1–B_4.

The squaring and square-root calculation operations could be easily implemented in digital hardware, but the cost would be high, in particular for a large dynamic range. For this reason, we use the approximation

$$B_j \approx \max\{I_j, Q_j\} + \left(\frac{1}{4} + \frac{1}{8}\right) \cdot \min\{I_j, Q_j\} \qquad (4.92)$$

which supplies equivalent results at a much lower cost. Note that the bigger correlation value I_j or Q_j of a signal component has to be added to the smaller, multiplied by (1/4 + 1/8). The simple arithmetic and logical operations required for this are implemented in the functional blocks G_ADD, switch 2, and the magnitude registers of Figure 4-61. After each chip interval, switch 1 goes into position II for exactly eight clocks, so that the eight correlation values reach block G_ADD. The first four values (I_1–I_4) are immediately shifted into the magnitude registers. Switch 1 goes back to position I, so that the correlation values for the next waveform can accumulate in the registers 1–8. Now switch 2 closes for further calculation of the amounts B_j based on the above calculation rule, so that the four values I_1–I_4 are fed back consecutively into functional block G_ADD, where each of them is compared with the pertaining quadrature values Q_1–Q_4, still stored in G_ADD. The higher value of each is determined and pushed onto the magnitude registers.

The smaller correlation values are now each divided by four or by eight, respectively, by shifting them by two or three bit positions to the right in G_ADD, and the results are added. For the closing addition based on the above calculation rule, the maximum values stored in the magnitude registers are fed back to functional block G_ADD via switch 2 for the purpose of adding them to the minimum values each scaled by (1/4 + 1/8), and the results are pushed onto the magnitude registers, where there are now the four desired amounts B_j. These amounts are used to execute the symbol decision.

According to Table 4-16, a symbol decision can be made after each four chip intervals. For this purpose, the amounts B_1–B_4 from the magnitude registers have to be added in four consecutive chip intervals based on the scheme shown in Table 4-18, and the results have to be placed into the four symbol registers SR_1–SR_4.

Table 4-18 Symbol value calculation.

Chip Interval Number					Symbol Register	Data Symbol
1	2	3	4			
B_1	B_2	B_3	B_4	$\Sigma \Rightarrow$	SR4	LL
B_2	B_3	B_4	B_1	$\Sigma \Rightarrow$	SR3	LH
B_3	B_4	B_1	B_2	$\Sigma \Rightarrow$	SR2	HL
B_4	B_1	B_2	B_3	$\Sigma \Rightarrow$	SR1	HH

The required operations are executed by means of the multiplexer MUX and the adder ADD 2 in connection with the magnitude registers and the symbol registers SR_1–SR_4. Assume that the symbol registers are filled with zeros at the end of the first chip interval. Now, the multiplexer supplies one by one the four amounts B_1–B_4, as given in column 1 of Table 4-18, to the adder ADD 2, while the second input of this adder receives zeros. After the second chip interval, the amounts given in column 2 of Table 4-18 arrive cyclically exchanged from the multiplexer, i.e., B_2, B_3, B_4, B_1, so that now, after two addition steps, the symbol registers contain partial sums according to the first two columns of Table 4-18, because the symbol register contents have been fed back to the adder ADD 2. If we continue this process, we obtain the final symbol results in the registers SR_1–SR_4 after the two remaining chip intervals. All the decision logic has to do now is to decide for the maximum of the four results to determine the relevant data bit combination, which is then fed to the data sink. However, we can, in fact, gain more information from the results of the symbol calculation, which can be used for autonomous correlative synchronization, i.e., independent from the mains voltage, which will be described below. First, we will explain the remaining blocks in Figure 4-61, which contain the most important transmitter functions.

The core component of the transmitter hardware is the waveform memory, containing the samples for the transmission waveforms, the majority of which can also be used as reference signal samples in receive mode. First, the data to be sent reach functional block SB, where they are combined into frequency sequences according to the symbol formation given in Table 4-16. Next, in the sequence shown in Table 4-16, the functional block SB addresses and reads from the address spaces of the waveform memory where the samples with the desired frequencies are stored. At a sampling frequency of 600 kHz, the address spaces would each have a maximum length of 125. Altogether, a maximum of 500 samples would have to be stored. To keep down the cost for the reconstruction lowpass after the digital/analog converter, it is a good idea to increase the clock frequency for transmission-signal synthesis, e.g., to 1.2 MHz or 2.4 MHz.[23] This means that the memory requirement grows proportionally. Another important detail in the design of the transmitter functions is that abrupt phase hops should be prevented both during a frequency change within a symbol and at the symbol borders; i.e., phase continuity should be ensured. Otherwise, the rules with regard to the out-of-band noise power specified in EN 50065 [NO3] would be difficult to meet.

This completes the functional description, and we will now introduce the autonomous synchronization, which is basically a side product of the special receiver architecture [D8].

In many cases, it is an advantage not to use the mains voltage as a synchronization reference. In some eastern European countries where the stability of the mains frequency is poorer than in the central European grid, a sufficiently precise synchronization is no longer achievable for chip rates of about 1000 s^{-1}. Otherwise, if a functioning communication is demanded even during power failure, then an autonomous receiver synchronization should be possible from the received signal. It is generally a good idea to apply a coarse synchronization by means of the mains voltage and then compensate continually for synchronization errors by use of the correlative method introduced below, so that eventually the synchronization is more exact than it would be on the basis of the mains voltage in the central European grid, even if the synchronization reference could be freed from jitter by means of a digital phase-locked loop (PLL).

In the known matched filter receivers based on correlator structures, autonomous synchronization from the received signal means generally higher costs than the correlation itself. Instead, the described ASIC hardware can be used to provide this synchronization at negligible additional cost. A stable time basis is available everywhere, because all transmitters and receivers basically use quartz oscillators for clock generation. Finally, the synchronization equipment matches the time raster present in the received signal exactly to the time raster generated locally in the receiver. This complex process is described below and shown in Figure 4-62, which is based on Table 4-16, and by use of the time frame given in Table 4-19.

Figure 4-62 represents the four information-carrying symbols and the corresponding data bit combinations. Starting from the bit combination "LL", we can see that the symbol belonging to "HL" is created by a cyclic right shift and the symbol belonging to

---

[23] As mentioned before, this oversampling also reduces the quantization noise of the D/A converter.

Examples of System Implementations

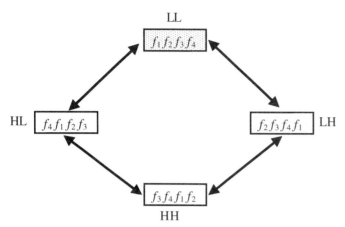

Figure 4-62 Explanation of the correlative synchronization.

Table 4-19 Explanations to the principle of correlative synchronization.

transmitted: LL	f_1	f_2	f_3	f_4	
LL $\longrightarrow$	f_1	f_2	f_3	f_4	
LH	f_2	f_3	f_4	f_1	
HL $\longrightarrow$	f_4	f_1	f_2	f_3	f_4
HH	f_3	f_4	f_1	f_2	

|←→| time lead

"LH" is created by a cyclic left shift. Relating to the synchronization, "right shift" means that the time raster of the local reference in the receiver leads the received signal; i.e., a delay would have to be executed for correction. "Left shift" means that the time raster of the local reference in the receiver lags behind the received signal.

If we had selected "LH" as our starting point, the cyclic right shift would result in "LL" and left shifting would result in "HH". Viewing each symbol in this respect, we can determine whether or not there is a lead or a lag in the receiver time raster, solely by looking at the highest and the second-highest correlation values. The difference between the two values also gives an idea about the size of the synchronization error; i.e., the higher it is, the better the synchronization matches, and the lesser the corrective actions required. Taking these facts into account, we see that the synchronization can be corrected quite easily.

Assume the symbol belonging to "LL" is received. For example, if leading is detected with a relatively large difference between the highest ("LL") and the second-highest ("HL") correlation values, then the receiver's local reference is slightly delayed. We can achieve this delay easily in the digital way by inserting additional clock periods. After the next correlation process, the highest and second-highest correlation values are studied again. Let us further assume that the symbol belonging to the bit combination "HH" was just transmitted, so that the highest correlation value is received for this symbol. After the previous lead correction, it is now expected that the second-highest correlation value for "LH" will arrive, and that the difference to the highest value will have increased, i.e., that the synchronization has improved. The described correction steps are continued, if necessary by constantly reducing the correction amounts, until the receiver reference is found to lag instead of leading. A stable optimum state is achieved when alternate leading and lagging of the receiver reference is detected after each correlation process, i.e., after each symbol duration. This ideal state will hardly occur in practice, because the difference to be observed is subject to fluctuations, even if the synchronization is ideal, due to inherently present noise. The problem is that these fluctuations cannot be distinguished from synchronization errors. This means that, in order to achieve the described stable state, the corrections will generally have to be bigger (more steps that insert or omit clock cycles) than theoretically expected.

Table 4-19 shows the described connections with regard to the time raster, where the symbol belonging to the bit combination "LL" is presumably transmitted (see top line: "transmitted").

We can see that there is a considerable lead of the reference time raster at the receiver side, with two effects: first, the correlation value for the transmitted symbol will be too small; second, contributions will result for symbols not transmitted. In this example, this is the case for only a single symbol, namely the symbol allocated to the bit combination "HL". The vertical dashed rectangles in Table 4-19 include contributions correlated for each of the four chips of a symbol in the wrong position due to the time shift. We can see that contributions can arise only for the symbol belonging to "HL", while there is no match of the frequencies in the rectangles for "LH" and "HH".

4.4.9 The Symbol-Processing Multicarrier Scheme—Summary

The symbol-processing multicarrier scheme presented above is a modification of the classical fast frequency hopping, where the drawback of a high chip rate is avoided. The implementation of this scheme in an integrated circuit with mixed analog and digital functional units (mixed-signal ASIC) and application of proven basic principles (e.g., ring architecture) was an important prerequisite for broad practical trials. The functions integrated in the ASIC cannot be realized at low cost by any other method, not even by the use of expensive digital signal processors (DSPs), let alone device volume and power consumption. A standard 8-bit microcontroller (80C51), also integrated in the form of a macro cell, handles signal evaluation and system control, and its integrated interfaces allow bidirectional communication with data sources and data sinks of any type. This microcontroller offers high flexibility, as it can be programmed. In addition, the memory mapping allows control of the entire modem operation, using the familiar instruction set of this commonly used microcontroller known by many developers in the industry. This basis, allowing complex modulation schemes, creates the main prerequisites to open up electric energy distribution networks as a versatile communication medium.

Meanwhile, this ASIC has been produced in large numbers, and the industrial series production of systems for energy-related value-added services has been started. Trials in a number of PSUs have been in progress for more than a year now, and excellent results have been observed, with regard to both the bit error rate and the availability of all modems used in a complete network supplied by a transformer station (up to 400 households). Not a single total connection failure was observed, even in extreme situations. However, high bit error rates have been reported, which will require suitable measures for error detection and correction. Although an FEC like the one described in connection with building automation systems is very effective (particularly good results were achieved with the modified (31, 25, 4) Hamming code), it has important drawbacks. These drawbacks become obvious especially whenever the channel enters bad states only briefly. The reason is that FEC requires redundancy bits to be always transmitted, even in totally errorless states, so that the net data rate is permanently reduced.

One option currently in the testing phase to avoid this drawback of FEC is to equip the receiver with very good error-detection capabilities and to request the transmitter to retransmit every packet found to be faulty. This method is called ARQ (automatic retransmission request) [D5]. ARQ reaches its limits when there are permanently high bit error rates, because the effective transmission speed can quickly drop to unacceptable values when many packets have to be retransmitted. This means that, for example, to use ARQ to overcome impulsive interference from a dimmer would be useless.

ARQ (and often FEC) cannot do without special measures for safe detection of uncorrectable errors. Codes found to be effective in FEC are not suitable to detect many different types of uncorrectable errors. The requirements with regard to the so-called "residual error probability"—the probability that an error slips through detection—are very high in PSU data transmission applications. For this purpose, three "integrity classes"—IP1, IP2, IP3—were defined for remote-control technology standards. The IP3 class allows a maximum residual error probability of 10^{-12}. This means that a maximum of one bit error in 26 years would get through undetected in a continual transmission at a data rate of 1200 bits/s. Note that achieving such low residual error probabilities does not require expensive methods; relatively simple calculation operations that can be easily implemented on a low-cost microcontroller are sufficient. One such proven method is the cyclic redundancy check (CRC). Based on special division algorithms, this method generates a few check bits and appends them to the data record [D5]. The receiver repeats the transmitter's calculation operations on the received data and compares the result with the received check bits. If deviations are found, then the transmission is faulty, and retransmission of the data packet is requested.

4.4.10 Modem Implementations for Energy-Related Value-Added Services

In closing this chapter, we will introduce implemented modems developed partly as prototypes within the scope of research work and in the industry as a preparation for small series production. We will make ample use of pictures to represent these modems and will show that, based on the achieved state of the art in implementing the new services of PSUs, at least as far as powerline communication is concerned, they are ready for practical use. Today's fast developments in the electric power market will force all parties involved to take quick action. All household meters that cannot yet be read electronically have proven to be a massive handicap. Bringing the modem and electronic power meter together seems to be an attractive and probably the only acceptable solution.

Initial trials for application of powerline communication in PSUs were started in 1995, using modems based on ASIC 4 from Figure 4-49. SFSK was the chosen modulation with a gross data rate of 1200 bits/s. The (31, 25, 4) Hamming code described in an earlier section was used for forward error correction. In addition, CRC was used to detect uncorrectable errors [D5, ST23].

Figure 4-63 shows a coupling into the low-voltage supply of the Westhochschule at the University of Karlsruhe, Germany. The transmission led over ground cable (approximately 400 m) to the author's faculty building. Roughly one-third of all wall plugs in the building could be connected over modems to the transformer station at an acceptable error rate. Due to the office and working hours, there were strong fluctuations in transmission quality, depending on the time of day. The main impact originated from the powering-on of the personal computers (PCs). However, this was due not to interference from switching power supplies of these computers, but to their EMI suppression equipment, which obviously dramatically reduced the powerline access impedance locally due to large parallel capacitors. This means that, with a modem operating near a PC connection, one must expect considerable additional attenuation for the signals to be received and overheating of the transmitter output stages due to low impedance.

The modem shown in Figure 4-64 was implemented for exhibition at the Interkama 1995 trade fair in Dusseldorf. Its core component is the ASIC 6 from Figure 4-49, which is easily visible at the top right of this picture. The modulation scheme was again SFSK with a gross data rate of 1200 bits/s. In addition, FEC on the basis of the well-known (31, 25, 4) Hamming code and CRC for error detection were used [D5].

All components, including power supply and two bistable relays with high switching capabilities, are accommodated in a standard housing of a ripple control receiver. The transmission quality of these devices is comparable to the setup in Figure 4-63, because the transmission scheme used is virtually identical in both cases. Only the hardware of the decision logic of the receiver branch was modified, and this did not have any influence on the system behavior.

The devices introduced at the Interkama 1995 trade fair were the first to awaken some interest in new PSU services over their supply networks. However, remote meter reading was still considered a totally uneconomical and useless application. The major argument was that one reading per year is sufficient. Still, the presentation triggered extensive trials, some of which were started immediately after the Interkama by several PSUs.

For this purpose, the structure shown in Figure 4-65 was manufactured in a small series by the industry. About 1000 units of ASIC 7 (see Figure 4-49) were eventually produced.

On the motherboard (on the right-hand side in the picture) we see in the top part the ASIC mentioned above, and to the left of it the sample memory. No change was made to modulation, data rate, and error correction and detection versus the predecessor version. This meant that the same transmission quality was achieved. However, one

Figure 4-63 Experimental system for data transmission over the last mile.

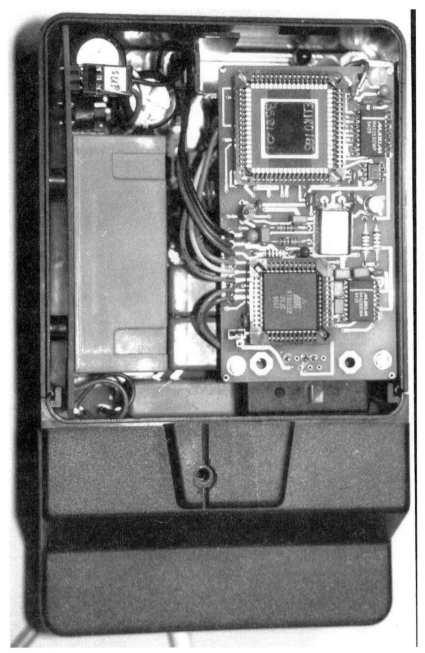
Figure 4-64 Modem exhibited at the Interkama 1995 trade fair.

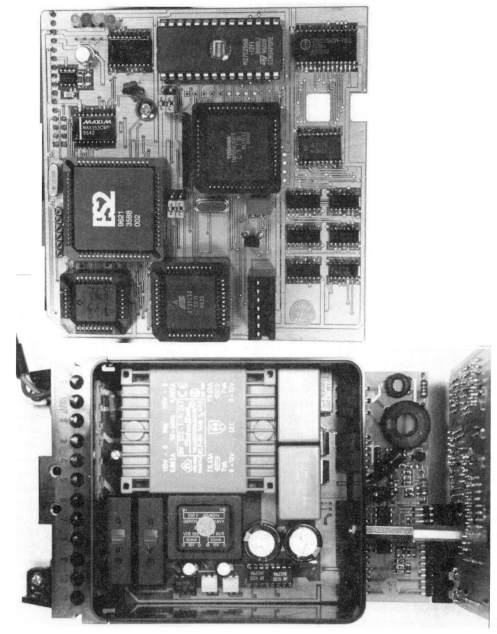

Figure 4-65 Industrial modem produced in a small series for extensive field trials in cooperation with PSUs.

Examples of System Implementations

addition was a second so-called "application" microcontroller (in the center of the motherboard), implementing higher layers of the transmission protocol, so that it supported a connection to standard interfaces customary in mains control engineering. This step was absolutely necessary to win PSUs as future customers in order to be able to demonstrate the integration of the new powerline communication into existing higher-level control structures.

The left half of the pictures shows again the standard ripple control receiver cabinet, housing the entire structure. We can see the power supply, the power relay, and signal coupling components.

The most important step, extensively discussed in this chapter, namely the modification of frequency hopping toward a special symbol-processing multicarrier system and its integration on a mixed-signal ASIC, is finally illustrated in Figures 4-66 and 4-67.

Figure 4-66 shows the first prototype of the symbol-processing multicarrier system based on the described ASIC, arranged in the left half of the picture (QFP 100 package). To the right of it, we see the program memory in the form of an EPROM. This relatively large DIL package was selected, because an EPROM simulator was used for program development. In addition to the program for the microcontroller integrated in the ASIC, this EPROM contains also the samples for the transmission and reference signals, which are loaded during booting (after a reset) into corresponding RAM areas of the ASIC, where they are available for signal synthesis. To the far right of the board, there is the RS232 interface socket with pertaining level converter (bottom). This makes the modem capable of universal communication. To the left and underneath the ASIC are a large number of discrete SMD components. These components are mainly capacitors and resistors, together with integrated operation amplifiers forming active filters,

Figure 4-66 MFH modem prototype with mixed-signal ASIC.

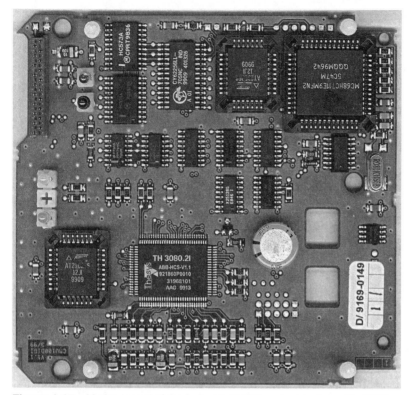

Figure 4-67 Motherboard of the industrial MFH series device with ASIC.

which have bandpass character at the receiver side and lowpass character at the transmitter side. In addition, the basic gain of the AGC is defined here over resistors.

Extensive testing of all ASIC functions revealed a series of defects, but they could all be removed by workarounds, so that they did not hinder the system trial. The entire volume of the defects—the most critical caused by the semiconductor manufacturer—required a redesign of the ASIC, which was available at the beginning of 1999. This is visible in the ASIC labeling: TH3080-II in Figure 4-66 versus TH3080-2I in Figure 4-67.

On a then-tested basis, the industry started the development of series devices, first aimed at implementing the energy-related value-added services in the A band according to the EN 50065 specifications. Of course, this ASIC is also excellently suited to building automation applications, e.g., by use of the B–D bands. It can be assumed that a significant increase of the noise immunity by MFH can be achieved versus the currently used SFSK, without risking an unacceptable rise in the overall costs.

Figure 4-67 shows the structure of the motherboard. Note certain obvious similarities with Figure 4-65.

The bottom part of the picture shows the new ASIC and the sample memory to the left of it. The application microcontroller system with a series of peripheral components is arranged in the top area of the picture. Again, this type implements the higher layers of the transmission protocol, so that the device connects to the standard interfaces used in mains control engineering. The board in Figure 4-67 is again intended for installation into a standard ripple control receiver cabinet. In addition, there are power supply, power relays, and signal coupling components.

This type of modem is currently in the trial phase in PSUs all over Europe. Current results suggest that several renowned PSUs will soon implement energy-related value-added services on the basis of this technology.

CHAPTER 5

Innovation Potential from Deregulation—Possibilities and Limits of Signal Transmission

5.1 Deregulation of the Telecommunication Markets

The development of alternative fixed network access within the last mile has gained highest significance since the telecommunication monopoly was lifted in 1998. Fast powerline communication will allow the distribution of a new high-quality communication technology in the form of Internet access. It will also support the use of telephony, telefax, and many different types of data services. The vision of the Internet over the wall plug is particularly attractive, because the world's largest knowledge and information repository will be at everyone's fingertips at low cost. Currently, high access costs represent a massive hindrance. The development of powerline offers a potential for dramatic change to this situation, making access to the Internet easy and inexpensive. Everybody can get online by simply plugging their PCs into the wall plug. From the view of costs, the situation could be such that, regardless of the data quantity to be transmitted, there will be the same fee or perhaps a monthly "flat rate." There will be no billing of online idle periods as occurs when accessing the Internet via conventional telephone lines. Powerline will make worldwide communication, procurement of information, purchasing and trade as commonplace and omnipresent as drawing electric power from the wall plug is today. There is no doubt that this development is of enormous significance and consequence for the future of a highly industrialized society with global interconnections on all levels.

Deregulation in telecommunications allows new market entrants to provide telecommunications services along electrical networks. Proprietary technology is being developed, which allows telecommunications services to be realized over the existing public and private mains power networks. As we will see in later sections and throughout the book, national regulatory bodies have specified and/or suggested regulations for use that would hinder the full exploitation of this new technology. In an effort to shape a common position, several alliances including members from both theory and practice have formed recently, such as the HomePlug Powerline Alliance.

5.2 Fast Data Transmission over Building Installations (Last Meter Solutions)—the HomePlug Powerline Alliance

Worldwide, there is an ever-growing interest to squeeze more and more data capacity out of indoor installations, i.e., to exploit the so-called "last meter." Here, in contrast to the last mile, there is even less competition. Other than wireless approaches such as DECT[1] or Bluetooth,[2] there are very few alternatives except additional wiring, which is normally not acceptable for existing buildings. Wireless services appear almost ideal at first glance but will face severe problems in practice, when several floors of a building have to be linked. Powerline is not challenged by walls and other obstacles because the signals mainly run along the wires and the coverage is perfect even for large buildings.

Powerline will become an issue mainly in the field of home automation and entertainment, not only in industrial or commercial environments, but also for the private customer. The following sections will show that there is a sufficiently high capacity at indoor installations to provide data rates of up to several tens of megabits per second. The following essential steps will have to be taken to exploit this potential:

- Channel analysis and modeling
- Regulation: definition of limits for electromagnetic compatibility
- Standardization: agreements of system manufacturers and service providers, e.g., concerning spectrum usage, modulation schemes, protocols, etc.
- Field trials for performance analysis and EMC verification
- Hardware and software development for high-volume series production

5.2.1 Mission and Vision of the HomePlug Powerline Alliance

HomePlug Powerline Alliance, Inc., was formed in March 2000 by 13 founding Members. The paragraph and list of purposes that follow are an excerpt from the bylaws of HomePlug—see [W5].

---

[1] Digital Enhanced Cordless Telephone.
[2] A system operating in the "microwave oven" frequency range (2.4 GHz) with spread-spectrum techniques.

HomePlug is a non-profit corporation formed to provide a forum for the creation of an open standard and specification for home powerline networking products and services and to accelerate the demand for products based on these standards worldwide through the sponsorship of market and user education programs. The purposes for which the corporation is organized are:

1. To define, establish and support a home networking specification which provides the basic networking capability but is also compatible with other uses of the home powerlines, and foster the rapid adoption of these specifications and standards by developers of related products and services.
2. To provide a forum and environment whereby the corporation's members may meet to approve suggested revisions and enhancements that evolve the initial specification; to make appropriate submissions to established agencies and bodies with the purpose of ratifying these specifications as an international standard; and, to provide a forum whereby users may meet with developers and providers of home networking products and services to identify requirements for interoperability and general usability.
3. To educate the business and consumer communities as of the value, benefits and applications for home networking products and services through public statements, publications, trade shows demonstrations, seminar sponsorships and other programs established by the corporation.
4. To protect the needs of consumers and increase competition among vendors by supporting the creation and implementation of uniform, industry-standard conformance test procedures and processes which assure the interoperability of home networking products and services.
5. To maintain relationships and liaison with educational institutions, government research institutes, other technology consortia, and other organizations that support and contribute to the development of specifications and standards for home networking.
6. To foster competition in the development of new products and services based on specifications developed by the corporation in conformance with all applicable antitrust laws and regulations.

In addition, ETSI (European Telecommunications Standards Institute) has initiated a PLT project to develop standards and specifications to cover the provision of voice and data services over the mains power transmission and distribution network and in-building electrical wiring. One of the main goals of this project is to harmonize ETSI standards with the relevant EU/EC Directives.

5.3 Telecommunication Access over the Low-Voltage Grid (Last-Mile Solutions)

Figure 5-1 summarizes the scenery with new alternative last-mile solutions enabled by powerline. The central node within a local area network is a transformer station, which supplies up to a maximum of 400 households via several—usually three to six—outgoing cables. Communication is always controlled from that node, also called "base station," so that a point-to-multipoint network structure is created.

The conventional fixed telecommunications network (TCN) is structured quite differently. Its general aim is to provide point-to-point connections efficiently, because for telephony, a participant usually needs a single, exclusive link to its partner. In contrast, the demands for Internet access are completely different; they can be satisfied more effectively through the point-to-multipoint structure of the power supply grid than by any TCN.

During a typical Internet session the participants contact a server over "uplinks," which supply the requested information via "downlinks." Two significant differences become apparent in comparison with a normal telephone connection:

- A server can be occupied by numerous participants at the same time.
- The traffic is highly "asymmetrical"; i.e., a participant normally sends very short messages, while the server is expected to be able to supply large data volumes very quickly. Within the conventional telephone network its deficiencies concerning this kind of traffic are currently partially reduced by the introduction of the ADSL[3] technology.

The typical network of a transformer station—as shown in Figure 5-1—represents a logical bus system. If several participants have to be served at the same time, a multiuser protocol is needed to handle the access. In a multiuser environment it is important to exploit the resources of the medium, i.e., the entire channel capacity, as efficiently as possible, so that even during high-traffic periods the remaining data rate for each user does not fall below acceptable limits.

Since the typical cable lengths for energy transportation are restricted to no more than a few hundred meters to keep energy losses low, there is a considerable channel capacity, which definitely will allow Internet access from the wall plug for numerous users. There are no physical limitations that would set insurmountable obstacles. Further transportation of data from concentrators (i.e., the base stations) within the trans-

---

[3] Asymmetric digital subscriber line.

Telecommunication Access over the Low-Voltage Grid (Last-Mile Solutions)

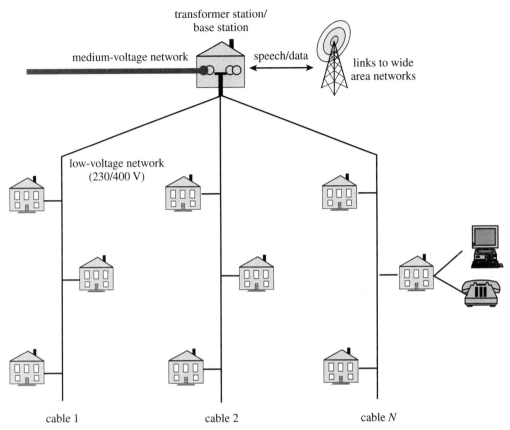

Figure 5-1 Powerline cell structure within the local loop.

former stations to wide area networks can be performed by conventional communication links such as optical fibers, broadband cables, wireless links (e.g., line-of-sight radios), or even medium-voltage powerlines.

Besides powerline communications and ADSL, other last-mile solutions have been proposed by the industry and different service providers. For example, cable television channels could be shared to a certain extent. Appropriate systems and devices have already been developed. However, such a solution has serious disadvantages in comparison with PLC. On the one hand, substantial portions of European cable TV networks still are owned by the former monopolists, and on the other hand, expensive changes will be necessary. For example, the majority of the relay amplifiers are not designed for bidirectional operation, so that a complete replacement will be unavoidable.

In contrast, power supply networks need only slight modifications, and these networks are ubiquitous all over the world, including sparsely populated rural areas where no cable TV is present.

An additional possibility for last-mile solutions was recently introduced by special short-distance line-of-sight radios. In August 1999, frequency dispatching took place in Germany through the RegTP.[4] Now two frequency bands are available: around 3.5 GHz and around 26 GHz. Since electromagnetic waves in these frequency ranges are almost totally blocked by all kinds of obstacles, in particular by buildings, it will always be necessary to have an open line-of-sight connection between base stations and participants. Base stations must therefore always be installed on elevated points, such as the roofs of skyscrapers or tops of towers. Open line-of-sight connection distances of up to eight km can be bridged under almost all weather conditions. The antennas of the participants are installed either on the roofs of their buildings or at the outer walls, if an orientation toward a base station is achievable. For further transportation of data to customers' terminals, i.e., computers or other telecommunication equipment, an additional indoor local area network is needed, the construction of which of course entails additional costs. Here again powerline communication offers highly interesting solutions.

For PLC the electrical power inlet of a building should be regarded as an interface from both a technical and a legal point of view, because the responsibility of the power supplier ends here. For the indoor installations the owner of the building is responsible. While the outdoor wiring usually is performed in accordance with certain rules and therefore is standardized to a large extent, and also well documented, building installations are very heterogeneous. Therefore a well-defined interface at the house connection makes sense in any case.

5.4 Energy Market Deregulation (Free Electricity Trade)

The process of deregulating the electricity markets in Europe is currently at its peak, although its activities began at a much slower pace than those following the deregulation of the telecommunication monopoly. This does not mean, however, that the subject is less explosive. Soon, electric power consumers will be able to freely select their power supplier, and change it as often as they want. Due to the difference in tariffs, frequent meter reading will be an inevitable consequence. Many supply companies will see themselves merely as so-called "power wheeling" enterprises in the future. This

---

[4] Regulierungsbehörde für Telekommunikation und Post = regulation authority for telecommunication and mail services (Germany).

will be tied to considerable losses in profits. So these companies have begun to devise options to keep their customers within the scope of additional services, which will by no means be easy. Powerline communication will undoubtedly play a central role in this endeavor.

With a bidirectional communication link between the transformer station and the customer premises, there is a way to exchange information quickly and directly between the PSU and the power consumer. In the future, a growing significance of such information flows is expected, because the power tariffs will be increasingly subject to a fine-grained structure, so that a fast and appropriate response to offer and demand will be possible. During peak load periods, e.g., around noon or in the evening in winter, electricity will be more expensive, compared to weak load periods, for example at night or on hot summer days or in the early morning on bank holidays, when scaled reduced tariffs could be offered. The new message channels can be used by PSUs to quickly determine the demand of a large number of customers, then calculate a current tariff on this basis and send their customers attractive offers. Basically, it would even be possible to prepare an individual proposal for customers. In connection with building automation, "smart" appliances, networked over the installation lines, could switch on when the power price fell below a certain limit. This would not only allow a cut in peak loads, which in itself would mean an enormous benefit, considering the high cost involved for the availability of power plant capacity, but also open up ways to offer tailored products to power customers. There are great possibilities for a PSU to not only deliver or put through electricity, but also to design and sell an individual service package. These possibilities can be realized at data rates of several kilobits per second in connection with electronic power meters and smart appliances. Of course, additional services not directly connected to the delivery of electric energy are conceivable, for example ordering and billing of pay-TV programs by means of remote control directly to the TV set connected over the mains (pay per view). Other, more intuitive services could be remote reading of water, gas, and heat meters, provided that these meters have appropriate data interfaces.

Chapter 4 described the qualified transmission technology that is already available for the tasks described above, and both the devices and the systems have been further developed, so that nothing would hinder a quick introduction of new services. In addition to the described MFH scheme, another more complex frequency-agile modulation method called OFDM has been implemented for powerline communication in the A to D bands in application-specific integrated circuits. While extreme robustness at data rates of up to several kilobits per second against relatively low development and devices costs had first priority in MFH, the OFDM approach is targeted to data rates of several

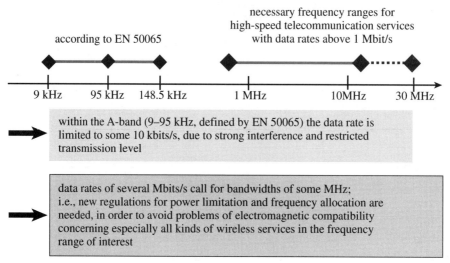

Figure 5-2 Spectral resources for powerline communications.

10 kbits/s at low robustness and clearly higher system cost. The market will certainly decide which one of these technologies will proliferate.

5.5 Bandwidth Requirements and Frequency Allocation

The realization of data rates up to several Mbits/s is possible at electrical energy distribution networks as well as at indoor installations. This has been proven by numerous studies, supported by extended measurement campaigns, and also in many field trials. Some significant results for indoor links will be presented later. The frequency range up to more than 30 MHz offers excellent possibilities, e.g., for fast Internet access and various indoor applications, such as digital audio or even video signal distribution, from the wall plug. Currently, however, only the use of the frequency range from 3–148.5 kHz for PLC is regulated (see Figure 5-2). As mentioned above, this is why PLC system development for this frequency range has reached a high degree of maturity and numerous applications related to building automation and energy information services have been operating in a legally safe environment. But for the higher frequencies neither standards nor rules exist for their use in PLC applications, and thus field trials need special experimental licensing, which can be obtained from the authorities when certain conditions concerning location, transmission power, and frequency use are met.

Since 1998, different work groups (PTF[5] and ATRT[6] of the RegTP) are concerned with regulation and standardization approaches. For Germany, a proposal was published by the RegTP in the form of the so-called NB 30.[7] This draft paper evoked fiercely controversial discussions at the beginning of 1999. The NB 30 involves not only powerline communications, but all kinds of wire-bound data transmission, including cable TV, the conventional telephone network and the new ADSL techniques, as well as all kinds of computer networks for commercial, industrial, and private applications. In March 1999, a hearing took place at the German federal ministry for economics and technology (BMWi), where all parties concerned by the NB 30 could express their views. In consequence of this hearing it became obvious that the current version of the NB 30 will need a rigorous revision before setting the rules into effect. Due to the general importance and sensitivity toward possible serious economic disadvantages for German industry the recommendations of the NB 30 draft paper were suspended for amendments and corrections until summer 2003. Details of the NB 30 in comparison with regulations of other countries will be discussed in Section 5.13.

5.6 Channel Characteristics, Coupling and Measuring Techniques at High Frequencies for PLC

Basic properties of power distribution lines used as waveguides at high frequencies have been analyzed in previous chapters. We might note that not only open wires but also underground cables are well suited for use as transmission media up to frequencies of more than 20 MHz. For indoor installations the usable frequency range exceeds 30 MHz. Investigations revealed that the characteristic impedance is relatively constant and independent of frequency in different wiring structures. We can conclude that typical powerlines are in fact only weakly lossy. Generally, however, a lowpass characteristic is found, which becomes particularly apparent at underground cables with PVC[8] as insulation material. Figure 5-3 provides an overview of the substantial characteristics of common cables. The diagrams refer to the four-sector cable depicted on the top right, which is the most common type in central European distribution grids. The lower left diagram shows the characteristic impedance over frequency. The attenuation for pure cables of 1 km length without taps is shown on the right.

---

[5] Power Line Telecommunications Forum.
[6] Arbeitskreis technische Regulierung in der Telekommunikation = working group of technical regulation for telecommunication.
[7] Nutzungsbestimmung = regulation for use.
[8] Polyvinyl chloride.

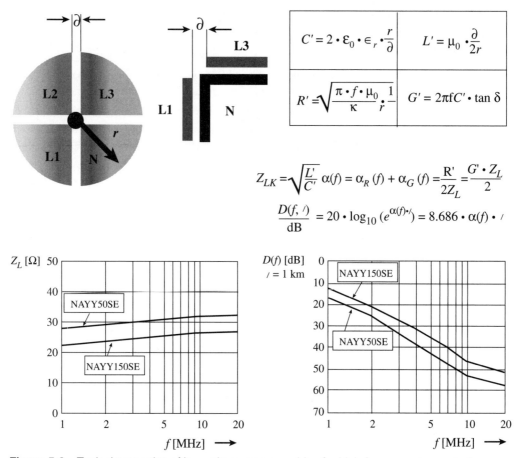

Figure 5-3 Typical properties of low-voltage energy cables for high-frequency transmission.

The curves marked NAYY50SE designate short supply taps of about 10 m bringing electrical energy directly to the house connections, while the curves marked NAYY150SE characterize the main supply cable. This cable's cross-sectional dimensions are about three times larger than those of the supply taps. The lowpass character is clearly visible and suggests a preferred use of frequencies below 20 MHz, or even below 10 MHz for long distances. This statement becomes clear when taking real supply structures, including numerous taps, into account. A general upper frequency limit, up to which a use for PLC is recommended, cannot be defined based solely on the attenuation effects discussed here. The reasons are that, on the one hand, attenuation values may change substantially, if a real network incorporating numerous taps is examined, and on the other hand the interference scenario will play an important role, because ultimately the signal-to-noise ratio at the receiver will always be the crucial figure for channel quality.

When indoor links are considered, however, the situation is quite different, because the lowpass character is hardly visible due to the short distances. Therefore, it will be advisable to use high frequencies here, e.g., more than 30 MHz. This may offer an additional advantage, as the background noise level usually diminishes with increasing frequency. Before presenting further details of the channel characteristics and channel modeling, we discuss some basic rules for appropriate RF signal coupling and measuring procedures.

5.6.1 Coupling of High-Frequency Signals

The coupling of signals into the powerline network in the form conditioned for the frequency range below 150 kHz cannot be used for higher frequencies, e.g., from 500 kHz up to more than 30 MHz. The basic principles, however, are still valid after some adjustments. The coupling circuit has to be modified as shown in Figure 5-4. In general, a high degree of symmetry is paramount for EMC (electromagnetic compatibility) reasons, as pointed out in detail later on. Toward this aim, two capacitors are inserted at the high-voltage side of the coupler, feeding the PLC signal as symmetrically as possible into the power supply network, e.g., over a standard wall plug.

Later sections will discuss in detail the fundamental importance of symmetry and appropriate choice of conductors for good EMC performance. For now, it suffices to note that the use of the neutral conductor within the access domain proved to be unfavorable with respect to EMC. In contrast, symmetric coupling on two phases offers much better results. Due to the higher voltage (400 V instead of 230 V) between the phases, using two capacitors is also an advantage because they are connected in series, so that each is stressed with only half the voltage, i.e., 200 V. Since the capacity for high-frequency applications can be brought down to only 1/10 of the values needed for the CENELEC bands A–D (< 150 kHz), there is a significant reduction of costs. We will see later that the situation is totally different for indoor networks, because here the neutral conductor proves to be the best choice to carry PLC signals with good EMC performance.

On the modem side the protection circuit against transients must also be changed, because conventional suppressor diodes have a relatively large capacity, which would attenuate substantial signal portions. The protection bridge shown in Figure 5-4 performs well up to frequencies of about 30 MHz. However, the four Schottky diodes must be carefully selected, since the capacities vary from one type to another. In general, low-power small-capacity devices are sufficient, because the typical forward bias volt-

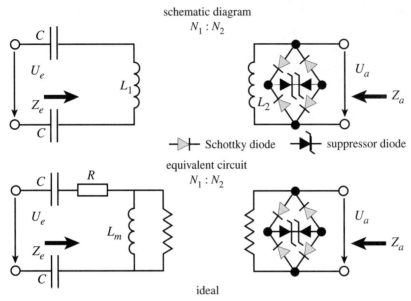

Figure 5-4 High-frequency coupling unit with protection circuitry.

age of a Schottky diode is approximately one-third of that for a usual silicon diode. It is well known that the capacity of a diode decreases with growing reverse voltage. Therefore, we need to make sure that each of the four Schottky diodes always remains reverse-biased for all signal amplitudes to be transmitted. In Figure 5-4, this is ensured by the bidirectional suppressor diode inserted diagonally into the bridge. Through its large capacity the diagonal voltage of the bridge does not follow the RF signal but remains almost unchanged, corresponding to the peak value of the transmitted signal. This is why the Schottky diodes are always reverse-biased. When the frequency range above 30 MHz has to be taken into account, the Schottky diodes should be replaced by PIN diodes, which have an even lower capacity. Although there is a larger forward bias voltage now, no damage problems are expected, because the main portion of spectral energy of dangerous transients is always located at relatively low frequencies.

Concerning the magnetic core materials there are also some changes necessary for the RF range. While toroidal cores are well proven for applications at frequencies of up to several hundred kHz, they cause significant degradation of performance for signals in the MHz-range. For high frequencies, "double-hole" ferrite cores turned out to be good solutions. The main inductance L_m (see Figure 5-4) can take on a rather small value now due to the high frequency, so that windings with very few turns are suitable, while a single turn is normally sufficient at the high-voltage side. Of course, this is an advantage with regard to core magnetization by the powerline current (50 or 60 Hz) passing the

Figure 5-5 Typical elements of an RF coupling unit.

coupling capacitors. On the other hand, as these capacitors are relatively small, core magnetization by the powerline current generally poses no problems at high transmission frequencies.

Figure 5-5 shows the construction of a powerline coupler suitable for frequencies up to more than 50 MHz. On the left-hand side are two coupling capacitors with typical dimensions (68 nF, 1000 V). We can see a typical core with primary and secondary windings (two turns each) in the center, and there is also a blank core. Twisting the wires proved to be very important with respect to EMC, because the area between the conductors must be kept as small as possible, so stretching of conductors should absolutely be avoided.

5.6.2 HF Measuring Systems and Measurement Results

During the transmission over a channel with the transfer function $H(f)$ and/or the impulse response $h(t)$, a signal $s(t)$ is attenuated on the one hand and corrupted with additive noise $n(t)$ on the other (see Figure 5-6). At the receiver, we eventually have a signal $r(t)$, which is a mixture of the attenuated transmission signal and added noise, i.e.,

$$r(t) = h(t) * s(t) + n(t) \tag{5.1}$$

For message detection based on the received signal $r(t)$, the signal-to-noise power ratio (S/N) is the crucial figure; it determines the bit error probability. The signal-to-noise ratio at the receiver can be calculated whenever the local interference scenario and the attenuation for the frequency range of interest are known. The only free variable is the transmission signal amplitude. Attenuation and interference have to be determined by measurements. This task turns out to be much more complex for high frequencies, compared to those below 150 kHz. One problem already mentioned is signal coupling, for instance to the crossbar system of a transformer station. Due to the large distances between the conductors, proper coupling of RF signals is difficult, so that fairly large

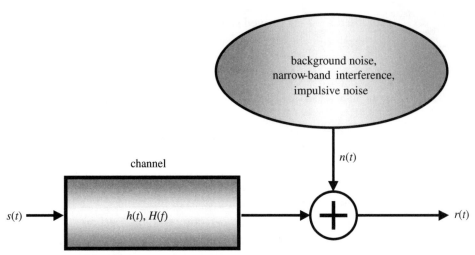

Figure 5-6 Principle of an information transmission system.

wiring loops will be inevitable. On the one hand, this will lead to signal losses, and on the other hand radiation will be produced, leading to potential EMC problems.

The actual transmission amplitude present at the powerlines must be determined over a special measuring channel to ensure clear and reproducible conditions to properly measure the transfer function. Note that the transmitter or signal generator output cannot immediately be taken as the transmission level. The attenuation of a powerline channel with the transfer function $H(f)$ can be determined by the ratio $r(t)/s(t)$ of the received and transmitted signals. $H(f)$ is generally a complex function of the frequency. Nevertheless, in order to determine $H(f)$, a real transmission signal $s(t)$ in the form of a sinusoid with frequency f and amplitude A can be applied without restricting the generality, i.e.,

$$s(t) = A \cdot \cos(2\pi f t) \qquad (5.2)$$

Normally, the received signal $r(t)$ is now complex, so that for $H(f)$ both a magnitude $|H(f)|$ and a phase response $j\varphi_H(f)$ have to be determined. In a first step $|H(f)|$ is of greater importance, because the results for $|H(f)|$ already allow rather clear estimations of the performance of a channel used for telecommunication. The measurement of $\varphi_H(f)$ requires substantially more effort, but will be inevitable for the selection and optimization of appropriate modulation schemes. In the meantime an extensive, almost complete measuring database concerning $|H(f)|$ for various power distribution grids is available. Measurements of the phase response $\varphi_H(f)$ and the impulse response $h(t)$, however, are still under way. The magnitude $|H(f)|$ of the transfer function and/or its reciprocal value, the attenuation $D(f)$ [in dB], are determined as follows:

$$|H(f)| = \frac{\text{amplitude of the received signal with frequency } f}{\text{amplitude of the signal with the frequency } f \text{ at the feeding point}} \qquad (5.3)$$

$$\frac{D(f)}{\text{dB}} = -20 \log_{10} |H(f)| \qquad (5.4)$$

Next, a transmission signal $s(t)$ is generated for measurement purposes, e.g., by use of a frequency synthesizer based on a DDS (direct digital synthesis) circuit, allowing a frequency variation over the range (for instance, 500 kHz to more than 30 MHz) we are interested in. The synthesizer output is amplified and fed into the powerline network over a matched coupling circuit. The received signal $r(t)$ is evaluated, e.g., by use of a spectrum analyzer (see Figure 5-7).

A special feature of the measurement setup in Figure 5-7 is the possibility of remotely controlling the transmission system by powerline communications within the CENELEC A band (9–95 kHz). The synthesizer can be triggered from a standard personal computer (PC) or a notebook to produce certain frequencies (single tones) with certain amplitudes, following a predefined timing pattern. Alternatively, continuous sweeps over certain frequency ranges can be programmed. Accurate information about the transmitted signal is immediately available for the spectrum analyzer, so that the received signal can be sampled synchronously and transferred to the PC or notebook for storage and offline evaluation. During transmission pauses, the spectrum analyzer is able to record the interference scenario. In numerous studies the electrical energy distribution grid turned out to be a "long-term stationary" channel; i.e., changes to the trans-

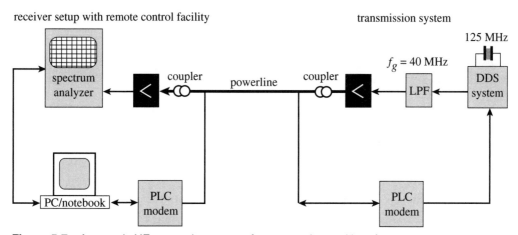

Figure 5-7 Automatic HF measuring system for attenuation and interference.

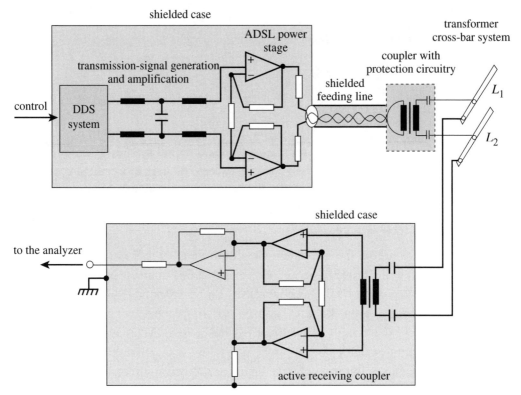

Figure 5-8 Details of appropriate powerline signal coupling at high frequencies.

fer function can be expected within minutes. Also, the typical interference scenario is generally subject to slow changes only, except for a few special cases. To gain the most important measurement parameter, i.e., the signal-to-noise ratio (S/N), two consecutive sweeps of the spectrum analyzer over the frequency range of interest are normally sufficient—one with the transmitter active (for S), and the other with the transmitter switched off (for N). The database mentioned above was acquired step by step in this way. In order to demonstrate the essential HF signal coupling problems, Figure 5-8 depicts the most important details.

Following these basic rules will facilitate the correct setup of the instrumentation for EMC analysis later. For universal and flexible signal generation the use of DDS circuits turned out to be a good choice. They can be easily operated from a PC or even a low-cost microcontroller. After filtering and amplification, the transmission signal is fed into a symmetric power stage with ungrounded outputs. Such components are inexpensive and are offered by different manufacturers in good quality for ADSL applications. The transmission signal is now conducted over a shielded twisted pair and symmetrically coupled into the powerline network—in this example to the crossbar system of a transformer station. An active receiving coupler is connected close to the feeding points in order to pick up the actual signal values at the entrance of the powerline wiring as exactly as possible. This information is mandatory to be able to determine the attenuation of a powerline channel correctly. The importance of symmetry and ungrounded signal coupling is revisited later. Note that this special way of coupling will avoid currents which usually spread along extended ground loops and significantly contribute to unwanted radiation. The reason is that, whenever the transmitter has a terminal connection to ground, the parasitic capacitance between primary and secondary winding of the coupler becomes effective.

This capacitance is no longer negligible in the frequency range of interest, because the windings must be placed close together to avoid large stray inductance. So with a ground-terminal connection of the transmitter, the transmitted signal may partially take its way over the coupler's parasitic capacitance and return over various more-or-less extended ground loops, causing the EMC problems mentioned earlier.

5.6.2.1 New Industry-Developed Measuring Systems

Although the measuring systems shown in Figure 5-7 have been automated, they are actually composed of standard devices and custom-made parts built by research institutions, such as universities. For parallel studies at different locations and in different power supply systems—which will be necessary for a sufficiently high statistical reliability of the results—numerous copies of the measurement equipment are desirable. For currently existing systems, however, this will not be possible for economic reasons. Furthermore, the operation is presently controlled by various self-developed (nonstandardized) software packages for different platforms such as PCs, notebooks, or microcontrollers. Besides these drawbacks, current measurement systems are not able to record important parameters, such as the phase of the transfer function, the impulse response, or the powerline access impedance. Most of these measurements can still be executed only by means of digital storage oscilloscopes and expensive network analyzers.

In such a measurement environment, large portions of acquired data have to be stored on fixed disks for subsequent offline analysis in PC programs. These drawbacks have encouraged universities and the industry to jointly develop a universal tool called "Power Line Analyzing TOol" (iPLATO). The concept of this tool is shown in Figure 5-9 and presented briefly in the following. This new system is planned to be available by the end of year 2001.

The core element of iPLATO is a fast digital signal processing (DSP) system with high computing power and an extended portion of high-speed memory, at least 128 MB.

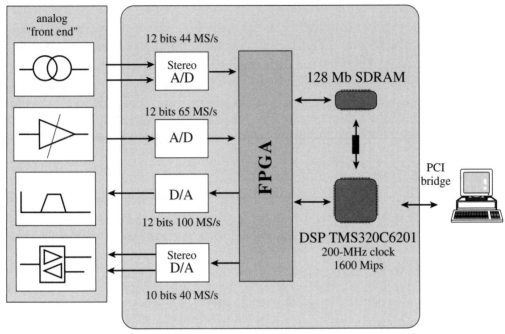

Figure 5-9 The universal powerline analyzing tool iPLATO.

The DSP system plugs into the PCI slot of a standard PC. Two pairs of A/D and D/A converters with different speeds and resolutions form the bridge between the "analog front end," which is equipped with connections to the powerline network to be analyzed, and the DSP system. A current prototype version of iPLATO allows signal recording at a sampling rate of 65 Mbits/s at a resolution of 12 bits.

Besides recording and analysis of powerline signals, the iPLATO measuring system will also allow signal synthesis for simulation and test and the implementation of complete PLC systems. The analog front end comprises not only filters and amplifiers, which are required to determine complex transfer functions and impulse responses as well as spectral and time-domain analysis of the interference scenario, but also directional couplers to determine the complex input impedance of powerline access points (see Figure 5-10). With this extension iPLATO includes all substantial functions of a network analyzer.

For the proper design of high-speed PLC systems, based for instance on OFDM schemes, we absolutely need to know the channel's impulse response to determine important parameters, such as the guard interval duration. This problem will be discussed in more detail in Chapter 6. The complex transfer function and the impulse response of a channel are connected over the Fourier transform. In practice, it will be

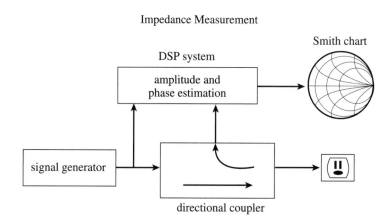

Figure 5-10 The powerline impedance measurement feature of iPLATO.

sufficient to measure either of the two parameters. Figure 5-11 illustrates the essential relations. The necessary measurements are not easy, as there are generally no loopback possibilities for outdoor powerline channels. This is why most of the currently published measurement results cover only attenuation, i.e., the reciprocal magnitude of the transfer function. Such results are usually yielded with a measurement setup similar to that shown in Figure 5-7. A few measurements of the complex transfer function are found for indoor networks in the literature, e.g., in [H9, H38]. They were performed with standard network analyzing equipment, where a synchronization link is necessary. Normally, this limits the use for powerline lengths below 100 m. Basically, it would also be possible to do time-domain measurements of the impulse response with short pulses. However, they are difficult to apply to "live" powerline networks, because non-linearity may become effective due to the large impulse amplitudes needed to investigate cable lengths of more than 300 m. Furthermore, there is also a potential risk to damage or interfere appliances connected to the mains.

To solve these problems, a new channel estimation method based on broadband (spread-spectrum) stimuli was developed; it works without special synchronization requirements. In addition, a fast and precise estimation of the desired results is possible in presence of noise. The new measurement method is based on periodic spread-spectrum signals stimulating the channel under test and an unbiased estimation algorithm to evaluate the results. Figure 5-12 gives a general overview of the architecture of the new measuring system, which can be regarded as an extension of the basic powerline link depicted in the lower part of Figure 5-11. Instead of detecting and processing data, the receiver has to calculate an estimate of the complex frequency response or impulse response, respectively, from the received signal $e(t)$.

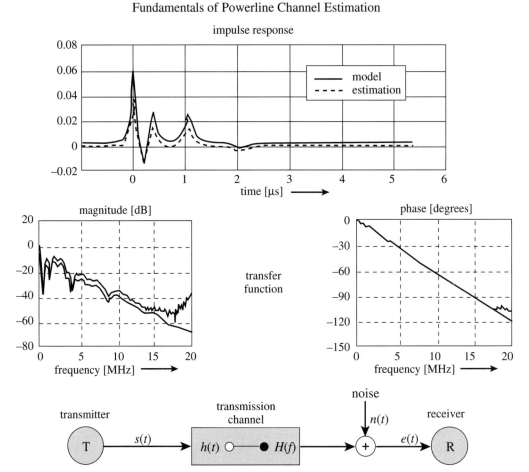

Figure 5-11 Characterizing the powerline channel by its impulse response and the complex transfer function.

As all the calculations are based on digital signal processing, a discrete system model is introduced, where the continuous time t is replaced by time steps k at intervals of the sampling rate $f_s = 1/T_s$.

The discrete samples are represented by a vector of appropriate length, designated in bold letters. For example, the stimulation signal vector is denoted by **q** and the received signal by **e**. The additive noise on the channel is represented by **n** and the discrete impulse response by **h**. Finally, calculations with the received vector **e** are performed in order to determine an estimate $\hat{\mathbf{h}}$ for the channel's impulse response.

Starting with a continuous mathematical description of the link in the lower part of Figure 5-11 in connection with Figure 5-12, we can denote the received signal as

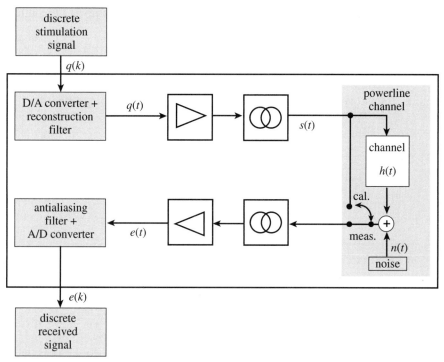

Figure 5-12 Broadband setup to estimate the powerline channel's impulse response and transfer function.

$$e(t) = q(t) * h(t) + n(t) = \int_{-\infty}^{\infty} q(t-\tau) h(\tau) \, d\tau + n(t) \quad (5.5)$$

i.e., a convolution of the stimulus $q(t)$ with the channel's impulse response $h(t)$ and addition of the noise $n(t)$. From (5.5) it is obvious that the essential task of the proposed estimator will be a deconvolution process in presence of noise. The discrete representation of (5.5) is as follows:

$$e(k) = q(k) * h(k) + n(k) = \sum_{\nu=-\infty}^{\infty} q(k-\nu) h(\nu) + n(k) \quad (5.6)$$

Of course, the sum from minus to plus infinity in (5.6) is not usable for a practical implementation. For the stimulation signal, a periodic digital pseudonoise sequence of length N is chosen with the total duration

$$T_P = N \cdot T_s \quad (5.7)$$

where T_s is the sampling rate as introduced above. Thus the discrete stimulation signal vector is

$$\mathbf{q} = \begin{bmatrix} q_1 & q_2 & \cdots & q_N \end{bmatrix} \tag{5.8}$$

and, based on the elements of this vector, we can build the transmission matrix

$$\mathbf{Q} = \begin{bmatrix} q_1 & q_N & q_{N-1} & \cdots & q_2 \\ q_2 & q_1 & q_N & \cdots & q_3 \\ q_3 & q_2 & q_1 & \cdots & q_4 \\ \vdots & \vdots & \vdots & \ddots & \vdots \\ q_N & q_{N-1} & q_{N-2} & \cdots & q_1 \end{bmatrix} \tag{5.9}$$

This matrix exhibits a so-called "right-circular" form, which means that it can be easily constructed by appropriate shifts of the vector $\mathbf{q}$ given by (5.8), and finally only this vector has to be stored in the transmitter.

With $\mathbf{Q}$ from (5.9) the discrete convolution according to (5.6) can be expressed very concisely as follows:

$$\mathbf{e} = \begin{bmatrix} e_1 \\ e_2 \\ e_3 \\ \vdots \\ e_N \end{bmatrix} = \begin{bmatrix} q_1 & q_N & q_{N-1} & \cdots & q_2 \\ q_2 & q_1 & q_N & \cdots & q_3 \\ q_3 & q_2 & q_1 & \cdots & q_4 \\ \vdots & \vdots & \vdots & \ddots & \vdots \\ q_N & q_{N-1} & q_{N-2} & \cdots & q_1 \end{bmatrix} \cdot \begin{bmatrix} h_1 \\ h_2 \\ h_3 \\ \vdots \\ h_N \end{bmatrix} + \begin{bmatrix} n_1 \\ n_2 \\ n_3 \\ \vdots \\ n_N \end{bmatrix} = \mathbf{Q} \cdot \mathbf{h} + \mathbf{n} \tag{5.10}$$

As already mentioned, it is now the estimator's task to determine the discrete impulse response $\mathbf{h}$ as precisely as possible from the received signal $\mathbf{e}$. As a first step, we can get an acceptable estimate

$$\hat{\mathbf{h}}_i = \mathbf{Q}^{-1} \mathbf{e} \tag{5.11}$$

by multiplying the received signal vector by the inverted matrix $\mathbf{Q}$ from (5.9). As the inversion of $\mathbf{Q}$ delivers $\mathbf{Q}^{-1}$, being a right-circular matrix again [F13], all the information to be stored is again contained in a single row of $\mathbf{Q}^{-1}$. For its calculations, the estimator uses a priori information about the length and the form of the stimulating signal. Equation (5.11), however, is generally not able to deliver exact results in presence of noise. But because the estimator is unbiased, it is guaranteed that the expected value of consecutive estimates will converge toward the true impulse response, so that, for a sufficiently large number Φ of estimates, we finally get

$$\mathbf{h} = \frac{1}{\Phi} \sum_{i=1}^{\Phi} \hat{\mathbf{h}}_i \tag{5.12}$$

which means that any desired precision can always be achieved despite the noise scenario.

In addition to the functions mentioned, the implementation of complete channel models for emulation purposes would also be a very attractive feature. Toward this aim a large FPGA[9] is included—see Figure 5-9. With the emulation capability, testing of modulation schemes and overall system performance can be accomplished in a laboratory environment to a great extent, significantly reducing the need for expensive and time-consuming field trials.

5.6.3 Transmission Characteristics at High Frequencies

5.6.3.1 Access-Domain Channel Characteristics

Currently an extended measuring database is available for the access network domain, i.e., the distribution grid between transformer station and house connections. The general result is that this domain can be exploited for telecommunication purposes up to frequencies of approximately 10 MHz. Among the critical channels we usually find links, which include both underground cables and open overhead wiring. Furthermore, some special filter effects are found in densely populated residential areas, where the supply cable has many taps and the length of the feeding cables to the houses is nearly constant. Such structures take on the characteristics of FIR (finite impulse response) filters, which can lead to complete suppression of the frequency range above a few MHz. In summary, however, the investigations clearly confirm that data rates up to several Mbits/s are realistic, even on critical links. Different cable types do not exhibit a completely different behavior, but their important characteristics can be described with few parameters. Another important result is that moderate transmission power levels will be sufficient for reliable operation, so that potential EMC problems can be avoided from the start. Figure 5-13 shows the result of a combined attenuation and noise measurement performed at a 300-m ground cable connection over the frequency range from 500 kHz to 20 MHz. Obviously the attenuation rises with the frequency. Such a lowpass characteristic can be observed at all ground cable connections and can, therefore, be

---

[9] Field programmable gate array.

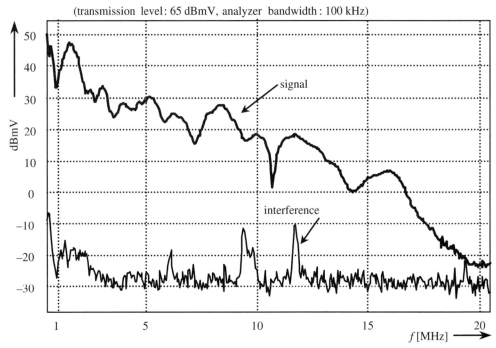

Figure 5-13 Results from attenuation and interference measurements over a 300-m ground cable link.

regarded as a substantial property, at least in the frequency range above 500 kHz. In addition, periodic fluctuations of the transfer function were noted. This observation is important, because it leads to the echo model [H16, H20, H28, D9] for the powerline channel discussed in Section 5.7.

Over extended ranges of the spectrum the interference in Figure 5-13 can be regarded as background noise, remaining clearly below the level of the received signal. The signal-to-noise ratio is of the order of 50 dB in the range from 500 kHz to 6 MHz. Also around 17 MHz, for instance, we still find more than 20 dB, which would be sufficient for undisturbed high-speed communication. In the range from 9 to 12 MHz, however, narrowband peaks appear in the noise spectrum, which can be easily identified as short-wave broadcast bands. While the applied transmission levels (65 dBmV ≡ 1.77 V) guarantee reliable powerline communications, the reception of weak short-wave broadcast stations could be affected.

Unfortunately, the conditions documented in Figure 5-13 are not representative; they describe a rather friendly channel. The channel quality varies enormously, and Figure 5-14 presents some typical examples covering the range of interest. Figure 5-15 ignores the ripples and details in order to give an overview of the essential properties.

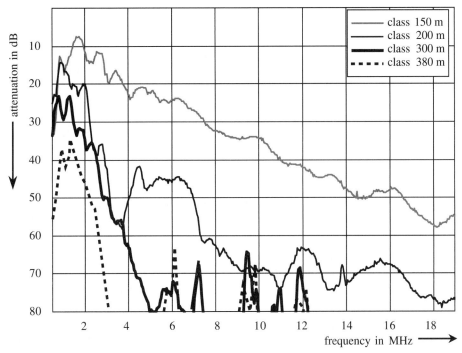

Figure 5-14 Dependence of attenuation and cable length.

The four different attenuation curves in both figures are assigned to length classes in the range from 150 to 380 m. An examination immediately reveals heavy variations within this range.

While the shortest link clearly exhibits a lowpass characteristic, its attenuation remains below 60 dB at up to 20 MHz. Furthermore, the flat course is remarkable. For such a link no communication problems are expected, even with simple modulation schemes and low transmission levels. For the 200-m class attenuation significantly increases, partially more than 25 dB. With the 200-m link, high density of house connections is most likely responsible for the significant difference in attenuation, rather than the relatively short distance difference of 50 m, while the 150-m cable probably supplies only a few customers. Quite extreme conditions can be noticed for the two additional classes, i.e., 300 m and 380 m. The lowpass behavior is no longer "graceful" but is a rather sharp transition from passband to stopband, similar to a filter. This leads us to assume that a high number of taps with regular geometry are present, forming a kind of FIR filter. It becomes evident from Figures 5-14 and 5-15 that powerline channels could exist in the access domain, and that the bandwidth available to these channels

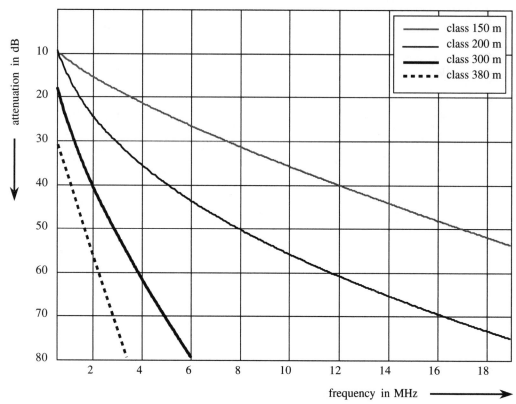

Figure 5-15 Dependence of attenuation and cable length (schematic).

can fall below 4 MHz. This observation suggests that the lowest available frequencies should preferably be used for long-distance links.

5.6.3.2 Indoor Channel Characteristics

Figure 5-16 shows some typical transfer functions of electric building installation networks. Ripples are visible here, too, and we find attenuation values in the 40–80-dB range. The results compare well with those for outdoor links, despite the rather short distances. A clear difference, however, can be recognized in the course of the attenuation curves over frequency. In particular, when considering the "1st floor" curve in Figure 5-16, we see that the lowpass character has almost disappeared. This is obviously due to the short distance.

Also, the longer link of the "2nd floor" curve does not exhibit significant lowpass effects, because the frequency range around 20 MHz still appears usable quite efficiently. These observations suggest extending the bandwidth used toward higher frequencies. According to the current state of the art, the frequency range from 10 MHz

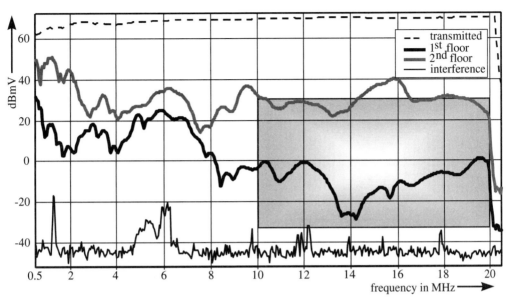

Figure 5-16 Typical transfer functions and interference in indoor installations (spectrum analyzer bandwidth: 30 kHz).

to more than 30 MHz appears very attractive for PLC within buildings. Thus, with respect to upcoming standardization, a separation of frequency bands for indoor and outdoor use can be recommended, i.e., about 1 to 10 MHz for the access domain (last mile), and 10 to more than 30 MHz for indoor applications.

Some initial indoor channel measurements were carried out up to 50 MHz to exploit the range of higher frequencies. Moreover, new signal possibilities between neutral conductor (N) and protection earth (PE) were included. The latter is of special interest for EMC reasons; this issue will be discussed in Section 5.11. Figure 5-17 shows typical results when signal coupling is used between a phase (L) and the neutral conductor (N) at a normal wall plug within a building. The upper left diagram shows both the transmitted and received signals recorded with a spectrum analyzer over a frequency range from 500 kHz to 50 MHz. The nominal transmitted voltage is 50 dBmV ($\equiv$ 316 mV). The signal is traced immediately at the feeding point and exhibits some fluctuations there, caused mainly by variations of the network input impedance found at the wall plug. In any case, it is important to record such effects in order to correctly calculate the transfer function, which is shown in the lower left diagram.

The transfer function shows attenuation values of up to 50 dB and significant fluctuations with sharp notches at several frequencies. Also a slight lowpass character is visible, but astonishingly there is almost no decrease over the 25–50-MHz frequency range. The fluctuations and notches are caused by taps and unmatched ends of lines.

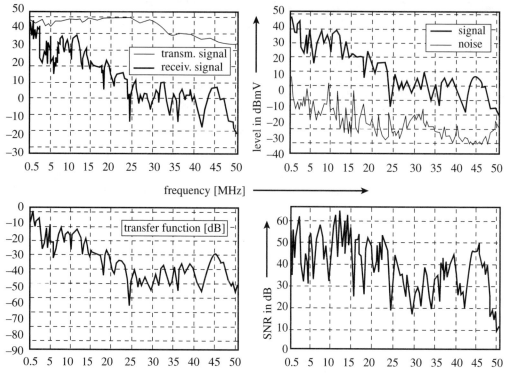

Figure 5-17 Transfer function and interference of an indoor link for L-N coupling.

Similar effects are also found in the access domain, so that the same echo-based model presented in Section 5.6 [H7, H19, H34] applies to both cases. In the upper right diagram, the noise is recorded together with the signal for comparison. The spacing of generally more than 20 dB over the entire frequency range is pretty good, which is clearly visible in the lower right SNR diagram. Note that the transmitted signal level is 15 dB below the value of Figure 5-16. In spite of this, reliable data transmission would be possible, even with simple modulation schemes.

For comparison, Figure 5-18 shows the results from signal coupling between neutral (N) and protection earth (PE). With regard to the transfer function there is almost no difference—the maximum attenuation is also around 50 dB with no significant increase toward higher frequencies. Obviously the feeding point exhibits lower impedance, compared to the L-N case. This can be explained by the direct galvanic connection between PE and N at the house's grounding point. A simple and inexpensive conditioning method to isolate the neutral conductor from the rest of the network for high-frequency signals will be outlined in Section 5.12.

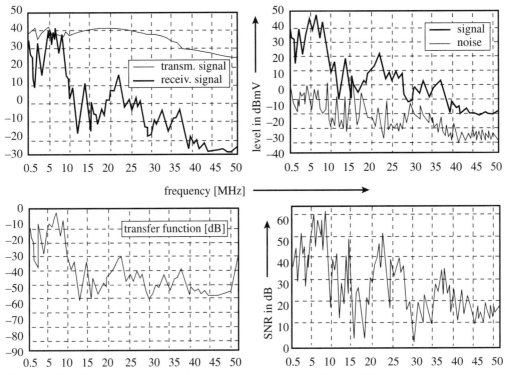

Figure 5-18 Transfer function and interference of an indoor link for PE-N coupling.

Due to the lower feeding impedance, the received signal appears additionally attenuated, so that the SNR diagram in the lower right of Figure 5-18 indicates less than 10 dB at several frequencies. In many extended parts of the spectrum, however, we find SNR values of more than 30 dB—in fact up to 50 dB, so that high-speed communication would be feasible, even without the conditioning mentioned above. On the other hand, conditioning would improve fast data-transmission possibilities in any case due to higher SNR, because attenuation decreases. In addition, conditioning is also important with respect to EMC because symmetry is improved, so that differential mode signaling takes place, especially at the feeding point and its near neighborhood.

5.7 The Powerline Channel Model

Attenuation increases with frequency in most long-distance powerline links. A lowpass characteristic can also be observed in a less significant form on indoor links. In addition, periodic fluctuations can be ubiquitously discerned, suggesting the following echo-based transfer-function model approach—see Figure 5-19.

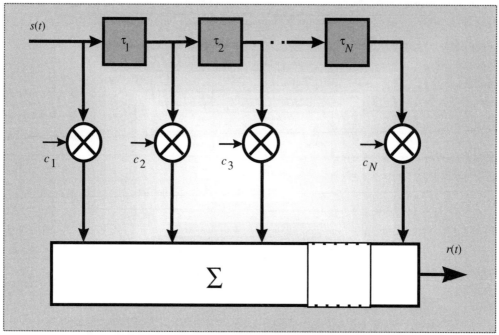

Figure 5-19 Basic structure for the echo-model definition.

The basic behavior of a transmission channel incorporating N echoes can be described by an impulse response

$$h(t) = \sum_{i=1}^{N} c_i \cdot \delta(t - \tau_i) \qquad (5.13)$$

where the coefficients τ_i denote the echo delays and the factors c_i stand for the echo attenuation. From (5.13) the transfer function

$$H(f) = \sum_{i=1}^{N} c_i \cdot e^{-j2\pi f \tau_i} \qquad (5.14)$$

can be calculated. Under real-world conditions, however, the coefficients c_i depend not only on the cable length, but also on frequency. Based on numerous channel investigations, the following expression is finally found:

$$c(f, \ell_i) = a_i \cdot e^{-\alpha(f) \cdot \ell_i} \qquad (5.15)$$

In (5.15) ℓ_i stands for the cable length and a_i is a specific weighting factor regarding network topology details; a_i represents the product of reflection and transmission factors along the *i*th echo path. Combining multipath propagation and frequency- and length-dependent attenuation, and introducing the phase velocity v_p, finally leads to the complete transfer function

$$H_E(f) = \sum_{i=1}^{N} a_i \cdot e^{-\alpha(f) \cdot \ell_i} e^{-j2\pi f \frac{\ell_i}{v_p}} \tag{5.16}$$

The coefficient $\alpha(f)$ needs further explanation: Based on an analysis using the physical cable parameters R',[10] G',[11] L',[12] and C',[13] we can derive the following result for the complex propagation factor:

$$\gamma = \sqrt{(R' + j\omega L')(G' + j\omega C')} = \alpha + j\beta \tag{5.17}$$

Under the assumption that $R' \ll \omega L'$ and $G' \ll \omega C'$ in the frequency range of interest, a cable can be considered as weakly lossy and hence the following approximation is valid:

$$\gamma = \underbrace{\frac{1}{2}\frac{R'}{Z_L} + \frac{1}{2}G' Z_L}_{\mathrm{Re}(\gamma) = \alpha} + \underbrace{j\omega\sqrt{L'C'}}_{\mathrm{Im}(\gamma) = \beta} \tag{5.18}$$

with

$$\alpha(f) = \frac{R'}{2Z_L} + \frac{G' \cdot Z_L}{2} \approx \vartheta_1 \cdot \sqrt{f} + \vartheta_2 \cdot f \quad \text{and} \quad \beta = \omega\sqrt{L'C'} = \frac{\omega}{v_p} \tag{5.19}$$

Here $R'/2Z_L$ describes the impact of the skin effect, and $G' Z_L/2$ denotes the dielectric losses within the insulation material. A further approximation step leads to the expression

$$\alpha(f) \approx \left(\eta_0 + \eta_1 \cdot f^\varepsilon\right) \tag{5.20}$$

which turns out to be advantageous for setup and handling of the echo model in practice, because the coefficients η_0, η_1, and ε are generally constant for a cable type, and ε stays in a quite narrow range between 0.5 and 0.7.

---

[10] Resistance per meter.
[11] Lateral conductivity per meter.
[12] Inductance per meter.
[13] Capacitance per meter.

The proposed channel model has been verified in numerous applications, delivering excellent compliance with the measured values. A very interesting conclusion is that generally only a small number of echoes must be considered. In most cases 3 to 5 will be sufficient, although there might be up to 40 reflecting taps along a cable. This astonishing result can be explained by the fact that distant echoes are normally extinguished by attenuation, so that only the most dominant reflections in the receiver's closer neighborhood must be taken into account.

The echo model presented above will serve as a basic reference for powerline channels and will be included in upcoming standards. Some characteristic examples follow to verify the model's validity. The first example, given in Figure 5-20, is based on a measurement documented in Figure 5-13. For simulation, Equation (5.16) was evaluated including four dominant echoes, after the parameters a_i, $\alpha(f)$, ℓ_i, and τ_i had been determined suitably. Considering the strong simplification, the agreement of the two curves is astonishing. The locations of the notches are met very exactly, and also the absolute values of the transfer function closely fit. Thus the model fulfills the essential requirements. Some minor deviations can be observed only at low frequencies. This is due to neglecting far echoes, which are generally subject to high attenuation, and for which therefore only insignificant errors are observed. It is important to note that the

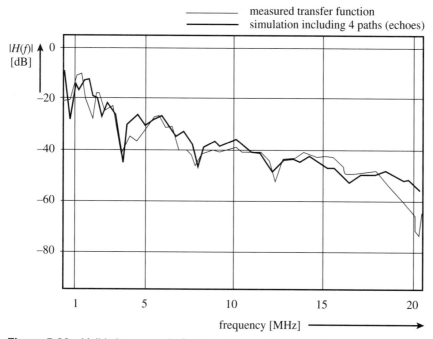

Figure 5-20 Validation example for the echo-model approach.

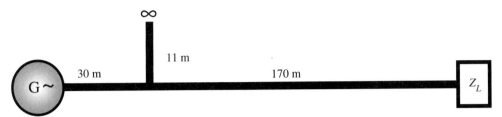

Figure 5-21 Experimental network for echo-model validation.

precision of the model is totally controllable by the number of echo paths. The precision improves with each additional path taken into account.

The results of the next example were gained under well-defined conditions within an experimental network, constructed especially for such purposes. A 200-m power cable of type NAYY150SE was installed as a "backbone" in the form of a loop in a large factory building, including a single branch of 11 m at a distance 30 m from one of the ends. Figure 5-21 illustrates the conditions.

During the measurements analyzed in the following, the cable was matched at the right end with its characteristic impedance Z_L, while the 11-m tap remained open, i.e., without any load. Signal injection took place from the left with an RF generator (G ~). A sweep with a sinusoidal signal was led through the frequency range for the transfer function, while very short impulses with high amplitude were fed in to determine the impulse response. The measurements were then taken each at the right line end terminated by Z_L.

With the constellation shown in Figure 5-21, only one main echo must be expected, provided that the feeding generator is well matched to the cable's characteristic impedance. This is obviously the case here, because the impulse response shows exactly two main peaks at a time distance of about 140 ns. The further fine structure of the impulse response comes from multiple reflections on the tap, because a signal reflected at the open end normally does not run fully toward the generator, or to the right line end, but will be partly reflected once more at the tap. This means that, even with the simple constellation in Figure 5-21, there is actually a very complex reflection scenario. However, multiple reflections decrease quickly in the amplitude, so that they are mostly negligible in practice. Due to the insulating material (PVC) with a dielectric constant $\varepsilon_r \approx 4$, the phase speed on the cable is about 150,000 km/s = 150 m/μs, corresponding to half the speed of light. The first peak in Figure 5-22 (bottom right) is the transmission impulse from the generator, which has traversed the entire cable length of 200 m, thus appearing after time 200/150 μs = 1.33 μs. The second peak forms from the fact that the signal entering the tap is reflected at the tap's open end, appearing at a path

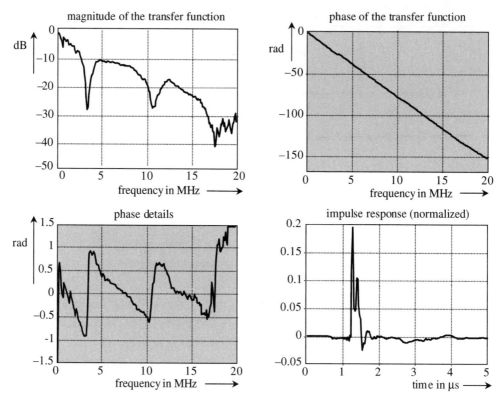

Figure 5-22 Measurement of transfer function and impulse response at the experimental setup from Figure 5-21.

difference of 2·11 m = 22 m from the main peak. The passage of 22 m takes about 140 ns.

The effects of the reflection appear in the transfer function in the form of notches with fixed frequency spacing. The first notch occurs where the direct and the reflected signal portions are shifted exactly a half wavelength against each other, which leads to subtraction. For the frequency f_{n1} belonging to the first notch the run-time difference of 140 ns corresponds to half a wavelength, which means that we have a period of 280 ns, and thus f_{n1} = 3.57 MHz results. The repetition of notches always occurs at odd multiples, of f_{n1}, and so we get f_{n2} = 10.7 MHz and f_{n3} = 17.8 MHz. In both Figures 5-22 and 5-23 these locations are met very exactly. Whenever a notch appears, we find strong fluctuations the corresponding phase curve—see the "phase details" diagram.

Figure 5.23 represents the results of simulations based on the echo model. Obviously the agreement of measurement and simulation is excellent for all representations.

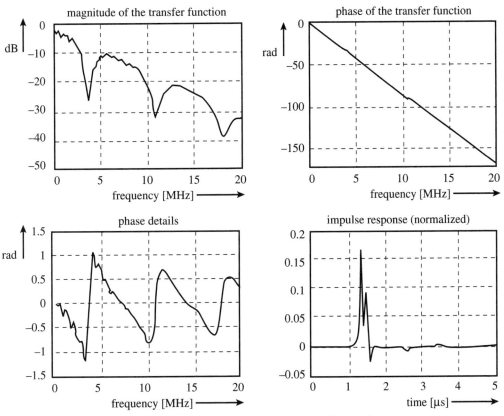

Figure 5-23 Simulation of the sample network based on the echo model.

Even details of the impulse response are identical. For simulation 6 echoes were considered; i.e., some of them passed the 11-m branch several times.

The examples shown above sufficiently demonstrate the performance and efficiency of the echo model. The model has been transferred into industry and will be part of upcoming standardization for powerline channel modeling. At present the integration into standard simulation tools for communication systems such as "COSSAP" from Synopsys is under way. Also the channel emulator, which is planned for integration into the iPLATO system, will include the echo model.

5.8 The High-Frequency Interference Scenario

Additive noise is another important degrading effect for data transmission, in addition to signal distortion. Unlike many other communication links, the powerline channel does not represent an additive white Gaussian noise (AWGN) environment. Referring to

Figure 2-14, this section takes a closer look at the typical powerline noise scenario in the range of high frequencies. First a classification of noise types similar to that in Section 2.1.3 is provided:

1. *Colored background noise* exhibits relatively low power spectral density (PSD), varying with frequency and consisting mainly of a sum of numerous low-power noise sources. The PSD is time-variant in terms of minutes up to hours.
2. *Narrowband noise* consists mostly of continuous wave signals with amplitude modulation. This type of noise is caused mainly by ingress of broadcast stations operating in the medium- and short-wave bands. Levels generally vary with time of day.
3. *Periodic impulsive noise asynchronous to the mains frequency.* Here in most cases impulses are found with a repetition rate between 50 and 200 kHz, producing a spectrum with discrete lines spaced according to the repetition rate. This type of noise is generally caused by switching power supplies.
4. *Periodic impulsive noise synchronous to the mains frequency.* Here impulses have a repetition rate of 50 or 100 Hz (60 or 120 Hz, respectively, in the United States) and are synchronous to the mains cycle. They are of short duration (10–100 µs), and the PSD decreases with frequency. This type of noise is caused mainly by appliances such as light dimmers.
5. *Asynchronous impulsive noise.* This type of impulsive noise is caused by transients due to switching within the network. Impulse duration ranges from several microseconds to a few milliseconds with arbitrary arrival times. The PSD can be more than 50 dB above the background noise level.

Noise types 1 through 3 are usually stationary over periods of seconds and minutes or sometimes even hours and can be regarded as background noise. In contrast, types 4 and 5 are time-variant in terms of microseconds and milliseconds. During the occurrence of impulses, the noise PSD rises perceptibly and may cause bit or burst errors. Some measurement results are briefly discussed below.

Background Noise As stated above, the background noise comprises colored noise, narrowband noise, and periodic impulsive noise with repetition rates much higher than the mains frequency. The dominant noise type is narrowband noise caused by ingress of broadcast stations. Especially the 49-m, 41-m, 32-m, and 25-m broadcast bands are quite obvious. But even in the frequency range below 5 MHz, most interference is narrowband noise. Normally, narrowband noise experiences the highest amplitudes in the evening hours, when propagation conditions for short-wave radios are pretty good, while it is much lower in daylight. Equally spaced lines with varying amplitudes can sometimes be detected at frequencies above 10 MHz. The typical spacing of such lines is in the range of 100 kHz, corresponding to periodic impulse noise with repetition periods around 10 µs.

Impulsive Noise While background noise is generally stationary, impulsive noise caused by switching transients introduces significant time-variance. Analysis and modeling of this type of noise are currently subjects of research activities. The following three parameters have been introduced for characterization:

- A_i—impulse amplitude
- t_w—impulse width
- t_a—interarrival time between impulses

A_i, t_w, and t_a are random variables, and their statistical properties have to be investigated by measurements. Currently work is under way toward this end. First results indicate that only about 1% of the overall time is affected by impulsive noise in typical powerline environments (indoor and outdoor), even under worst-case conditions. The typical noise PSD during the occurrence of impulses is 20–60 dB above the background noise level. About 1% of the detected impulses exceed a width of 200 µs, and 0.1% have a width of more than 1 ms. Maxima of the impulse width are in the range of 5 ms.

Modeling the interference scenario at high frequencies is currently still the subject of scientific research. Severe problems still exist for the detailed description of the different kinds of impulse noise. In the HF range and for high-speed data transmission different approaches are needed than for low data rates within the CENELEC bands A–D. For the high-speed communication considered here, waveforms of very short duration are used, which may be shorter than the duration of a typical noise impulse. Thus multiple (burst) errors may be introduced on the symbol or bit level, respectively. Simple forward error-correction measures, examined previously for low data rates, will usually fail. For effective countermeasures substantially more detailed knowledge of the typical impulse-noise parameters is needed. Therefore the quantities "impulse amplitude" (A_i), "impulse width" (t_w), and "interarrival time" t_a have been introduced as relevant figures. The measuring system iPLATO—see Figure 5-9—will be useful for both automated acquisition and statistical evaluation of these figures for impulse-noise modeling, and for impulse-noise synthesis in further steps toward powerline channel emulation.

The latter will be of high importance for realistic transmission-system simulation. Figure 5-24 gives an idea of the concept for impulse-noise synthesis as it is under way for implementation into iPLATO. The mentioned relevant figures—amplitude A_i, dura-

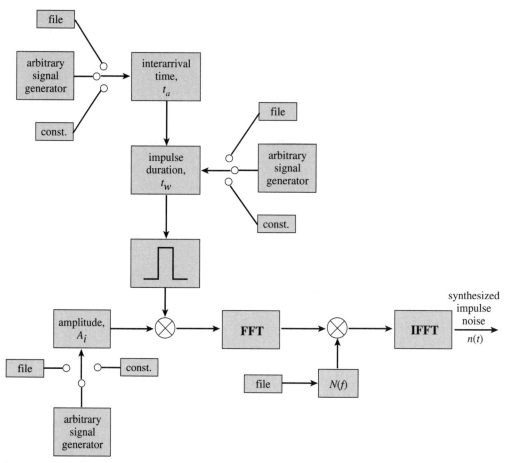

Figure 5-24 Concept for universal impulse-noise synthesis within iPLATO.

tion t_w, and inter-arrival time t_a—can be defined in different ways, for example from a file, by an arbitrary signal generator, or as constants. An impulse train synthesized by appropriate specification of these three describing quantities can now undergo spectral forming, which is digitally implemented in the frequency domain for convenience. Toward this aim first a fast Fourier transform (FFT) takes place. The computed Fourier coefficients are now weighted by multiplication with a spectral forming function $N(f)$, which is usually stored in an appropriate file. In this way the desired spectral shaping is achieved. Finally, the result is transformed back into the time domain by means of an inverse fast Fourier transform (IFFT), and the synthesized impulse noise is available.

A short overview is now given of additional interference, which occurs besides impulse noise in the frequency range from 500 kHz to 20 MHz. In this frequency range it can be assumed that most of the interference caused by the normal use of electricity

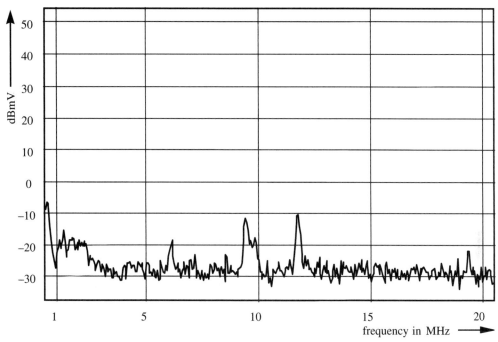

Figure 5-25 Noise level at a 300-m ground cable path (analyzer bandwidth: 100 kHz).

(except, of course, the impulse noise) has been attenuated to negligible residual values. On the other hand, however, other signals appear, which in fact do not represent noise, because they come from intended and allowable transmissions of different radio services. The fact that certain voltages caused by these services have always been present at the power distribution grid did not affect anybody in the past. If, however, in upcoming PLC applications the frequency range of such wireless services is affected, there will be an urgent need for coordination. Just as broadcast reception in long-, medium-, and short-wave radio bands may be impaired by powerline communications, the received signals from broadcast stations can impair PLC.

Figure 5-25 shows a typical record, where some signals from broadcasting stations are clearly recognizable as peaks above the relatively low background noise, which, on the average, is found around −28 dBmV ($\approx 40\ \mu V$) in a bandwidth of 100 kHz. Below 1 MHz a medium-wave transmitter is visible, while at 6, 9, and 12 MHz the signals of HF bands clearly appear.

Also at building installation networks similar ingress is observed, as the records depicted in Figure 5-26 indicate. When analyzing the levels quantitatively, the spectrum-analyzer bandwidth, which was set to 30 kHz here, must be considered, being approximately one-third in comparison with the settings for Figure 5-25. For represen-

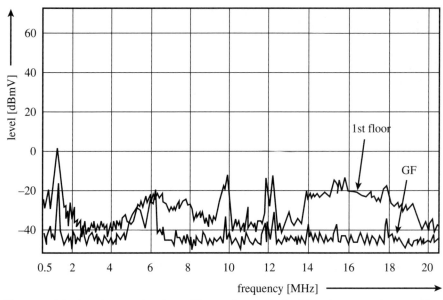

Figure 5-26 Interference level on different floors of a building (analyzer bandwidth: 30 kHz).

tation of the broadcast bands this is of almost no significance, because here we have narrowband signals. The background noise, however, appears attenuated now, due to its broadband character. Furthermore, it is remarkable that both background noise and received broadcast signals on the ground floor (GF) are clearly weaker than, e.g., on the first floor. Concerning the broadcasting stations this can be easily explained: for antenna efficiency of the powerline wiring the elevation above ground plays an important role.

5.9 Access Impedance

Usually in conventional communication systems matching of both the transmitter's output and the receiver's input impedance to the medium's characteristic impedance is accomplished. This is generally well realizable, e.g., for an antenna or a coaxial cable, because such devices exhibit a characteristic impedance which is time-invariant and frequency-independent. At power lines, however, we find completely different conditions. Within the CENELEC A-band, for example, the access impedance is heavily frequency dependent, so that, e.g., at 10 kHz several hundred watts of power would be needed to achieve the maximum permissible transmission amplitude, while at 90 kHz some wattages would already be sufficient. Usually powerline access impedance rises constantly with increasing frequency. For the high-frequency range this impedance converges for outdoor networks toward approximately 50 Ω, which represents the typical cable's characteristic impedance. At indoor installations the typical values are higher, i.e., in the range of 10 Ω, but may be significantly influenced by the proximity of certain appliances containing interference-suppression equipment.

Furthermore at low frequencies (< 300 kHz) we also recognize a time-variance of the access impedance besides the mentioned frequency dependence. Also the location of measurement has substantial influence on the results. At the crossbar system of a transformer station, where up to 10 ground cables start, the access impedance is usually about ten times lower than at a typical house connection. In principle this is also true for high frequencies; however, the typical values stay an order of magnitude higher than, e.g., for the A-band. Thus typical access impedance values below 10 Ω must be taken into account in a transformer station. The connection of a transmitter power stage with a standard output impedance of 50 Ω would thus not be recommendable, as the reflection coefficient would be about 44%. Also at the house connection point the access impedance can be astonishingly low, generally substantially below the values found at a wall plug. This can be explained by the fact that the house connection point behaves like the root of an extended tree, from which numerous wires start in parallel. Even if the characteristic impedance of a typical pair of indoor installation wires is about 2 times larger than that of the outdoor feeding cable, the high degree of parallelism causes very low resulting values.

With the measurement equipment used in the past, recording of powerline access impedance has always been a difficult and also risky task, since the measurements generally have to be performed under tension. The new concept based on directional couplers—as mentioned with iPLATO—will make access impedance recording much simpler, faster and safer. Thus a database can be rapidly created to give a more detailed picture of access impedance statistics for different locations, including also the time dependence.

5.10 Estimating the Powerline Channel Capacity

5.10.1 Shannon's Theory for the Powerline Channel

The ability to transfer data can be estimated for any channel with knowledge of the essential parameters used in Shannon's formula

$$C = B \cdot \log_2 \left(1 + \frac{S}{N}\right) \quad (5.21)$$

C indicates the maximum data rate (in bits/s) for which a theoretically error-free transmission is possible. C normally cannot be achieved by technically realizable and also profitable systems in practice. Therefore in the following a distinction is made between the terms "theoretical channel capacity C" and "realizable data rate r_d." In (5.21) B is the available bandwidth and S/N the signal-to-noise ratio at the receiver's input. The use of (5.21) is not immediately feasible for powerline channels, since the signal-to-noise ratio is not constant within the bandwidth B, but may vary substantially. In practice therefore the transmitted signal-power density spectrum $S_{rr}(f)$ and a frequency-dependent noise-power spectral density $S_{nn}(f)$ must be taken into account, so that (5.21) has to be modified as follows:

$$C = \int_{f_u}^{f_o} \log_2\left(1 + \frac{S_{rr}(f)}{S_{nn}(f)}\right) df \quad \text{with } B = f_o - f_u \quad (5.22)$$

In order to be able to use (5.22), we must know the transmission-power density spectrum $S_{tt}(f)$, the channel's transfer function $H(f)$ as well as the noise-power density spectrum $S_{nn}(f)$ at the receiver. The received signal-power density spectrum is then

$$S_{rr}(f) = S_{tt}(f) \cdot |H(f)|^2 \quad (5.23)$$

The transfer function $H(f)$ and the noise-power density spectrum $S_{nn}(f)$ are fixed channel characteristics, which can be influenced only by measures such as network conditioning. Thus the only remaining variable is the transmitted signal power density spectrum $S_{tt}(f)$, which is determined mainly by the applied modulation scheme. From (5.22) and (5.23) it is immediately evident that arbitrarily high data rates could be achieved even for bad channels, if the transmitted power spectral density were not limited. However, the requirement of electromagnetic compatibility (EMC) between PLC and different kinds of wireless services calls for power limitation over the entire frequency range of interest for high-speed PLC, i.e., 1 MHz to more than 30 MHz. Such restrictions may be made straightforward without regarding the typical conditions of powerline channels, defining a limiting curve which is either frequency-independent or in a very simple way frequency-dependent—see Section 5.11.

On the other hand, detailed knowledge of typical courses of the transfer function $H(f)$ and the noise-power density spectrum $S_{nn}(f)$ allows an optimization of the transmission-power density spectrum, whereby—particularly for bad channels—a significant increase of the channel capacity can be achieved. This, however, presupposes that in certain frequency ranges rather high power density is allowable—see Section 5.11.

Capacity estimations for numerous measured channels, through evaluation of (5.22) and (5.23), clearly indicate that the available bandwidth B represents the most important resource for a high data rate. As the usable bandwidth B will in practice not be contiguous but more or less fragmented, so that "slots" of different width will be distributed over the range from 1 MHz to more than 30 MHz, modulation schemes are needed, which can exploit such a fissured spectrum as well as possible. Besides bandwidth reduction by regulation, the usable frequency range can also be restricted through high attenuation. An increase of the transmitted power density can hardly compensate bandwidth restrictions, because B directly and proportionally affects the obtainable data rate, whereas transmitted power—according to (5.21) and (5.22)—only takes effect over a logarithmic function.

Furthermore, channel capacity generally decreases with distance due to the lowpass characteristic of a powerline channel. For short distances, e.g., 100 m, the theoretical upper limit exceeds 250 Mbits/s. Even for distances of 300 m and restrictions of the usable bandwidth to only 5 MHz, still more than 14 Mbits/s are obtained. For access channels of more than 200 m, however, it must be noted that in most cases the frequency range above 10 MHz contributes almost nothing to the channel capacity.

Regarding realizable transmission systems, it is obvious that the theoretical channel capacity cannot be reached with reasonable effort. Modulation schemes, however, differ substantially in their ability to exploit the capacity. For excellent channels, with theoretical capacities in the range of 250 Mbits/s, data rates around 100 Mbits/s are quite realistic. Even for most channels classified as very bad, with lengths in the 300-m range, nevertheless data rates of at least 5 Mbits/s seem feasible. Toward this aim it will be necessary to include complex modulation schemes such as QAM (quadrature amplitude modulation). Best results are generally obtained by adaptive and optimizing modulation schemes, capable of dividing the available spectrum into sufficiently narrow subchannels, which can transfer suitable portions of the data stream, individually adapted to their quality.

Detailed evaluations of an extended measurement database of powerline access channels between transformer station and house connections performed in [D9] revealed that including frequencies above 10 MHz increased the data rate only for approximately 40% of the involved links. For only 20% of all examined links an increase of more than 25% for the data rate could be obtained by the frequency range above 10 MHz. This leads to the conclusion that a separation of frequency ranges for access and indoor use makes sense at 10 MHz; i.e., upcoming standards should dedicate the range below 10 MHz to PLC access use and the range above 10 MHz to indoor applications.

Concerning EMC, the following results are available for the access domain. These results are based on a limit of the electrical field strength of $E_1 = 40$ dBµV/m (see Section 5.11, Figure 5-28, at 1 MHz), using a receiver with the EMC measuring bandwidth

$B_M = 9$ kHz and referring to empirical values of the electromagnetic decoupling factor $K(f,d)$.[14] For the access domain $K(f,d)$ typically lies between 10^{-2}/m and 10^{-3}/m. Now—for best- and worst-case scenarios—the transmitted power spectral density

$$S_{tt}(f) = \frac{E_l(f)^2}{|K(f,d)|^2 \cdot B_M} \tag{5.24}$$

can be estimated, delivering

$$S_{tt} \approx -59 \text{ dBV}^2/\text{Hz to } -79 \text{ dBV}^2/\text{Hz} \tag{5.25}$$

Thus, if the decoupling factor is small ($K = 10^{-3}$/m), the transmitted power density can be 20 dB higher than in case of strong coupling. According to (5.21) this means a more than six times higher data rate in case of weak coupling.

For indoor PLC channels $K(f,d)$ may take on values which are even larger than 10^{-2}/m if no conditioning measures are provided. Obviously a significant lack of symmetry is leading to common-mode propagation of PLC signals here.

5.10.2 Access-Channel Capacity Estimation

The following analysis is based on the situation depicted in Figure 5-13, where the noise spectrum is only insignificantly stressed by narrowband interference. In most portions low-level background noise is observed, with a spectral amplitude around -28 dBmV ($\equiv 40$ μV) within a spectrum analyzer filter bandwidth of 100 kHz. From these data, we can calculate the single-sided noise-power density as follows:

$$N_0 \approx \frac{(40 \text{ μV})^2}{100 \text{ kHz}} = 1.6 \cdot 10^{-14} \frac{\text{V}^2}{\text{Hz}} \equiv -138 \text{ dBV}^2/\text{Hz} \tag{5.26}$$

In order to get a basic impression of the theoretical capacity of this powerline channel, let us consider the frequency range from 500 kHz to 9 MHz in Figure 5-13. We find an excellent signal-to-noise ratio of approximately 40 dB. According to (5.21) this leads to a spectral efficiency of

$$\log_2\left(1 + 10^4\right) > 13 \text{ bits/s/Hz} \tag{5.27}$$

which means that a data rate of 110 Mbits/s could theoretically be achieved within an 8.5-MHz bandwidth. The noise floor is also well below the received signal over other

---

[14] The decoupling factor will be defined in detail in Section 5.11, but it can be used here in an introductory overview.

extended frequency ranges. An SNR of more than 20 dB can be observed even up to 17 MHz.

More sophisticated evaluations of a measurement database carried out in [D9], including more than 200 links by means of (5.22) and (5.23), revealed that for short distances of approximately 100 m within the access domain the theoretical upper limit of the data rate exceeds 250 Mbits/s. As already mentioned, more than 14 Mbits/s can still be obtained even over distances of 300 m and under worst-case conditions.

Of course, the theoretical channel capacity cannot be completely exhausted for technically and economically feasible PLC systems. Applicable modulation schemes, however, differ substantially in their ability to use the capacity. Therefore, a careful selection is recommended. For good channels with theoretical capacities in the range of about 250 Mbits/s, data rates around 100 Mbits/s are feasible. Even on most of the channels exceeding 300 m, which often represent the worst case, data rates above 5 Mbits/s are still possible. Here, however, it is necessary to use modulation schemes with adaptive optimizing features, as mentioned above and discussed in more detail in Chapter 6. Table 5-1 gives an overview of the channel capacity for the powerline access domain (last mile).

Table 5-1 Estimation of the powerline channel capacity over the last mile.

	Data Rate		Link Length
	Best Case	Worst Case	
Theoretical	250 Mbits/s	14 Mbits/s	100 to >300 m
Realizable	100 Mbits/s	5 Mbits/s	100 to >300 m

5.10.3 Indoor Channel-Capacity Estimation

Until now the available measurement database for indoor powerline channels has been extremely small, especially with respect to frequencies exceeding 30 MHz. Nevertheless, a first rough estimation based on the results from Figure 5-16 makes sense. Let us consider the shaded area in Figure 5-16. This shaded block covers the frequency range from 10 to 20 MHz, i.e., a bandwidth of 10 MHz. Taking the first-floor curve for a transmission link where transmitter and receiver are located on the same floor, we find a signal-to-noise ratio of around 60 dB over almost the entire frequency range of interest. From (5.21), we can calculate a spectral efficiency of 20 bits/s/Hz, which results in the following impressive value for the channel capacity:

$$C \approx 200 \text{ Mbits/s} \tag{5.28}$$

Of course this channel does not represent a worst-case scenario, but on the other hand there might be even better ones. Currently further evaluations do not seem sensible, because the basis is too small for statistically reliable results. Toward this aim various measurements have to be performed at indoor links, including frequency ranges in excess of 30 MHz and different coupling alternatives. Here the use of PE–N coupling in connection with the conditioning procedures discussed in Section 5.12 will be of utmost interest in order to keep the limits of radiation currently given in Figure 5-28. However, a measuring bandwidth $B_M = 120$ kHz must be used now, and a maximum electrical field strength of 27 dBµV/m is defined for frequencies above 30 MHz. Assuming a decoupling factor $K(f,d) \approx 10^{-3}$/m, which appears realistic when network conditioning is applied, the allowed transmitted power spectral density will be $S_{tt} \approx -83.8$ dBV^2/Hz, according to (5.24). Let us first determine the indoor noise-power spectral density from Figure 5-16 to estimate the channel capacity achievable with such a level. Since the noise-level is around –45 dBmV here for frequencies above 10 MHz, a straightforward calculation similar to (5.26) delivers the noise-power spectral density:

$$N_{0i} \approx \frac{(5.62\,\mu V)^2}{30\,\text{kHz}} = 1.05 \cdot 10^{-15}\,\frac{V^2}{Hz} \equiv -150\,\text{dBV}^2/\text{Hz} \qquad (5.29)$$

Assuming an attenuation of 50 dB leads to the received signal spectral power density $S_{rr} = -133.7$ dBV^2/Hz; see also (5.23). Thus the signal-to-noise ratio will be 16.3 dB, allowing a spectral efficiency of 5.43 bits/s/Hz. For a usable frequency range from 30–70 MHz the channel capacity would reach about 217 Mbits/s. Exploiting only one-tenth of this value would mean a success for PLC-based high-speed indoor digital networks with data rates of some 10 Mbits/s.

In concluding this section, let us finally demonstrate how to realize Internet access over the wall plug with a data rate of 2 Mbits/s in a relatively narrow bandwidth, with low transmitting power and a very simple modulation scheme. We start with a channel according to Figure 5-13 in the frequency range around 15 MHz. The attenuation is about 60 dB, and we find no interference, except the background noise with the power spectral density $N_0 \approx 1.6 \cdot 10^{-14}$ V^2/Hz given by (5.26). The system example is based on the following parameters:

Transmission signal amplitude: 1 V
Data rate: 2 Mbits/s
Modulation: BPSK
Carrier frequency: 15 MHz
Total bandwidth: approximately 4 MHz

With the normalized transmission power 1 V^2 and the data bit duration $T_b = 0.5$ μs we have the transmitted bit energy

$$E_{bs} = 0.5\,\mu V^2 s \tag{5.30}$$

which reaches the receiver with an attenuation of 60 dB, i.e., decreased by a factor of 10^6. With the result from (5.26) the ratio of bit energy to noise-power spectral density at the receiver can now be determined, i.e.,

$$E_b/N_0 = (0.5 \cdot 10^{-12}\,V^2 s)/(1.6 \cdot 10^{-14}\,V^2/Hz) = 31.25 \equiv 15\,dB \tag{5.31}$$

According to (5.21) we have a spectral efficiency of approximately 5 bits/s/Hz within a bandwidth of more than 4 MHz around the center frequency of 15 MHz—see Figure 5-13. Thus the theoretical capacity of this portion of bandwidth would be 20 Mbits/s, only 10% of which is actually used in the system example. Therefore, in any case it would be worthwhile to apply a more sophisticated modulation scheme, such as QPSK or QAM, in order to exploit more of the channel capacity. Here again the potential of the powerline channel for future enhancement of data rates becomes clearly obvious. When today the first generation of PLC equipment is aiming at data rates of 1–2 Mbits/s, there is still a "clearance" of several orders of magnitude until theoretical limits are reached. This knowledge should make the system designer confident that he will be able to master his current job with moderate effort. Based on increasing experience with operating PLC systems, and the general progress of technology, it can be expected that within a short time span it will be attractive and possible to launch further approaches toward the capacity limits. For all those tasks it is advisable, however, to have the following possible restrictions in mind:

- The limits of permissible radiation, which at present are not yet definitely fixed, must be kept by any PLC system.
- Frequency allocation is still under discussion, and the exemption of certain frequency bands for powerline communications must be taken into account.
- Many links exceed 300 m and/or may exhibit transmission properties substantially worse than depicted in Figure 5-13.
- In the above considerations only background noise was included as interference; in practice, however, also narrowband noise and various kinds of impulsive noise may be dominant.

Despite possible restrictions, this section has impressively demonstrated the enormous data-transmission capacity for both PLC-based local loop and in-home digital network solutions. Further investigations of indoor powerlines should include transfer-function and interference measurements as well as radiated fields. The next section provides an introduction to the current problems of defining limits for radiation from PLC as well as frequency allocation. A preview of possible regulation and standardization outcomes is also presented.

5.11 Electromagnetic Compatibility: Problems and Solutions

When using electrical energy distribution networks as shared media for different telecommunication services and fast data-transmission purposes, it must be kept in mind that they represent "electromagnetically open" structures, insufficiently protected against both ingress and emission of signals with high frequencies. Thus PLC transmission signals may affect on one hand other PLC systems in the same network segment directly over the connecting wires, and on the other hand different wireless services via radiated fields. Electromagnetic compatibility means both the coexistence of different PLC systems in close proximity and coexistence with wireless services.

5.11.1 Compatibility with Wireless Services

Electromagnetic compatibility of radio services and PLC is basically a bidirectional problem. On the one hand the electromagnetic fields from radio transmitters produce corresponding voltages or currents in the power distribution network, because the unshielded lines operate as more-or-less efficient "spread" antennas, and on the other hand PLC signals radiate electromagnetic fields, which can impair radio receivers. While received radio signals on powerlines have to be mastered by appropriate PLC techniques, it will be necessary to limit radiation from PLC, so that the primary users of wireless services, to whom the shared frequencies have been assigned, remain unaffected. As Figure 5-27 indicates, for the frequency range up to 30 MHz the scenario is rather complex. According to statements of the RegTP, at present a total spectrum of approximately 7.5 MHz in the frequency range up to 30 MHz would be usable in principle for PLC. An assignment for PLC, however, appears currently infeasible for reasons which we now discuss.

The spectrum is not contiguous, as Figure 5-27 clearly demonstrates. Some gaps of different width, distributed arbitrarily in the depicted frequency band, appear usable for PLC. It should be noted that the representation in Figure 5-27 is only schematic; i.e., an accurate location of usable frequencies cannot be taken from this picture.

At first glance, it appears feasible to assign the open gaps to PLC services, permitting an increased transmitting-power spectral density within these "chimneys." From such a solution most modulation schemes could profit, particularly those which are able

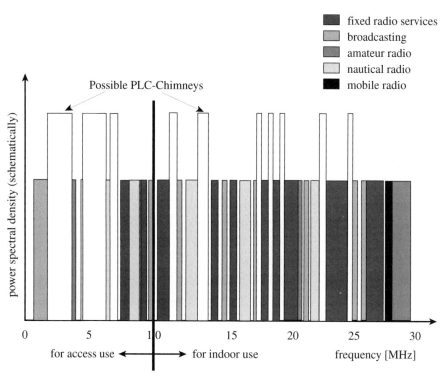

Figure 5-27 Possibilities of frequency allocation and the "chimney approach."

to use narrow gaps with high spectral efficiency. Here primarily OFDM will be an ideal candidate, followed by broadband multicarrier schemes (see Chapter 6). Unfortunately, the chimney approach sketched in Figure 5-27 encounters some serious problems, so that a corresponding regulation for PLC appears questionable for the following reasons:

- The frequency gaps in Figure 5-27 are not really free, but are dedicated to certain primary users for wireless services through worldwide treaties and agreements. These users do not currently use them, or a shared use would be possible due to large distances. An assignment for PLC, however, would mean that the current owners would have to give up some of their rights, at least partially. The necessary procedures for such changes usually take years of time and are loaded with enormous costs, which would have to be imposed on prospective PLC service providers.

- If the chimneys are very narrow, it will be impossible in practice to make appropriate use of the enhanced power spectral density within them, because the power density must decrease rapidly with extremely steep slopes at the edges. For example, an OFDM system would have to leave several subchannels unused at the edges or implement complex filters. In many cases the achievable advantages can thereby disappear very quickly.

Another possible way to accomplish EMC between PLC and wireless services is a general limitation of radiated fields from powerlines. At the beginning of 1999 in Germany the RegTP suggested the limiting curve labeled D in Figure 5-28. As the figure clearly indicates, the frequency range of interest for high-speed PLC, i.e., 1 MHz to >30 MHz, is strongly affected by the decreasing slope from 40 dBµV/m ($\equiv$ 100 µV/m) at 1 MHz down to 27 dBµV/m ($\equiv$ 22.3 µV/m) at 30 MHz. For frequencies above 30 MHz up to 1 GHz the limit remains constant at 27 dBµV/m. In comparison with the upper curve denoted US, which is valid for the United States (FCC Part 15), the disadvantages imposed by possible German regulations become clearly obvious. Around 2 MHz, for example, the allowed American limits are about 30 dB above the RegTP's suggestion. This means a factor of 1000 in terms of transmission power or a factor of approximately 10 concerning the possible data rate. With these prerequisites in mind, it is not astonishing that we find enormous progress in developing high-speed PLC equipment in the United States, heading for data rates above 10Mbits/s for indoor applications [W11]. Also the recent foundation of the HomePlug Alliance [W5] must be considered against this background. Due to the current lack of standardization HomePlug has set up a very tight roadmap. By December 2000, the full draft version 1.0 of the specifications was completed. The baseline technology chosen by HomePlug for high-speed powerline networking will be OFDM (see also Chapter 6), especially the concept presently provided by Intellon [W11]. For the beginning of 2001 HomePlug is planning field tests in more than 500 households worldwide. Here, for the first time, the 10Mbits/s technology will be deployed and evaluated. Within 2001 the corresponding specifications shall be published; subsequently, a certification lab shall be established to guarantee compliance.

Figure 5-29 gives an impression of the excellent market opportunity for fast powerline networking during the next four years. The curves are based on a study by Frost Sullivan for the United States. For the rest of the world, of course, the numbers will be different, but the overall trend of growth can be assumed to be very similar. While currently the number of homes on line is in the range of 35 million, roughly a doubling is predicted by 2004. Regarding home networks, however, an exponential increase is expected from about 6 million to 60 million by 2004. From this point of view the efforts of the HomePlug alliance can be regarded as very rewarding in any case.

With the suggestions currently discussed in Germany, however, such a rapid progress for high-speed networking seems questionable. Even when data rates in the range of only 1–2 Mbits/s are encountered, the limits are hard to meet. Therefore

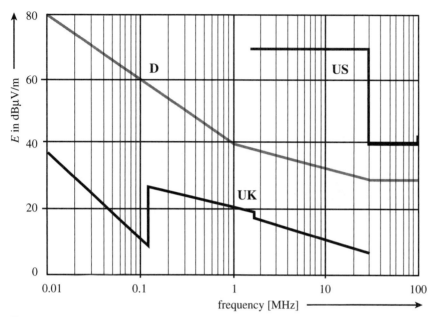

Figure 5-28 Electrical field level limits for radiation from wire-bound communication for three different countries.

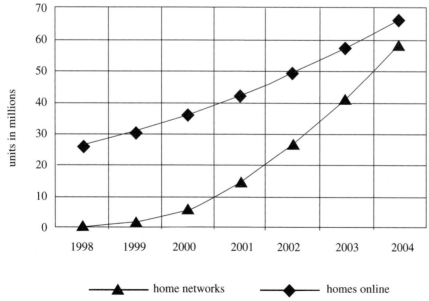

Figure 5-29 The market opportunity for high-speed powerline networking.
Source: Frost & Sullivan, Sept. 1999.

network conditioning toward more symmetry and common-mode rejection will be an important issue for further exploitation of the channel capacity if the *D*-curve from Figure 5-28 becomes effective without changes.

The lower curve in Figure 5-28, labeled UK, which is defined by the "MPT 1570" document for England, can obviously be regarded as a "killer regulation" for PLC. The level is generally about 20 dB below the RegTP's suggestion, which roughly means a sixfold decrease of possible data rates. Under such circumstances, industry and utilities will not be interested in investing manpower and money in PLC technology. This kind of hard limitation looks much like a campaign against powerline communications, because the technical sense is extremely doubtful. Allowed limits for unintended radiation, e.g., from various appliances, are more than 60 dB above the UK curve, and also radiation from computer networks or telephone lines generally exceeds these limits, so that such wire-bound communication networks will require expensive additional shielding or have to be switched off, as soon as the limitation becomes law. Enforcing such a law would call for close examination of all kinds of radiation from wires and thereby lead to a tremendous effort concerning measurements and shielding procedures and even to the shutting down of various services, entailing considerable damage for each country where technical and economical progress essentially depend on the development and extension of information technology.

Generally, the intensity of electromagnetic fields decreases with increasing distance from signal-carrying lines. Coexistence of PLC with radio services therefore means keeping the field strength caused by PLC at locations where radio reception takes place so low that receiving of the wanted services will not be noticeably impaired. Thus two crucial influences determine the permissible PLC transmission level: the frequency range used for PLC and the radio receiver location.

While broadcast and amateur radio services will typically be of interest in populated areas, i.e., in the near neighborhood of PLC systems, this is hardly the case for nautical or aeronautical radios. Thus within broadcast bands of interest a very low field strength must already be reached in close proximity to the lines. Investigations revealed that without network conditioning only relatively low transmission-voltage levels are permissible for PLC, in order to keep the limits given in Figure 5-28 (*D*-curve). Simple modulation schemes will therefore generally be bad candidates for PLC. Furthermore, it makes sense to leave such frequency ranges, which are occupied by strong radio transmitters, completely unused by PLC, in order to avoid serious effects upon weak PLC signals. Concerning nautical and aeronautical radios, however, it is usually sufficient when the fields radiated by PLC fall below certain limits at a distance of some 100 m, in contrast to 3 m, which is currently specified in the measuring procedure belonging to Figure 5-28. The idea outlined here represents some kind of modified chimney approach, allowing PLC transmission levels significantly above the limits in Figure 5-28, without affecting sensitive wireless services, and it could therefore become an important approach toward agreements of coexistence.

Besides distance, the transmission level generally determines the strength of a radiated field linearly. This dependence can be quantitatively described by a "mains decoupling factor" $K(f,d)$, which is introduced by means of Figure 5-30.

For PLC equipment manufacturers it will be important to have clear and easy-to-measure limits of the voltage they are allowed to feed into the power supply network. Regulation, however, due to the expected low value, must not concern itself with this voltage, but only with the possible radiation. Therefore, the decoupling factor, similar to a definition in CISPR/A [NO5], is introduced as an instrument that is easy to handle. The decoupling factor $K(f,d)$ must in principle be determined for each configuration separately, e.g., for each cable and substation or each indoor PLC link. Whenever $K(f,d)$ of a certain geometry is known, it is possible to calculate the radiated fields for arbitrary signals. Such results also offer the possibility of comparing measurements for different signals. $K(f,d)$ can be regarded as a transfer function between the communication signals on the lines as input and the inadvertently radiated field as output, summarizing the impact of all parameters that influence this transfer function. For determining $K(f,d)$ a sinusoidal continuous wave signal is injected symmetrically into the powerline wiring. The signal generator is swept over a certain frequency interval, and the radiated field is measured with an H-field loop antenna. According to EMC standards for the frequency range below 30 MHz, such a loop antenna has to be used. During the sweep, the H-field amplitudes are determined.

When measurements with an H-field loop antenna according to Figure 5-30 are performed, exclusively magnetic fields are captured. The associated electrical field strength can be calculated by multiplying the magnetic field strength with the characteristic impedance $Z_H = 377\ \Omega$ of the free space, i.e.,

$$E = Z_H \cdot H \qquad (5.32)$$

Note that (5.32) is correct only under far-field conditions, i.e., if the E-field and H-field components are orthogonal and no components in the direction of propagation exist. If results from (5.32) are acquired under near-field conditions, they can nevertheless be useful for comparative investigations and finally for the definition of limits for radiation.

It is important to note that an electrical field strength due to radiation from PLC determined this way cannot be compared with the field strength of a broadcasting sta-

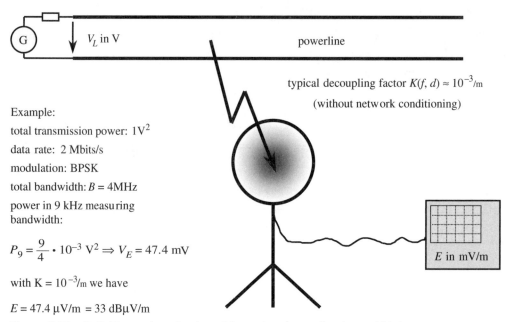

Figure 5-30 Definition and application of the mains decoupling factor $K(f,d)$.

tion measured in the same place. Under the assumption that for reliable PLC a certain transmission level V_L (in volts) is necessary on the powerline, the question is, which electrical field strength E (in volts/meter) results from that at a certain distance? The answer is anything but simple, because E depends substantially on the network structure, the method of signal coupling, and various environmental conditions within a building, such as the presence of large metal parts. Figure 5-31 shows some first-measurement results of the decoupling factor $K(f,d)$ in dB (m^{-1}) of a typical building installation. The measurements were performed at three distinct frequencies, 4, 7.8, and 11 MHz, where nothing else could be received, in order to get clear results. The curves in both diagrams of Figure 5-31 clearly indicate the near-field problem; i.e., at a distance up to 10 m we find a strong decrease of the K-factor. Furthermore, the antenna efficiency of the wiring grows with frequency in the range considered here. For future investigations, however, the frequency range above 10 MHz up to more than 50 MHz will be of major interest. Especially for the higher frequencies a decrease of the decoupling factor is expected. Although the results from Figure 5-31 cannot be regarded as representative, because they were recorded for a single building, they are nevertheless very encouraging, because at a distance of only 15 m we generally have a decoupling factor below –60 dB (m^{-1}), which confirms the former assumption that $K \approx 10^{-3}$/m is a valid approximation for many real-world scenarios.

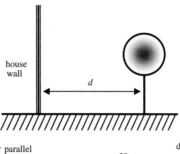

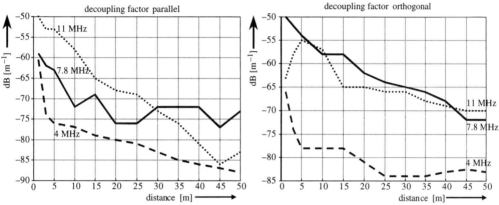

Figure 5-31 Decoupling factor measurement results for building installations.

This statement is also supported by various published results [H17, H22]. $K \approx 10^{-3}$/m means that for a transmission level $V_L = 1$ V on the line (unmodulated carrier) there is an electrical field strength of approximately $E = 1$ mV/m ($\equiv 60$ dBµV/m). In order to estimate which values of the electrical field strength are to be expected from real PLC signals, the following example is considered—see also Figure 5-28:

The carrier amplitude is assumed to be 1 V. For analysis a resistance of 1 Ω is chosen, which simplifies the calculations. With this notation the total transmitted power is $P_{tot} = 1/\sqrt{2}$ V^2. For a data rate of 2 Mbits/s and BPSK (binary phase-shift keying) modulation the transmitted power is approximately distributed over a frequency band of 4 MHz. The distribution is in fact not even, but for conciseness, the details are neglected here. Then we have the power spectral density of the transmitted signal:

$$S_{ss}(f) = \frac{1 \, V^2}{\sqrt{2} \cdot 4 \, \text{MHz}} = 17.67 \cdot 10^{-8} \, \frac{V^2}{\text{Hz}} \quad (5.33)$$

Applying the measuring standard with a 9-kHz bandpass filter, only the portion of power which falls into this bandwidth is received, i.e.,

$$P_9 = S_{ss}(f) \cdot 9 \, \text{kHz} = 1.59 \cdot 10^{-3} \, V^2 \quad (5.34)$$

where P_9 corresponds to a line voltage $V_L = 47$ mV. In case of a network without conditioning, the radiated field expected from such a voltage can now be estimated by the decoupling factor defined above. For $K(f,d) \approx 10^{-3}$/m, we have

$$E = K(f,d) \cdot V_L = 10^{-3} \cdot 47 \cdot 10^{-3} \text{ V/m} \equiv 33 \text{ dB}\mu\text{V/m} \tag{5.35}$$

which is obviously very close to the limits depicted in Figure 5-28, and clearly indicates the feasibility of high-speed PLC even under severe restrictions. In addition, however, the decoupling factor can be substantially lowered by network conditioning measures, which introduce a higher degree of symmetry. For very high speed PLC with data rates exceeding 10 Mbits/s, network conditioning can be considered as a must. For each measure, however, efficiency and costs have to be carefully checked. Generally the following steps appear suitable for reduced radiation:

- Insertion of HF barriers to prevent PLC signals from propagating into unwanted directions
- Network modifications for common-mode rejection
- Symmetrical signal coupling

5.11.2 Compatibility of Different PLC Systems

For the overall success of PLC, compatibility of different systems will be a crucial requirement. While within the access domain a PLC service provider will definitely take care of the implementation of compatible systems only within the range of a base station; this cannot be easily ensured for the indoor environment. The coexistence of both the PLC access systems and different indoor systems are concerned. In principle coexistence could be ensured by separation through appropriate filtering. However, in practice a sufficient separation is not possible, because stop-band attenuation of more than 50 dB will not be feasible. Besides, every house connection reached by a base station would have to be conditioned in order to avoid possible collisions. Finally, frequency separation remains as a favorable solution, which is currently discussed in the PLCforum.[15] It is proposed to dedicate the frequency range below 10 MHz to the access domain, and assign frequencies above 10 MHz with no upper limit to indoor use. For the access domain this limitation represents a restriction of possible data rates only for very short distances. Since building installations exhibit no significant low-pass characteristics, there will be sufficient bandwidth above 10 MHz for many kinds of high-speed applications.

---

[15] The new international body of the powerline community.

Adaptive frequency allocation depending upon actual use of the spectrum could be a further alternative. Although such a solution can generally guarantee optimal exploitation of the available bandwidth, it requires a very efficient frequency-assignment algorithm in each network node, which will of course be a matter of cost.

For undisturbed operation of PLC in neighboring flats, precisely controlled access to the spectral resources will in any case be paramount. Due to various ways of signal coupling, separation can hardly be achieved by measures of network conditioning here. A fixed assignment of frequencies to individual PLC users also does not make sense. Currently no solution of the problem is at hand. Finally, the access to the frequency resources must be performed somehow adaptively by a procedure which is yet to be developed.

5.11.3 What to Expect from Regulations in the Near Future

In Germany the RegTP's draft paper NB 30 and the associated measuring procedures have been suspended for revision and amendments until summer of 2003. Comparing these German limits (see Figure 5-28) with the corresponding values from the United States and England reveals that the German curve is roughly centered between the US and the UK curves. On the one hand the powerline community strives to enhance the limits toward the US values—at least within some chimneys to be defined—while on the other hand there are efforts to lower the limits, initiated, e.g., by certain police and military departments, where scanning of the HF frequency range is still an issue. Also, astonishingly, several broadcasting corporations have recently stressed the "enormous" importance of long-, medium-, and short-wave broadcasting. With this constellation in mind, a more generous definition of limits cannot be expected at the moment. Also lowering the limits is not probable, since most European governments, i.e., their ministries of economics, meanwhile are well aware of the innovation potential of PLC technology, and industry has launched a series of investments for PLC system and service development. Furthermore, regarding the situation in the United States, nobody would seriously fear an electromagnetic disaster from PLC with a more generous definition of radiation limits.

An important task for the European PLC community, however, will be to point out that the limitations imposed by the *D*-curve in Figure 5-28 do not leave much clearance for the success of PLC technology, and that lowering the limits at this time would mean a substantial threat also for complex transmission techniques. Future PLC system generations with higher data rates will probably not get along without network conditioning. This will also be true for the United States, despite the higher limits of radiation, because the problem of overall coexistence has to be solved as soon as PLC applications become ubiquitous. The margins attainable with such measures will clearly exceed every achievement which would be possible by sophisticated signal processing and adaptation mechanisms. Especially indoor PLC networks are concerned here, because they are considered to be most critical, as confirmed by investigations performed under contract of the RegTP [H33]. Until now, however, generally worst-case conditions were considered, assuming that indoor powerline network conditioning is not feasible. Meanwhile the situation has significantly changed, as the next chapter will demonstrate.

5.12 Measures of Network Conditioning—Analysis of Feasibility and Efficiency

In this section new and known suggestions of powerline network conditioning will be evaluated. Network conditioning mainly includes measures to reduce unintended radiation of PLC signals, and at the same time tries to improve their propagation along the lines. Generally the critical points of unintended radiation from PLC are found at the feeding locations. The radiation can usually be reduced by two measures:

- Symmetrical coupling, so that pure differential-mode propagation is ensured over a distance which should be as long as possible. If, then, at a significant distance, due to inevitable asymmetry, a differential to common-mode conversion takes place, this will have minor impact on radiation, because then the PLC signal level has already sharply decreased.
- Avoiding PLC signal coupling on unused conductors and network sections, because they are not involved in useful signal transportation, but carry the PLC signals into unwanted directions and mostly also exhibit antenna effects, so that not only are PLC signals radiated, but also broadcast signals are received, which may disturb PLC.

Considering the real-world scenario within a typical transformer station as shown in Figure 5-32 makes the practical difficulties very obvious. Signal coupling will generally involve the complete crossbar system, so that usually only a small fraction of the provided RF power enters the cable of interest. The greatest portion—in practice often more than 90%—will spread in unintended directions, whereby radiation through the metal parts of the crossbar system is most critical. Therefore, the general aim of conditioning within a transformer station is to feed the PLC signals as symmetrically as possible solely into the selected supply cables for PLC customers, and to prevent propagation toward the crossbar system.

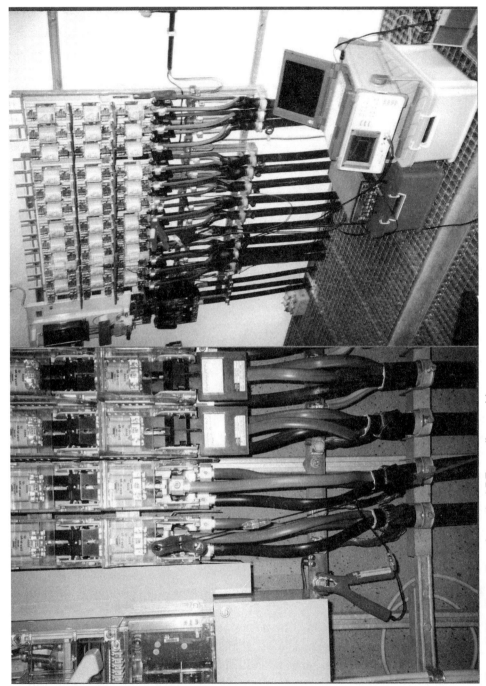

Figure 5-32 Illustration of practical RF coupling problems.

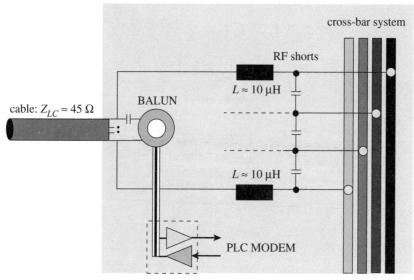

Figure 5-33 Network conditioning in a transformer station.

Appropriate solutions will not be easy, but a promising approach is presented in Figure 5-33. An inductor of approximately 10 µH is inserted for RF attenuation into each of the two conductors selected to carry the PLC signals. The capacitors at the crossbar system will reduce the remaining RF energy which eventually passes the inductors.

Applying this kind of conditioning to each of the supply cables will put two barriers between the cables, yielding a high degree of decoupling. Also for the opposite direction this kind of conditioning will be advantageous, because received broadcast signals, for which the crossbar system has mainly an antenna function, are kept away from the cable. The input/output impedance of a modem can now be matched to the characteristic impedance of the cable by means of a balun.

Investigations revealed that the two remaining conductors of the supply cable do not need further conditioning, because possible crosstalk is cancelled by the RF shorts at the crossbar system. Crosstalk coming up later at a certain distance from the feeding location is no longer critical, because the PLC signal level has usually dropped significantly, so that possible radiation will be negligible. The construction of the inductors mentioned, however, turns out to be a severe challenge, although their inductance is not very large. Since they have to carry the full load current, which may exceed 100 amperes, the use of magnetic materials is problematic due to saturation. Solutions in form of coreless coils will normally fail because of their size. Appropriate core materials and constructions are currently being tested in close cooperation with industry and

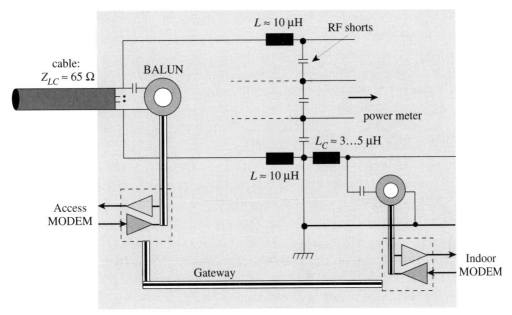

Figure 5-34 Network conditioning at a house connection point.

power supply utilities with encouraging results: isolation values in the range of 20 dB seem feasible. Regarding size and insertion techniques some details must still be worked out. An isolation of 20 dB would represent enormous progress, enabling a six-times-higher data rate, or if necessary for EMC reasons, a corresponding lowering of the transmission level.

The solution for transformer stations can be applied in a simplified form also for the house connections as shown in Figure 5-34. Especially the construction of the inductors causes less effort due to the lower current loads. After the inductors, we again find three capacitors to provide RF-shorts, which are additionally connected with the protective earth conductor. This way we have a clean and reliable termination of the access domain, which guarantees that the access network does not suffer from possible disturbances outside of building installations. Furthermore, this kind of decoupling also enables different kinds of indoor PLC applications, the signals of which must not spread into the outdoor network. Toward this aim it will also be advisable to build a gateway, e.g., realized by optical fibers, to connect the PLC domains (outdoor and indoor) without the risk of unwanted electromagnetic coupling.

In connection with high-speed indoor PLC, electromagnetic compatibility is regarded as a special challenge, because until now no measures of conditioning were known. Both RF isolation and differential-mode signaling were considered impossible to a large extent. Thus relatively strong coupling between the PLC signal voltage V_L on the line and the strength of a corresponding radiated field had to be assumed; i.e., the decoupling factor was supposed to be significantly larger than 10^{-3}/m [H33]. On the other hand, the relatively small distances to be bridged can generally be regarded as an advantage, because low transmission levels are supported by this fact. An obvious disadvantage, however, for indoor PLC is the close proximity of the PLC equipment and the sources of interference, such as dimmers, CD players, and all kinds of appliances with switched power supplies. So the advantage of small distances disappears at least partially.

A new approach for indoor network conditioning, the principle of which is already introduced in the right part of Figure 5-34, will now be outlined in detail by means of Figure 5-35.

The feeding point for PLC signals in buildings is typically a 230-V standard wall plug, which is, according to worldwide standards, connected with three wires, i.e., one phase (L1, L2, or L3), the neutral, and the protective earth conductor. In the past, PLC signal coupling was usually performed between phase and neutral conductor. While one terminal of all wall plugs is always galvanically connected with the neutral conductor, running through the whole building and finally being grounded at the house connection point, the phases are normally evenly distributed to the wall plugs. Thus a galvanic connection of the phases is not present before the transformer. With conditioning, however,

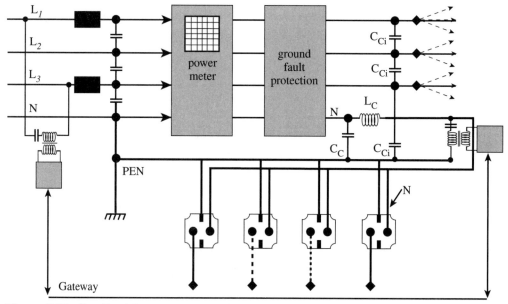

Figure 5-35 Details of the new approach for indoor mains conditioning.

a connection for high frequencies and simultaneous grounding takes place at the house connection, according to Figure 5-35. For indoor PLC applications the described feeding method between phase and neutral is not very favorable, because a sending modem may feed its signals into a wall plug which is, e.g., supplied by phase L1, while the receiving modem is connected to L2. Without any conditioning the received signals will usually arrive very weak via crosstalk between L1 and L2. Therefore, very high attenuation values may be observed at building installations, even for relatively small distances between transmitter and receiver. Also, when coupling PLC signals between a phase and the neutral, we usually find a high degree of asymmetry, which causes strong common-mode propagation immediately at the feeding point, so that even for low transmission levels a noticeable electromagnetic radiation may occur.

The basic idea of the new approach is to configure the phases and the protective earth conductor together as a ground plane for PLC signals, and to feed the neutral conductor against this ground. Toward this aim an inductor L_C of approximately 3–5 µH is inserted into the neutral conductor as shown in Figure 5-34. This way the neutral conductor is isolated from the rest of the powerline wiring for high frequencies, while nothing is changed for the frequency range where power supply takes place. The construction of the inductor is very simple, because no magnetic material is needed here, due to the low inductance needed to make it effective for frequencies above 10 MHz.

Figure 5-35 outlines the new idea in more detail. The conditioning inductor L_C will usually be placed after the power meter and the ground fault protection device. A small capacitor C_C of some nanofarads may be placed at the terminals of this device to prevent PLC signals from entering it, or from propagating to conductors not involved in signal transmission. Furthermore, Figure 5-35 suggests distributing capacitors C_{Ci} over the powerline wiring, in order to keep phases and protective earth grounded for high frequencies throughout the whole building. According to Figures 5-34 and 5-35, PLC signals can now be coupled between protective earth and neutral. The phases are not directly involved in signal transmission. Furthermore, as they are grounded for high frequencies, they provide an additional shielding effect which is advantageous with respect to EMC. The use of the neutral conductor as the "hot wire" for PLC signals leads to the following advantages, regarding both the transmission quality and electromagnetic compatibility:

- Protective earth and neutral are normally at the same electrical potential. Because these conductors are galvanically connected at the house's grounding point, interference resulting, e.g., from the operation of various appliances within the house is reduced to a great extent, so that the noise level between these conductors is generally low.

- Moreover these two conductors usually run in parallel from wall plug to wall plug throughout a building. Therefore we have a two-wire network with a relatively high degree of symmetry. So, on the one hand we observe fair transmission properties and low interference levels, and on the other hand we can expect low radiation, because—due to the symmetry—the PLC signals can propagate in differential mode over significant distances.
- Unwanted differential- to common-mode conversion can take place only at some distance from a PLC feeding point, where the signal levels have already strongly decreased. Therefore significantly lower decoupling factors can be assumed than those determined, e.g., in [H33], or depicted in Figure 5-31.

CHAPTER 6

Appropriate Modulation Schemes for PLC and Communication System Concepts

6.1 Introduction

Unfortunately, none of the various proven modulation techniques and media access procedures known from standard telecommunication applications can be used in PLC without major modifications. The following modulation schemes are basically applicable to powerline communication (PLC), when taking data rates above 1 Mbit/s into account:

1. Spread-spectrum modulation, in particular "direct sequence spread spectrum" (DSSS)
2. Broadband single-carrier modulation without equalizing
3. Broadband single-carrier modulation with broadband equalizing
4. Broadband multicarrier modulation with adaptive decision feedback equalizing
5. Multicarrier modulation in the form of "Orthogonal Frequency Division Multiplexing" (OFDM)

For schemes 4 and 5 the transmitted data stream does not have to be concentrated in a contiguous portion of the spectrum but may be distributed over numerous subchannels, optionally with gaps between them. For OFDM we generally have a high number of such equally wide subchannels, each of which can be loaded with data bits, according to their quality. Scheme 4 usually operates with larger "spectral portions," i.e., relatively few subchannels of different width. Both schemes 4 and 5 are basically able to deal with the expected unequally distributed spectral gaps of different width, which will be available for PLC. OFDM, however, clearly offers the highest degree of flexibility of all the schemes listed above and described in later sections.

6.1.1 Basic Consideration of Broadband Techniques

Band-spreading modulation techniques were originally developed solely for military communication purposes, in order to obtain robustness against intentional disturbers and eavesdropping. In the past the technology was characterized by high effort and considerable costs. The rapid development of integrated-circuit complexity, however, has made spread-spectrum techniques (SST) available for almost any application. Due to the resistance of SST against all kinds of narrowband interference and selective attenuation, such techniques are generally not a bad choice for PLC. An additional feature which makes SST very interesting, especially with regard to EMC, is the low spectral power density which characterizes the transmitted signals. Also, media access is elegantly accomplished by code-division multiple-access (CDMA) schemes.

6.2 Single-Carrier Modulation and CDMA

A particular feature of CDMA is multiple access without the need for global coordination or synchronization. As illustrated in Figure 6-1, a single carrier is used within a communication cell, and an individual spreading code $p_i(t)$ is assigned to each participant, being orthogonal to the codes of all others. First, the message signal $s_t(t)$ conventionally modulates the carrier which has the frequency f_0. Thereby a spectrum $S_k(f)$ of approximately the double bandwidth of the message is produced around f_0. In a further modulator (multiplier) fast 0/180° phase hops are inserted, according to the binary pseudonoise sequence $p_1(t)$. Now we have a transmission signal exhibiting a bandwidth that corresponds approximately to twice the clock frequency of the pseudonoise sequence.

At the receiver the same sequence $p_1(t - \tau)$ must be available, synchronized to the received signal, i.e., delayed by the signal propagation time τ between transmitter and receiver. In a first mixer the rapid phase hops are then removed and the spectrum $S_k(f)$ is restored; a conventional demodulation follows for recovery of the message. Another participant, to whom the orthogonal spreading code $p_2(t)$ has been assigned, cannot perform the mentioned spectral compression, i.e., the received spectrum $S_R(f)$ remains almost unchanged—see the shaded box in the lower part of Figure 6-1. If a narrowband interferer, e.g., in the form of a broadcast radio station, appears at the receiver, then it is subjected to the spreading process, so that only a small portion corresponding to the message bandwidth $S_k(f)$ can impair the desired signal.

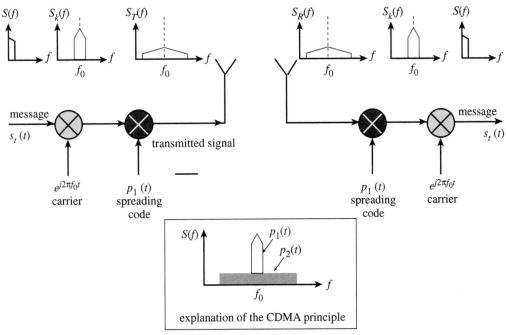

Figure 6-1 Direct sequencing and CDMA.

CDMA assigns the entire frequency band to each participant alternately, so that the access does not have to be coordinated. Each active participant, however, introduces some kind of background noise for all others. The more participants become active, the higher the probability of mutual disturbance. The situation is comparable with a room where numerous people are talking to each other at the same time but in different languages. It will work well only up to a certain limit. Therefore a general adjustment of channel capacity and the permissible number of active participants is needed. The crucial figure in this context is the so-called processing gain (*PG*), which is obtained through spreading.

If B_M is the message bandwidth after conventional modulation (spectrum $S_k(f)$ in Figure 6-1) and B_{SP} is the bandwidth of the transmitted signal (spectrum $S_T(f)$), then we have

$$P_G = \frac{B_{SP}}{B_N} \qquad (6.1)$$

PG should be between 10 and 100 to obtain an efficient system for practical applications. The higher *PG* is, the more participants can access the channel at the same time. However, as a general rule, the number of participants must always remain smaller than

PG; otherwise robustness against interference is lost and the signal quality may deteriorate to an unacceptable level for all users, even without additional interference.

With *PG* being sufficiently high, however, each user who becomes active contributes only a small portion of interference. This particular feature of a properly designed CDMA system is also denoted "graceful degradation." Not least for this feature CDMA has been applied to the American mobile phone standard IS-95 and for the upcoming UMTS (Universal Mobile Telecommunication System). Unfortunately, the main advantages of CDMA cannot be exploited in PLC for various reasons, including the lack of large contiguous spectrum portions (see Figure 5-27). The resulting fissured spectrum is bad for CDMA.

Apart from CDMA, two further aspects must also be mentioned which will rule out most kinds of wideband single-carrier schemes for PLC:

- Single-carrier broadband schemes do not exhibit high spectral efficiency; i.e., the transmission of a digital data stream with the symbol rate r_s requires a bandwidth of at least $2r_s$, which means that the following total bandwidth must be available for *N* users:

$$B_{SP} > 2 \cdot N \cdot r_s \qquad (6.2)$$

- The powerline channel does not provide a flat transmission characteristic over large bandwidth portions. We generally find lowpass and strong frequency-selective fading effects.

The second statement means that rather complex equalization would be required at the receiver. A thorough investigation of single-carrier and multicarrier modulation schemes for PLC is given in [D8]. As a result, clear disadvantages for single-carrier schemes may arise under the following circumstances:

- Poor transmission characteristics of the powerline channel
- Interference scenario, in particular when considering all kinds of impulsive noise
- Expected frequency-allocation and transmission-level limitation

For these reasons, experts in the field have concentrated on multicarrier techniques, in particular OFDM.

Table 6-1 presents a brief evaluation of the most important properties of modulation schemes that could be candidates for PLC. As a result, OFDM clearly turns out to be the most qualified scheme to cope with present and future conditions most likely found in PLC—see also [D8].

Table 6-1 Comparison of PLC modulation schemes (++ excellent; + good; 0 fair; – bad; – – very bad)

	Spectral efficiency	Max. data rate in Mbits/s	Robustness against channel distortions	Robustness against impulsive noise	Flexibility and adaptive features	System costs (incl. equalizers and repeaters)	EMC aspects, regulation
Spread-spectrum techniques	< 0.1 bits/s/Hz	≈ 0.5	–	0	– –	– –	++
Single-carrier broadband, no equalizer	1–2 bits/s/Hz	< 1	– –	+	– –	++	– –
Single-carrier broadband with equalizer	1–2 bits/s/Hz	≈ 2	+	+	0	–	–
Multicarrier broadband with equalizer	1–4 bits/s/Hz	≈ 3	+	0	0	–	0
OFDM	>> 1 bits/s/Hz	> 10	++	0	++	–	+

6.3 PLC Signal Characteristics and Level-Limit Measurements

The studies of the decoupling factor $K(f,d)$ in Section 5.11.1, Figures 5-30 and 5-31, used unmodulated carriers with zero bandwidth. A bandwidth of 9 kHz is suggested for the measuring procedures currently in the process of being defined within regulation NB 30. This bandwidth corresponds to the standard bandwidth of AM broadcasting stations, so that the measurement results are representative for the impairment of AM reception from interfering radiation caused by PLC. In the frequency range up to 30 MHz the maximum value of the magnetic field strength has to be recorded by use of a standardized loop antenna approximately 53 cm in diameter, as shown in Figure 5-30. The measured signal passes the 9-kHz bandpass filter, which is swept over the entire frequency range to be examined. The filter output can be evaluated in two ways: by determining either a peak value or quasi-peak value—see Figure 6-2. There have been lengthy discussions about the way to detect peak values, because things are rather complicated in practice, as soon as real modulated PLC signals instead of pure sinusoids are taken into account. If a real peak measurement is specified, this means substantial disadvantages for PLC signaling, because sophisticated digital transmission techniques are

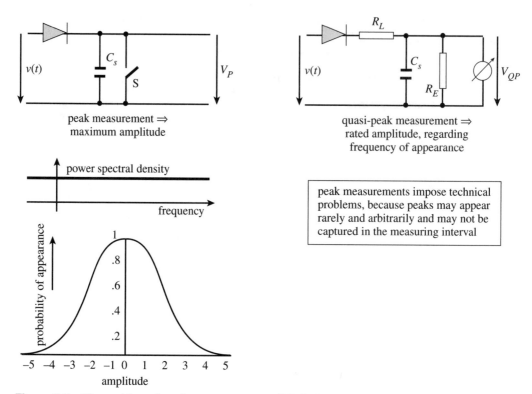

Figure 6-2 The problem of peak measurements of PLC signals.

normally used, producing typically wideband signals with an arbitrary distribution of amplitudes. In the lower left of Figure 6-2, power spectral density and amplitude distribution for typical PLC signals are sketched schematically. Note that PLC techniques can be optimized, so that low spectral power density will show an almost flat course, while the distribution of amplitudes becomes more and more Gaussian. It is well known that amplitudes of any size may occur in such a distribution, but the probability that very large peaks will appear is almost zero. This means in practice that, in order to truly capture the absolute peak values, the measurement time would have to be infinite, and the interference effect of such rare peaks will usually be negligible.

This means that the peak values have to be thought of as statistically distributed and rarely occurring noise and should, therefore, be treated like unintended interference from appliances. With the above considerations in mind, it would not make sense to specify level limitations based on absolute peak measurements. Meanwhile the discussion has moved to more sensitive quasi-peak detection methods with suitable time constants to be defined by appropriate choice of R_L, C_s, and R_E in the circuit diagram depicted in the upper left of Figure 6-2.

6.4 OFDM—A Multicarrier Scheme for High-Speed PLC

OFDM is a well-proven technique in applications such as digital audio broadcasting (DAB) and asymmetric digital subscriber line (ADSL). Furthermore, OFDM is a candidate for the upcoming digital video broadcasting (DVB). OFDM is strongly related to frequency hopping (FH), a spread-spectrum technique. Therefore OFDM exhibits robustness against various kinds of interference and enables multiple access. In OFDM, the available spectrum B_t is segmented into numerous narrowband subchannels. A data stream is transmitted by frequency-division multiplexing (FDM) using N carriers with the frequencies $f_1, f_2, ..., f_N$ in parallel [F12]. As shown in Figure 6-3, each of the subchannels has the bandwidth

$$\Delta f = \frac{B_g}{N} \tag{6.3}$$

The narrowband property of the subchannels justifies the assumption that attenuation and group delay are constant within each channel. Therefore equalization becomes an easy task which can be performed by a so-called "1-tap" technique. This is a substantial advantage of multicarrier signaling over single-carrier broadband approaches. A typical OFDM signal in the time domain can be described as follows:

$$s_{OF}(t) = A \cdot \text{rect}\left(\frac{t}{T}\right) \sum_{i=1}^{N} \sin\left(2\pi\left(f_0 + \left[i - \frac{N+1}{2}\right] \cdot \Delta f\right)t\right) \tag{6.4}$$

As (6.4) denotes, the minimum frequency spacing is $\Delta f = 1/T$, where T is the duration of a waveform. Equation (6.4) describes a set of N orthogonal waveforms, which cover the frequency range from

$$f_0 - [(N-1)/2] \cdot \Delta f = f_0 - (B_t - \Delta f)/2 \quad \text{up to} \quad f_0 + [(N-1)/2] \cdot \Delta f = f_0 + (B_t - \Delta f)/2$$

Figure 6-4 illustrates the spectral conditions for $N = 7$ waveforms. Orthogonality allows spectral overlapping, leading to an outstanding efficiency, which is about two times better in comparison with single-carrier techniques. With the symbol rate r_s for N subchannels we have an overall bandwidth of

$$B_t = (N+1) \cdot r_s \approx N \cdot r_s \tag{6.5}$$

Excellent spectrum utilization will turn out to be a key element for the success of high-speed PLC, because on one hand we have lowpass characteristics and filter effects limiting the usable frequency range, and on the other hand the use of certain portions of

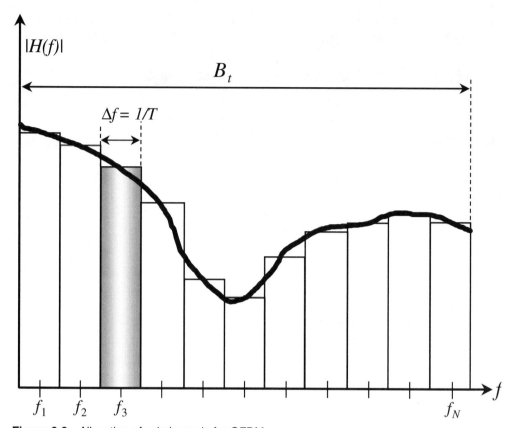
Figure 6-3 Allocation of subchannels for OFDM.

the spectrum may be excluded by regulation. In a classical FH system, information is contained in the sequence of frequencies; i.e., the carriers are transmitted sequentially. In OFDM, the substantial differences are that each carrier is now modulated and takes a part of the data stream, and that a large number of carriers—typically several hundred—are transmitted in parallel. As the transmitted signal is now the sum of many modulated carriers, its time-domain representation appears complicated, while the magnitude of the spectrum remains unchanged. Figure 6-5 shows the problem somewhat simplified with only $N = 32$ carriers, and using real-value (cosine-shaped) signals in the time domain. The complex relations (6.6) and (6.7) will be presented later.

The top chart in Figure 6-5 clearly outlines the constant spectral amplitudes over the entire frequency range, with a spacing $\Delta f = 1/T$. In the time domain the peak-to-rms (root mean square) ratio, i.e., the "CREST factor," represents an important figure, which should be kept as low as possible.

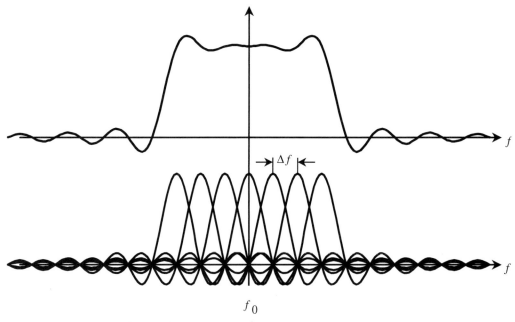

Figure 6-4 Example of seven OFDM waveforms in the frequency domain.

This is not an easy task due to the random nature of the data stream. The center chart in Figure 6-5 shows the worst case, where all 32 carriers are added with equal phase, which is zero in this example. Then we have a high peak amplitude of 32 only for the first sample, and all others are zero, which means that the CREST factor takes its maximum value. Of course, such a constellation is not desirable and must therefore be avoided in any case. A low CREST factor is not only paramount for keeping the cost of the transmitter power stage low, but also for EMC reasons—see Section 6.3. High peak values can cause significant levels of radiation. The lower chart in Figure 6-5 suggests how to achieve a significant improvement. In this example, the phases of the 32 carriers have been randomized, with the astonishing result that the CREST factor has dropped dramatically. Such a procedure may, however, not always lead to satisfying results. Due to the nature of a random process, it cannot be excluded that many or even all of the carriers have equal phases, so that a maximum peak as depicted in the center of Figure 6-5 may occur from time to time. In practice, the data stream to be transmitted will determine the carrier phases, which will usually also be a random process. In contrast to a purely random process, the so-called "bit loading" of the carriers can be controlled by the transmitter. This means that a sophisticated modulation method can be used to avoid constellations where the majority of carriers have equal phases to bring the CREST factor down.

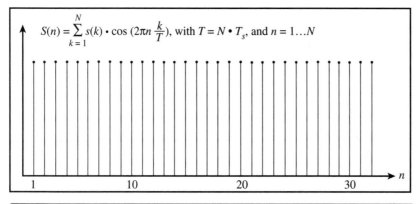
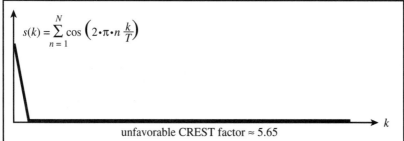
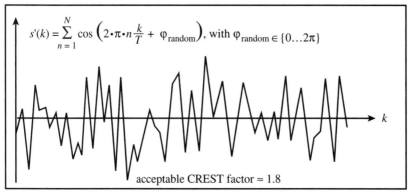

Figure 6-5 Illustration of the "CREST factor" problem with OFDM in the time domain for a 32-carrier example.

6.5 OFDM Signal Synthesis, Carrier Modulation and Demodulation

OFDM signal synthesis can be performed by an inverse discrete Fourier transform (IDFT), and at the receiver the complementary operation, namely a DFT, has to be carried out before the demodulation. Equations (6.6) and (6.7) describe these procedures for the sampling rate $1/T_s$.

$$s(kT_s) = \frac{1}{N} \sum_{n=0}^{N-1} S\left(\frac{n}{NT_s}\right) \cdot e^{j2\pi\frac{n}{N}\cdot k} \quad \text{with } k = 0, ..., N-1 \quad (6.6)$$

$$S\left(\frac{n}{NT_s}\right) = \sum_{k=0}^{N-1} s(kT_s) \cdot e^{-j2\pi\frac{n}{N}\cdot k} \quad \text{with } n = 0, ..., N-1 \quad (6.7)$$

The IDFT described by (6.6) supplies the transmission vector $s(k)$ of length N, which corresponds to the waveform duration $T = N \cdot T_s$. $S(n)$ denotes the vector of spectral lines spaced at $\Delta f = 1/T$ and carrying the information to be transmitted in phase and amplitude. At the receiver a vector $s(k)$, consisting of N discrete values of the received signal, is processed according to (6.7). The result of this DFT retrieves the spectrum $S(n)$, containing the necessary phase and amplitude information for data detection. In practice, calculations according to (6.6) and (6.7) can be efficiently performed by digital signal processors (DSPs) with optimized architectures for the fast Fourier transform (FFT) algorithm.

When comparing (6.6) and (6.7), the high degree of similarity becomes obvious immediately. The only formal difference, except factor $1/N$, is the sign of the exponential function. Introducing the following simple mathematical manipulations into (6.6) will allow the use of the same DSP program for FFT and IFFT:

$$s(kT_s) = \frac{1}{N} \left[\sum_{n=0}^{N-1} S^*\left(\frac{n}{NT_s}\right) \cdot e^{-j2\pi\frac{n}{N}\cdot k} \right]^* \quad (6.8)$$

As (6.8) indicates, the spectral coefficients $S(n)$ are represented in their conjugated complex form now, which is a very simple procedure, as only a change of the sign of the imaginary part has to be performed. And finally, after a complete computation of the sum in brackets, the sign of its imaginary part has to be changed.

As already mentioned, each carrier in an OFDM system carries part of the data stream. However, data need not be evenly distributed over the carriers, nor do equal modulation schemes have to be applied. Hereby considerable advantages can be yielded, whenever sufficient knowledge of the channel properties is available. If certain parts of the transmission band have low attenuation and low levels of interference, the subcarriers in this range can be loaded with complex modulation schemes, such as QAM. In other parts, where only small SNR values are expected, BPSK or a similar method may be used. Extremely bad portions or sections, which are excluded by regulation, can be faded by zeroing the corresponding Fourier coefficients. A simple carrier modulation scheme often used in PLC applications is discussed in the following. It is important to understand that no type of carrier modulation influences the width of an OFDM signal spectrum. Because each carrier represents a pure sinusoid during its entire duration T, the spectrum shape is determined only by T according to the Fourier relation

$$\text{rect}\left(\frac{t}{T}\right) \circ\!\!-\!\!\bullet T \cdot \text{si}\left(\pi \cdot T \cdot f\right) \tag{6.9}$$

Figure 6-6 gives an example that uses QPSK for carrier modulation. The lower half of the figure shows four sinusoidal carriers with the same frequency and the duration T. Note that always an integer number of periods has to fit into T (only two periods are drawn here for clarity). They represent a symbol element, occupying a single OFDM subchannel—see Figures 6-3–6-5. The zero-phase of the carriers in the time domain carries the information to be transmitted. In our case there are four positions: 0°, 90°, 180°, and 270°. The upper part of Figure 6-6 shows the QPSK signal space chart. The four dots can be thought of as terminations of vectors of equal length, pointing in orthogonal directions. As there are four points, a 2-bit information can be assigned to each one. In other words, one of the dots or one waveform in the time domain represents a symbol. In this example, each symbol carries 2 bits. (Note the difference between a symbol and a data bit.) A possible symbol-to-bit correspondence could look as follows:

0° ⇒ 00
90° ⇒ 01
180° ⇒ 10
270° ⇒ 11

In the case of 16-QAM we would have 16 dots in the signal space, representing the 16 symbols, each now carrying 4 bits of information. In addition to 12 discrete phase locations, there would now also be three amplitude steps. The system example in Figure 6-7 summarizes the key features of OFDM for high-speed PLC applications:

Data rate: $r_D = 2$ Mbits/s
Number of carriers: $N = 1000$
Modulation: QPSK ⇒ symbol rate: $r_s = 1$ MHz

Each carrier is equally loaded with information, and the waveform duration is $T = 1$ ms. So IFFT and FFT—our key operations—have to be performed in less than 1 ms. It takes modern DSPs, such as the TMS320C6001, about 70 μs to do a complex

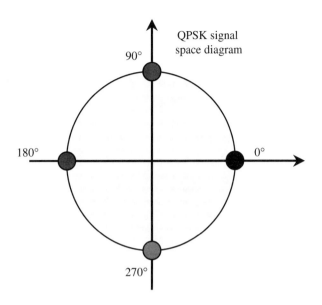

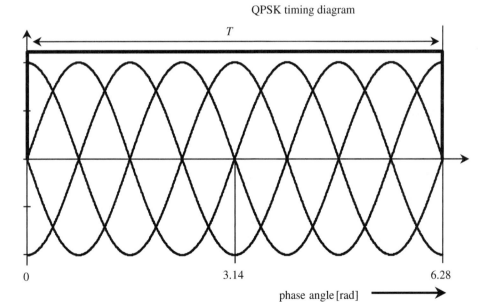

Figure 6-6 QPSK carrier modulation for OFDM.

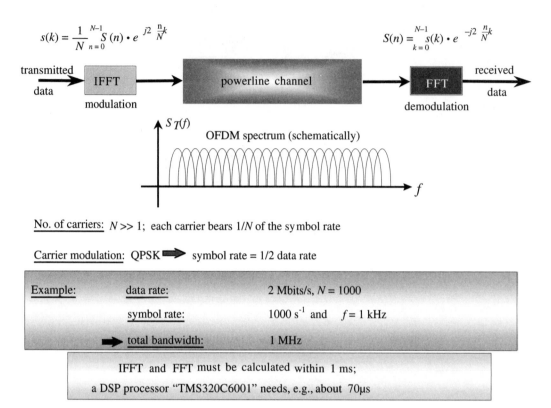

Figure 6-7 A simplified OFDM system example with QPSK carrier modulation.

1024-point FFT. This example clearly indicates that current DSP technologies can support a tenfold enhancement of the data rate.

6.5.1 The OFDM Transmitter

Figure 6-8 shows the block diagram of a typical OFDM transmitter. The source data stream is channel-coded, generally with interleaving. Then the coded bit stream is serial/parallel converted and divided into N groups of ν bits. Each of these groups represents a symbol, which is assigned to one of the subchannels. For the above example ($N = 1000$ and QPSK) we have $\nu = 2$, so that a data block of 2000 bits is fed in parallel to the symbol generators. Each symbol generator produces a Fourier coefficient $S_i(n)$, which maps the associated symbol onto signal space.

As already mentioned, it is also possible to perform the bit mapping individually for each subchannel. Of course, such a procedure makes sense only if channel quality information is available. This information can be gained in an initialization phase, during which a training sequence is transmitted. As the powerline channel parameters

OFDM Signal Synthesis, Carrier Modulation and Demodulation

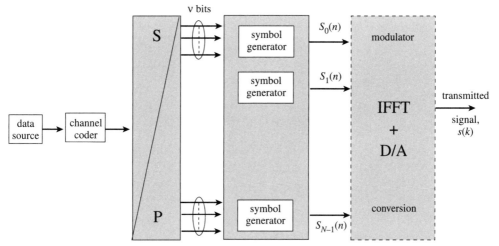

Figure 6-8 Typical OFDM transmitter setup.

(except impulsive noise) are highly stationary, a single training sequence at the beginning of a session will normally be sufficient. The complex symbols $S_i(n)$ in Figure 6-8 are now undergoing an IFFT, which delivers N samples of the transmission signal $s(k)$ in the time domain. After D/A conversion and lowpass filtering the signal $s(t)$ is fed to a power amplifier and coupled into the mains.

6.5.2 The OFDM Receiver

Figure 6-9 shows the channel with the impulse response $h(t)$, the addition of interference $n(t)$, and the typical structure of an OFDM receiver. The received signal $r(t)$ is a distorted version of $s(t)$ affected by $n(t)$. N samples of $r(t)$ are taken at the receiver input and A/D converted. With a sampling rate $1/T_s$ we get the received and digitized vector $r(k)$, which is then Fourier transformed. The result is a vector $R(n)$, representing the Fourier coefficients of the received signal.

In many cases, it makes sense to perform some kind of equalization to eliminate the influence of the channel transfer function, before symbol and bit detection take place. Equalization is amazingly simple due to the narrowband nature of the subchannels, as attenuation and group delay are constant. Thus each subchannel can be described by its transfer function

$$H_i = \alpha_i \cdot e^{j\varphi_i}, \quad \text{with} \quad n = 0, ..., N-1 \tag{6.10}$$

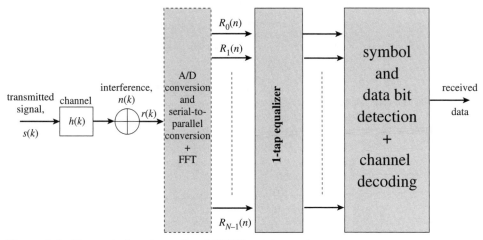

Figure 6-9 Transmission channel and OFDM receiver setup.

During an initialization phase, where a training sequence is transmitted (which is well known to the receiver) H_i is estimated and inverted, delivering

$$H_i^{-1} = \frac{1}{H_i} = \frac{1}{\alpha_i} \cdot e^{-j\varphi_i} \quad \text{with } n = 0, ..., N-1 \qquad (6.11)$$

Multiplying each Fourier coefficient R_i in Figure 6-9 by the corresponding H_i^{-1} will provide the desired equalization (1-tap equalizer). A training sequence will normally be necessary only in very large time intervals, because powerline channels are highly stationary. In practice, a single training sequence at the beginning of a communication session will normally be sufficient. Details of the equalizing procedure are shown in Figure 6-10.

6.6 OFDM for High-Speed PLC—Summary

OFDM promises robustness concerning both distortions by lowpass effects and strong fluctuations of the transfer function of powerline channels. The main advantage is obtained by the fact that the channel is divided into many narrow subchannels (see Figure 6-3). Therefore, as shown above, equalizing in OFDM is a simple procedure, in contrast to wideband equalizing, which would be needed, for instance, for multicarrier broadband systems using other modulation schemes such as GMSK.[1] In that case, equalizing would usually consume the entire computation capability of a fast DSP processor.

---

[1] Gaussian minimum shift keying.

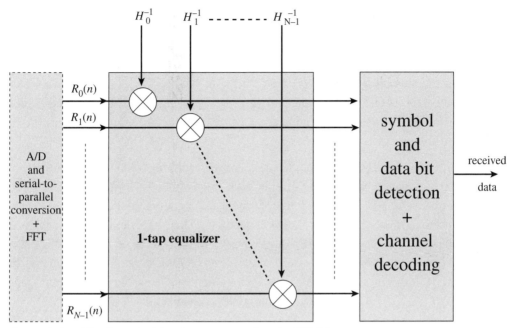

Figure 6-10 Working principle of a 1-tap equalizer for OFDM.

OFDM inherently solves one essential problem associated with high-speed PLC not mentioned previously: intersymbol interference (ISI) caused by multipath delay spread, an effect which is visible in principle in Figures 5-22 and 5-23. Hereby the desired signal and one or more delayed copies arrive at the receiver. The channel's impulse response clearly maps length, number, and attenuation of the signal paths. The overall duration of the impulse response is the crucial figure for a modulation scheme. Narrowband systems generally quickly run into problems with multipath. For example, a 1 Mbits/s BPSK symbol is only 1 μs wide. With 1 μs of delay spread, which easily occurs in powerlines, two symbols will interfere, so that the receiver cannot read them. If, however, the selected symbol duration can be much longer than the delayed paths, i.e., significantly longer than the impulse response, reception will always be correct. OFDM employs this approach in combating multipath, as the example in Figure 6-7 impressively demonstrates. We have a symbol duration of 1 ms here, so that only small portions can be affected by delay spreads up to 10 μs, which are typical for the access domain, and around 1 μs, which are typical for indoor channels. For higher data rates, however, OFDM symbols become shorter, and as they approach 100 μs, for example, it is a good idea to take steps to eliminate ISI. Fortunately, this is relatively easy by introducing a "guard interval," which is filled with a cyclic prefix. This prefix copies a sufficiently high number of samples—covering the duration of the channel's impulse response—from the end of an OFDM symbol and adds the copy to the beginning of the symbol. The long OFDM symbol time, generally many microseconds, as well as the cyclic prefix, which is usually a small percentage of the OFDM symbol time, are key factors that enable performance in a time-dispersive channel. For example, if we select a cyclic prefix of about 5 µs for a channel described in Figure 5-22, all received copies of the transmitted original will contain the same OFDM symbol (but shifted in time), and there will be no ISI.

As already stated, OFDM subchannels can be faded individually to counter the impact of narrowband interference or frequency-selective fading. Furthermore, very fissured spectra can be efficiently exploited, because the locations of the carriers and thus of the subchannels are almost arbitrarily selectable. This high flexibility is achieved without changes to hardware; usually only a DSP program has to be modified. The use of a DSP can be regarded as the "entrance fee" to OFDM technology. Compared to competitive techniques, this should not be seen as an obstacle, because such techniques also require DSP-like functions, e.g., for broadband equalization.

Since bandwidth efficiency must be considered an essential issue for the future of high-speed PLC, OFDM will be the ideal candidate. Even if certain bandwidth portions are lost, e.g., due to interference or fading, OFDM is able to omit the corresponding subcarriers and to continue working error-free with the remainder. As investigations performed in [D8] revealed, OFDM is able to use even heavily distorted channels where comparable modulation schemes generally fail. This can be achieved by repeated transmission of the information over different subchannels. This way, reliable transmission—although at reduced data rate—can be sustained even under extreme conditions, so that a 100% powerline link reliability can be guaranteed. No other modulation scheme which appears suitable and affordable for PLC exhibits comparable features.

OFDM-based demonstration systems for Internet access over powerlines have been in operation for some time. The feasibility of reliable high-speed PLC with OFDM has been proven. However, there is some work to be done toward cost-effective devices for series production. Considering the current efforts in industry, research institutions, and PSUs, it can be expected that consumer products will be in the market during the year 2001. Last but not least, the activities of HomePlug will play an essential role for the promotion of OFDM in this context. As already mentioned in Section 5.11.1, HomePlug has chosen OFDM as the baseline technology for high-speed powerline networking, and as the founding members of this alliance can be regarded as key players in the field, the results will undoubtedly be successful.

CHAPTER 7

Conclusions and Further Work

With the beginning of ubiquitous electricity supply approximately 80 years ago, communications over power lines began also. At a very early stage, energy and information transfer could be combined satisfyingly. In the past, however, only the supply utilities could make use of such communication links for their own purposes. Recently the situation has changed fundamentally. In 1998 telecommunications and energy markets were opened by cessation of the previous monopolies. As an immediate consequence, alternative fast communication links are needed to bridge the last mile. Powerline communication will be an ideal candidate, because the electrical power distribution system represents a perfect local area network, present everywhere and immediately usable. In the past not much information was available about the possibilities and the technology of powerline communications; obviously there was little need for such information. During the era of telecommunications and energy monopolies the topic interested only a few specialists. Currently, however, Internet access via the standard wall plug at data rates around 1–2 Mbits/s has initiated enormous activities and is becoming reality step by step. Furthermore, recently a very new approach was launched toward high-speed PLC for in-home digital networks (IHDN). In this context it will obviously be necessary to open the spectrum toward higher frequencies. No sensible upper limit can be specified at the moment, as typical indoor links, for example, are not expected to exhibit significant low-pass characteristics. Toward closer examination of the possibilities and limitations of high-speed indoor PLC at data rates in excess of 10 Mbits/s the following steps are planned:

1. Impulse-response and transfer-function measurements at different indoor installations in a frequency range from 10 to 100MHz, in order to establish a statistically reliable database:

    - in unconditioned networks with signal coupling between a phase and the N conductor.
    - in networks conditioned in the way suggested in Figures 5-34 and 5-35.

2. Investigation of the interference scenario in the mentioned frequency range with special emphasis on ingress of wireless services and the impact of impulsive noise:

    - in unconditioned networks with signal coupling between a phase and the N conductor.
    - in networks conditioned in the way suggested in Figures 5-34 and 5-35.

3. Statistical estimation of the channel capacity under the assumption of a decoupling factor in the range of 10^{-3}/m (allowable radiation level: $E_{max} = 27$ dBµV/m, according to Figure 5-28).
4. Measurements of the decoupling factor at different indoor installations in the mentioned frequency range from 10 to 100 MHz to establish a database:

    - in unconditioned networks with signal coupling between a phase and the N conductor.
    - in networks conditioned in the way suggested in Figures 5-34 and 5-35.

5. Definition and standardization of appropriate frequency ranges to achieve coexistence between the Internet from the wall plug and IHDN, e.g., using the spectrum from 10 to 20 MHz for the Internet and the spectrum above 20 MHz for IHDN.
6. Building experimental high-speed PLC systems based on OFDM, using fast standard DSP processors, microcontrollers, and FPGAs, followed gradually by higher integration, until the PLC system on silicon eventually becomes reality, enabling cost-effective consumer products for ubiquitous applications all over the world.

After it has turned out that the electrical power distribution grid, including building installations, has theoretical capacities of several hundred Mbits/s, major industrial companies have engaged in the development of various PLC systems and services. The innovation potential is regarded to be enormous, creating considerable economic values, from which, due to the nature of the powerline medium, everybody may benefit. Toward this goal, it is a primary intention of this book to help to estimate the possibilities and limitations of powerline communication with respect to everyone's individual needs. For a clear and complete illustration of the various facets of PLC, historical background information and recent results from research and development are presented. Concerning high-speed communication, many details are still under construction; the essential problems are discussed and the developments expected for the near future are analyzed, in order to prepare the potential user thoroughly for the upcoming technology.

The first part of the book (Chapters 1 and 2) presented the motivation for PLC and summarized the basic characteristics of the power supply system from the viewpoint of communication technologies. After a brief description of historical communication methods, such as carrier-frequency signaling on the high-voltage level and ripple-carrier signaling on the medium- and low-voltage levels, current possibilities based on existing standards were proposed in Chapters 3 and 4. In this context, new so-called "energy information services" were introduced which will play an important role in the deregulated electrical energy market. For numerous PSUs, it will become a matter of "do or die" to provide such services in the very near future.

After an analysis of the limits imposed by the laws of physics, Chapter 5 discussed problems of electromagnetic compatibility. Because existing wireless services, such as medium-wave and HF broadcasting, as well as amateur radio might be affected by upcoming ubiquitous PLC systems, it will be important to find appropriate frequency-allocation schemes for PLC and to define clear and sensible limits for allowable levels of unwanted radiation. For the powerline community, including the recently established international "PLC Forum," it is currently a task of utmost importance to set up sensitive fundamentals for radiation limits and appropriate measurement procedures. Also, network conditioning toward more symmetry and shielding will be an essential issue, especially for coexistence of high-speed indoor PLC with radio service reception, due to the proximity of both.

Chapter 6 dealt with the state of the art concerning realization of high-speed PLC systems. Different signaling schemes were analyzed and compared. The OFDM multicarrier modulation scheme turned out to be most favorable. Most of the key players in the field of high-speed PLC consider OFDM the ideal candidate to cope with both the problems of the rather hostile medium and the expected restrictions by regulations. The commitment of those key players from industry, PSUs, and research institutions to intensive cooperation for the development of sophisticated PLC equipment has engendered optimistic outlooks that worldwide communication over the electric power supply system will become as familiar to everybody in the near future as the use of electricity. In contrast to last-mile solutions, e.g., for Internet access, the exploitation of indoor powerline networks is of global interest, especially in the United States and Canada. Therefore, it will be rewarding to free international synergies and set up worldwide standards as soon as possible.

CHAPTER 8

Reading List and Bibliography by Topics

8.1 Carrier-Frequency Modulation in High-Voltage Lines

[T1] Podszech, H. K.: *Trägerfrequenz-Nachrichtenübertragung über Hochspannungsleitungen.* Springer-Verlag, Berlin, Heidelberg, New York, 1971.

[T2] Stevenson, W. D.: *Elements of Power System Analysis.* McGraw-Hill Kogagusha, Ltd., ISBN 0-07-061285-4, 1975.

[T3] Gommlich, H.: Messungen an TFH-Systemen für die Übertragung von Datensignalen. *bits*, Wandel & Goltermann Customer Information No. 42, 1987, pp. 26–28.

[T4] Hosemann, G. (ed.): *Elektrische Energietechnik.* Part 3. *Netze.* Hütte Taschenbücher der Technik, Vol. 29, Springer-Verlag, 1988.

8.2 Ripple Carrier Signaling

[R1] Zur Mengede: Telenerg System. *Siemens-Zeitung*, Vol. 7, 1933, p. 165.

[R2] Jacob, Heyl: Transkommando-Fernschaltung. *etz*, Vol. 57, 1936, p. 575.

[R3] Dennhardt, A.: *Grundzüge der Tonfrequenz-Rundsteuertechnik und ihre Anwendung.* Verlags- und Wirtschaftsgesellschaft der Elektrizitätswerke mbH, Frankfurt, Germany, 1971.

[R4] Deutsche Elektrotechnische Kommission im DIN und VDE: Rundsteuerempfänger mit Hintergrundschaltuhr. DIN draft, *Deutsche Normen*, March 1990.

8.3 Standards and Regulation Issues

[NO1] Deutsche Bundespost: Richtlinie für die technische Prüfung von TF Industriefunkanlagen auf Niederspannungsleitungen (bis 380V). *FTZ Guideline 446R2022*, 1974.

[NO2] Deutsche Bundespost: Technische Richtlinien für TF-Funkanlagen mit einer Nutzleistung von maximal 5mW und einer DBP-Zulassungsnummer der Kennbuchstabenreihe "TWF". *FTZ Guideline 17R2040*, 1985.

[NO3] EN 50065-1: Signalling on Low Voltage Electrical Installations in the Frequency Range 3 kHz to 148.5 kHz, CENELEC, Brussels, 1991.

[NO4] CISPR-16: Specification for Radio Interference Measuring Apparatus and Measuring Methods. *CISPR Publication 16*, Geneva, Switzerland, 1993.

[NO5] CISPR/A (Secretariat) 67, Report 60 (7/1985): Methods of measurement of mains decoupling factors.

8.4 Spread-Spectrum Techniques

[S1] Dostert, K.: Ein neues Spread-Spectrum Empfängerkonzept auf der Basis angezapfter Verzögerungsleitungen für akustische Oberflächenwellen. Dissertation im Fachbereich Elektrotechnik der Universität Kaiserslautern, 1980.

[S2] Baier, P. W., Dostert, K., and Pandit, M.: A novel spread-spectrum receiver synchronization scheme using a SAW tapped delay line. *IEEE Trans. on Comm.*, Vol. COM-30, No. 5, May 1982, pp. 1037-1047.

[S3] *Spread Spectrum Communications*. C. E. Cook, F. W. Ellersick, L. B. Milstein, D. L. Schilling (eds.). IEEE Press and John Wiley & Sons Inc., New York, 1983.

[S4] Dixon, R. C.: *Spread Spectrum Systems*. John Wiley & Sons, New York, 1984.

[S5] Simon, M. K., Omura, J. K., Scholtz, R. A., and Levitt, B. K.: *Spread Spectrum Communications*. Vols. I, II, III. Computer Science Press Inc., Rockville, Maryland, 1985.

[S6] Langewellpott, U.: Anwendung der Spread-Spectrum-Technik im Mobilfunk. *Frequenz 40*, 1986, pp. 249-254.

[S7]   Friederichs, K.-J.: Die Übertragungsqualität von Spread-Spectrum-Systemen mit M-ärer orthogonaler Modulation. Dissertation im Fachbereich Elektrotechnik der Universität Kaiserslautern, 1986.

[S8]   Baier, P. W., and Dostert, K.: Signalverarbeitung in Spread-Spectrum-Systemen. *Kleinheubacher Berichte*, Vol. 31, ISSN 0343-5725, 1988, pp. 161–170.

[S9]   Schilling, D. L., Pickholtz, R. L., and Milstein, L. B.: Spread spectrum goes commercial. *IEEE Spectrum*, Aug. 1990, pp. 40–45.

[S10] Dixon, R.: *Spread Spectrum Systems with Commercial Applications*. John Wiley & Sons, 3d ed., 1994.

8.5 Fundamentals of Communications, Systems Theory, and RF Technology

[F1]   Shannon, C. E.: A mathematical theory of communication. *Bell Systems Technical Journal*, 27, 1947, pp. 379–423 and 623–656.

[F2]   Shannon, C. E.: Communications in presence of noise. *Proc. IRE*, Vol. 37, 1949, pp. 10–21.

[F3]   Stein S., and Jones J.: *Modern Communication Principles*. McGraw Hill Book Company, New York, 1967.

[F4]   Viterbi, A. J., and Omura, J. K.: *Principles of Digital Communications and Coding*. McGraw Hill Book Company, New York, 1979.

[F5]   Unger, H.-G.: *Elektromagnetische Wellen auf Leitungen*. Dr. Alfred Hüthig-Verlag, Heidelberg, 1980.

[F6]   Spaniol, O.: *Computer Arithmetic*. John Wiley & Sons, New York, 1981.

[F7]   Furrer, F. J.: *Fehlerkorrigierende Block-Codierung für die Datenübertragung*. Birkhäuser Verlag, Basel, 1981.

[F8]   Brigham, E. O.: *The Fast Fourier Transform*. Prentice-Hall Inc., Englewood Cliffs, NJ, 1974.

[F9]   McCanny, J. V., and McWhriter, J. G.: Some systolic array developments in the United Kingdom. *IEEE Computer*, Vol. 20, No. 7, 1987, pp. 51–53.

[F10] Lüke, H. D.: *Signalübertragung*. Springer-Verlag, Berlin, 1991.

[F11] Lüke, H. D.: *Korrelationssignale*. Springer Verlag, 1992.

[F12] Proakis, J. G.: *Digital Communications*. McGraw-Hill, 1995.

[F13]   Yarlagadda, R., and Suresh Babu, B. N.: A note on the application of the FFT to the solution of a system of Toeplitz normal equations. *IEEE Transactions on Circuits and Systems*, Vol. 27, 1980, No. 2, February, pp. 151–154.

8.6 PLC-Related Electronic Circuit Design, Application Notes

[A1]   Tietze, U., and Schenk, Ch.: *Halbleiterschaltungstechnik*. 5th ed., Springer-Verlag, Berlin, Heidelberg, New York, 1980.

[A2]   Busch-Jaeger Elektro GmbH: Der Schalter für die Schalter. Zum Verständnis und zur Installation von Busch Timac. BJE478/11.85/0502, Lüdenscheid, Germany, 1985.

[A3]   Lock-in Amplifier. NF Circuit Design Block Co., Ltd., Yokohama, Japan, 1985.

[A4]   Sedayao, M. J., and Fenger, C. K.: NE5050 Powerline Modem Application Board Cookbook. Signetics Corporation, 1987. Published by Valvo as circuit description No. 3582 entitled: ASK-Modem-IC NE5050.

[A5]   Analog Devices: Monolithic Synchronous Voltage-to-Frequency Converter AD 652. *Data Conversion Databook*, Norwood, Mass., 1988, pp. 4–25 through 4–40.

[A6]   Busch-Jaeger Elektro GmbH: Hausleittechnik Busch Timac X-10. *Technisches Handbuch für Planung und Installation*. 1st ed., Lüdenscheid, Germany, 1988.

[A7]   Lasarray, S. A.: *Design Manual*. Bienne, Switzerland, 1989.

[A8]   Digital RF Solutions Corporation: Data Sheet DRFS-3250 NCMO, Number Controlled Modulated Oscillator. Santa Clara, 1990.

[A9]   Digital RF Solutions Corporation: Application Note AN 1003: Basic analog NCMO applications. Santa Clara, 1990.

[A10]   Digital RF Solutions Corporation: Application Note AN 1004: Control of spurious signals in direct digital synthesizers. Santa Clara, 1990.

[A11]   Digital RF Solutions Corporation: Application Note AN 1008: Numeric synthesis of square waves (NSS). Santa Clara, 1990.

8.7 Communication over the Electric Power Distribution Grid

8.7.1 Channel Analysis and Modeling at Low Frequencies

[N1]   Nicholson, and Malack J. A.: RF impedance of power lines and line impedance stabilization networks in conducted interferences measurement. *IEEE Trans. on Electromagnetic Compatibility*, Vol. EMC–15, No. 2, 1973, pp. 84–86.

[N2] Malack, J. A., and Engstrom, J. R.: RF impedance of United States and European power lines. *IEEE Trans. on Electromagnetic Compatibility*, Vol. EMC–18, No. 1, 1976, pp. 36–38.

[N3] Dvorak, T., and Ochsner, H.: Low tension power line as a fast digital data transmission channel. *Proc. of the 4th EMC-Symposium*, Zurich, 1981, pp. 1–6.

[N4] Vines, R. M., Trussell, H. J., Gale, L. J., and O'Neal, J. B., Jr.: Noise on residential power distribution circuits. *IEEE Trans. on Electromagnetic Compatibility*, Vol. EMC–26, No. 4, Nov. 1984, pp. 161–168.

[N5] Dostert, K.: EMC Problems in data transmission over indoor power lines using spread spectrum techniques. *Proc. of the 6th EMC-Symposium*, Zurich, 1985, pp. 453–456.

[N6] Vines, R. M., et al.: Impedance of the residential power distribution circuit. *IEEE Trans. on Electromagnetic Compatibility*, Feb. 1985, pp. 6–13.

[N7] Chan, M. H. L., and Donaldson, R. W.: Attenuation of communication signals on residential and commercial intrabuilding power-distribution circuits. *IEEE Trans. on Electromagnetic Compatibility*, Vol. EMC–28, No. 4, Nov. 1986, pp. 220–230.

[N8] Chan, M. H. L., and Donaldson, R. W.: Amplitude, width, and interarrival distributions for noise impulses on intrabuilding power line communication networks. *IEEE Trans. on Electromagnetic Compatibility*, Vol. EMC–31, No. 4, Aug. 1989, pp. 320–323.

[N9] Dostert, K.: Der Einfluß von Frequenzfehlern in Vielfachzugriffsystemen mit Frequenzsprungmodulation. *Int. Journal of Electronics and Communications (AEÜ)*, Vol. 43, No. 3, 1989, pp. 144–148.

[N10] Baier, P. W., and Dostert, K.: Synchronization error analysis in a mainsborne frequency hopping spread spectrum system. *Proceedings of the 1989 URSI International Symposium on Signals, Systems and Electronics, ISSSE '89*, Erlangen, Germany, 1989, pp. 544–547.

[N11] Dostert, K., and Ruse, H.: Modellierung von Niederspannungs-Stromversorgungsnetzen als digitale Übertragungskanäle im Frequenzbereich 30kHz bis 146kHz. *Kleinheubacher Berichte*, Vol. 34, 1991, pp. 79–88.

[N12] Dostert, K., and Karl, M.: Übertragungseigenschaften des Niederspannungs-Energieversorgungsnetzes zur digitalen Datenübertragung im Frequenzbereich von 10kHz bis 150kHz. *Kleinheubacher Berichte*, Vol. 39, 1996, pp. 333–342.

[N13] Arzberger, M., and Dostert, K.: Das Stromnetz als Kommunikationsmedium. *Funkschau*, No. 14, 1996, pp. 70–73.

[N14] Dalby, A.: Signal transmission on power lines—analysis of power line circuits. *Proceedings of the 1997 International Symposium on Power Line Communications and Its Applications*, Essen, Germany, April 2–4, 1998, pp. 37–44.

[N15] Hooijen, O: A channel model for the residential power circuit used as a digital communications medium. *IEEE Transactions on Electromagn. Compat.*, Vol. 40, 1998, pp. 331–336.

[N16] Barnes, J.: A physical multi-path model for power distribution network propagation. *Proceedings of the 1998 International Symposium on Power Line Communications and its Applications*, Tokyo, Japan, March 24–26, 1998, pp. 76–89.

[N17] Dostert, K., et al.: Fundamental properties of the low voltage power distribution grid used as a data channel. *European Transactions on Telecommunications (ETT)*, Vol. 11, No. 3, May/June 2000, pp. 297–306.

8.7.2 System Concepts, Experimental Modems, and Test Results for Low Bit-Rate PLC

[ST1] Ochsner, H.: Data transmission on low voltage power distribution lines using spread spectrum techniques. *Proc. Can. Commun. Power Conf.*, Montreal, Canada, 1980, pp. 236–239.

[ST2] Van der Gracht, P. K., and Donaldson, R. W.: Communication using pseudo-noise modulation on electric power distribution circuits. *IEEE Trans. on Com.*, Vol. COM–33, No. 9, 1985, pp. 964–974.

[ST3] Hagmann, W., and Braun, W.: Digitale Datenübertragung über das elektrische Verteilnetz mittels FH/PSK. *Frequenz 40*, 1986, 9/10, pp. 260–265.

[ST4] Dostert, K., and Multer, H.: Vielkanal-Datenübertragung über Installationsnetze: Gleichzeitig störsicher. *Elektrische Energie-Technik*, Vol. 31, No. 5, 1986, pp. 48–54.

[ST5] Becker, T., and Dostert, K.: Datenübertragung über das Niederspannungsnetz mittels Spread-Spectrum Technik. *Abschlußbericht vom 22.12.1986 zu einer BMFT-Fördermaßnahme "Forschungskooperation zwischen Industrie und Wissenschaft."*

[ST6] Piety, R. A.: Intrabuilding data transmission using power line wiring. *Hewlett-Packard Journal*, May 1987, pp. 35–40.

[ST7] Braun, W. R.: ROBCOM: Ein neues Konzept für die Datenübertragung über das elektrische Verteilnetz. *Bulletin SEV/VSE 13/1988*, Vol. 79, pp. 730–735.

[ST8] Dostert, K.: Multi-channel transmission of measurement and control data over indoor power lines using spread spectrum techniques. *ACTA IMEKO '88*, ISBN 1-55617-144-7, 1988, pp. 151–159.

[ST9] Onunga, J. O., and Donaldson, R. W.: Personal computer communications on intrabuilding power line LAN's using CSMA with priority acknowledgements. *IEEE Journal on Selected Areas in Comm.*, Vol. 7, No. 2, 1989, pp. 180–191.

[ST10] Dostert, K.: Optimierung der Sendeleistung in einem stromnetzgebundenen Verbrauchszähler-Fernabfragesystem. *ntzArchiv*, Vol. 11, No. 2, 1989, pp. 85–89.

[ST11] Dostert, K.: Design of a digital signal processor as a matched filter for frequency hopping spread spectrum signals. *Proceedings of the 1989 URSI International Symposium on Signals, Systems and Electronics, ISSSE '89*, Erlangen, Germany, Sept. 1989, pp. 764–767.

[ST12] Dostert, K., Ruse, H., and Threin, G.: Experimentelles Spread-Spectrum-System zur Zählerfernabfrage über Niederspannungsnetze. *Beitrag E4 im Konferenzband zum Thema 3 "Ablese- und Zählersysteme" des 4. Internationalen Kongresses über Zählersysteme und Stromverrechnung (CIFACES)*, Strasbourg, May 9–11, 1990.

[ST13] Vynckier, N. V.: Upwards to a reliable bi-directional communication link on the L power supplies for utility services: field tests in Belgium. *IEE 6th International Conference on Metering, Apparatus and Tariffs for Electricity Supply. IEE Conf. Pub.*, 317, 1990, pp. 168–172.

[ST14] King, M. C., Adame, J., Schaub, T., Rossi, G., and Ziglioli, F.: Experimental systems for tele-reading over the low voltage network. *IEE 6th International Conference on Metering, Apparatus and Tariffs for Electricity Supply. IEE Conf. Pub.*, 317, 1990, pp. 154–157.

[ST15] Dostert, K.: Frequency hopping spread spectrum modulation for digital communications over electrical power lines. *IEEE Journal on Selected Areas in Communications (JSAC)*, Vol. 8, No. 4, May 1990, pp. 700–710.

[ST16] Dostert, K.: Störsichere Meßdatenübertragung auf Stromversorgungsnetzen. *Tagungsband der Tagung des VDEW-Sonderausschusses "Zählerprüfwesen,"* Hamburg, Germany, Sept. 11–12, 1990.

[ST17] Dostert, K.: A novel frequency hopping spread spectrum scheme for reliable power line communications. *Proceedings of the 2nd IEEE International*

Symposium on Spread Spectrum Techniques and Applications, ISSSTA '92, Yokohama, Japan, Dec. 1992, pp. 183–186.

[ST18] Dostert, K.: Datenübertragung auf Stromnetzen—Stand der Technik in Europa. *Proceedings of the Workshop on "Communications over Power Lines,"* Institut für Experimentelle Mathematik der Universität GH Essen, Preprint No. 12, 1994, Part IV.

[ST19] Dostert, K.: A signal processing ASIC for an all digital spread spectrum modem for power line communications. *Proceedings of the 3rd IEEE International Symposium on Spread Spectrum Techniques and Applications, ISSSTA '94*, Oulu, Finland, July 1994, pp. 357–361.

[ST20] Dostert, K.: Signal processing ASICs for power line communication systems. *Proceedings of the International Symposium on Information Theory and Its Applications (ISITA '94)*, Sydney, Australia, Nov. 1994, pp. 1139–1144.

[ST21] Dostert, K.: Error control coding for mainsborne communications. *Proceedings of the 1996 IEEE International Symposium on Information Theory and Its Applications (ISITA '96)*, Victoria, Canada, Sept. 1996, pp. 135–138.

[ST22] Dostert, K., and Waldeck, T.: Comparison of modulation schemes with frequency agility for application in power line communication systems. *Proceedings of the 4th IEEE International Symposium on Spread Spectrum Techniques and Applications (ISSSTA '96)*, Mainz, Germany, Sept. 1996, pp. 821–825.

[ST23] Dostert, K.: All digital spread spectrum modems for power line communications. *European Transactions on Telecommunications (ETT)*, Vol. 7, No. 6, 1996, pp. 507–514.

[ST24] Goodenough, F.: Chip set puts 100kbits/s data on noisy power lines. *Electronic Design*, March 18, 1996, pp. 177–184.

[ST25] Dostert, K., Lehmann, K., Rosch, R., and Zapp, R.: Gebäudesystemtechnik: Datenübertragung auf dem 230-V-Netz. *Bibliothek der Technik*, Vol. 161, *Verlag moderne Industrie*, Landsberg/Lech, Germany, 1998, ISBN 3-478-93185-1.

[ST26] Hooijen, O.: On the channel capacity of the residential power circuit used as a digital communications medium. *IEEE Communications Letter*, Vol. 2, Oct. 1998, pp. 267–268.

8.7.3 Doctoral and Habilitation Dissertations Concerning PLC

[D1] Threin, G.: Datenübertragung über Niederspannungsnetze mit Bandspreizverfahren. *VDI Fortschritt-Berichte*, Series 10, No. 156, VDI Verlag, Düsseldorf, Germany, 1991.

[D2] Ruse, H.: Spread-Spectrum-Technik zur störsicheren digitalen Datenübertragung über Stromnetze. *VDI Fortschritt-Berichte*, Series 10, No. 185, VDI Verlag, Düsseldorf, Germany, 1991.

[D3] Dostert, K.: Prinzipien der Informationsübertragung über elektrische Energieversorgungsnetze. Habilitationsschrift. *Deutsche Hochschulschrift* DHS 414, Verlag Hänsel-Hohenhausen, Egelsbach, Germany, Dec. 1991.

[D4] Karl, M.: Möglichkeiten der Nachrichtenübertragung über elektrische Energieverteilnetze auf der Grundlage europäischer Normen. *VDI Fortschritt-Berichte*, Series 10, No. 500, VDI Verlag, Düsseldorf, Germany, 1997.

[D5] Arzberger, M.: *Datenkommunikation auf elektrischen Verteilnetzen für erweiterte Energiedienstleistungen*. Logos-Verlag, Berlin, Germany, 1998.

[D6] Hooijen, O.: *Aspects of Residential Power Line Communications*. Shaker-Verlag, Aachen, Germany, 1998.

[D7] Hensen, C.: *Mehrnutzer-Datenübertragung über Niederspannungsleitungen mit hoher Summendatenrate*. Shaker Verlag, Aachen, Germany, 1999.

[D8] Waldeck, T.: *Einzel- und Mehrträgerverfahren für die störresistente Kommunikation auf Energieverteilnetzen*. Logos-Verlag, Berlin, Germany, 2000.

[D9] Zimmermann, M.: *Energieverteilnetze als Zugangsmedium für Telekommunikationsdienste*. Shaker Verlag, Aachen, Germany, 2000.

8.7.4 Channel Analysis, Modeling, and System Concepts for High-Speed PLC

[H1] Arzberger, M., et al.: Fundamental properties of the low voltage power distribution grid. *Proceedings of the 1997 International Symposium on Power Line Communications and Its Applications*, Essen, Germany, April 1997, pp. 45–50.

[H2] Dostert, K.: Telecommunications over the power distribution grid—possibilities and limitations. *Proceedings of the 1997 International Symposium on Power Line Communications and Its Applications*, Essen, Germany, April 1997, pp. 1–9.

[H3] Dostert, K., et al.: Modellierung elektrischer Energieverteilnetze als schnelle Nachrichtenkanäle—ein praxisorientierter Ansatz. *Tagungsband der Konfer-*

enz: "Power Line Telecommunications—Forschung trifft Markt," EUTELIS PTF e.V., Nov. 25, 1997.

[H4] Dostert, K.: Telekommunikation über Energieverteilnetze—eine Standortbestimmung. *Tagungsband der Fachkonferenz "Powerline—Daten- und Telekommunikation über das Stromnetz,"* Institute for International Research (IIR), Düsseldorf, Germany, June 2–3, 1997.

[H5] Dostert, K.: Telekommunikation über das elektrische Energieverteilnetz. *Proceedings des Telekom-Anwender-Kongresses '97*, Cologne, Germany, Dec. 2–3, 1997, pp. 289–308.

[H6] Dostert, K., and Zimmermann, M.: Sprache über die Stromleitung. *Funkschau*, No. 4, 1998, pp. 22–27.

[H7] Dostert, K.: RF-Models of the electrical power distribution grid. *Proceedings of the 1998 International Symposium on Power Line Communications and Its Applications*, Tokyo, Japan, March 1998, pp. 105–114, ISBN 90-74294-18-3.

[H8] Brown, P. A.: Some key factors influencing data transmission rates in the power line environment when utilizing carrier frequencies above 1 MHz. *Proceedings of the 1998 International Symposium on Power Line Communications and Its Applications*, Tokyo, Japan, March 1998, pp. 67–75.

[H9] Philipps, H.: Performance measurements of powerline channels at high frequencies. *Proceedings of the 1998 International Symposium on Power Line Communications and Its Applications*, Tokyo, Japan, March 1998, pp. 229–237.

[H10] Dostert, K., and Halldorsson, U.: Modulation für Powerline. *Funkschau*, No. 6, 1998, pp. 56–61.

[H11] Dostert, K.: Neue Dienste auf Energieverteilnetzen: Technische Machbarkeit und wirtschaftlicher Nutzen. *Kongreßband der Konferenz "Powerline für EVU,"* Ueberreuter Manager-Akademie, Berlin, Germany, April 27–28, 1998.

[H12] Dostert, K.: Powerline Telecommunications—Chancen und Risiken im deregulierten Markt. *Tagungsband des Funkschau-Seminars "Powerline,"* Munich, Germany, June 29, 1998.

[H13] Dostert, K., et al.: Konzepte für Powerline-Kommunikationssysteme. *Funkschau*, No. 14, 1998, pp. 40–43.

[H14 Dostert, K.: Telekommunikation über das elektrische Energieverteilnetz. *Handbuch der Telekommunikation*, Vol. 1, Contribution 3.5.3.0. Verlagsgruppe Deutscher Wirtschaftsdienst, Cologne, Germany, ISBN 3-87156-096-0.

[H15] Dostert, K., et al.: Telecommunication applications over the low voltage power distribution grid. *Proceedings of the 5th IEEE International Symposium on Spread Spectrum Techniques and Applications (ISSSTA '98)*, Sun City, South Africa, Sept. 1998, ISBN 0-7803-4281-X, Vol. 1/3, pp. 73–77.

[H16] Dostert, K.: Power lines as high speed data transmission channels—modeling the physical limits. *Proceedings of the 5th IEEE International Symposium on Spread Spectrum Techniques and Applications (ISSSTA '98)*, Sun City, South Africa, Sept. 1998, ISBN 0-7803-4281-X, Vol. 2/3, pp. 585–589.

[H17] Simmons, R. D., Tournadre, C. V., and Womersley, R. V.: Final report on a study to investigate PLT radiation. *Smith report on project AY 3062*. The Smith Group Limited, UK, Nov. 20, 1998.

[H18] Dostert, K.: Power Line—Neue Dienste auf Energieverteilungsleitungen. *ETG-Fachbericht 73*, VDE-Verlag, Berlin, Germany, 1998, pp. 121–129, ISBN 3-8007-2391-3.

[H19] Dostert, K.: Das elektrische Verteilnetz als TK-Medium für die letzte Meile. *Congreßband I der ONLINE '99*, Düsseldorf, Germany, Feb. 1–4, 1999, pp. C141.2–C141.17, ISBN 3-89077-192-0.

[H20] Dostert, K., and Zimmermann, M.: A multipath signal propagation model for the power line channel in the high frequency range. *Proceedings of the 3rd International Symposium on Power-Line Communications and Its Applications (ISPLC '99)*, Lancaster, UK, March/April, 1999, pp. 45–51, ISBN 90-74249-22-1.

[H21] Philipps, H.: Modelling of powerline communication channels. *Proceedings of the 3rd International Symposium on Power-Line Communications and Its Applications (ISPLC '99)*, Lancaster, UK, March/April, 1999, pp. 14–21.

[H22] Rickard, J.: A pragmatic approach to setting limits to radiation from powerline communications systems. *Proceedings of the 3rd International Symposium on Power-Line Communications and Its Applications (ISPLC '99)*, Lancaster, UK, March/April, 1999.

[H23] Dostert, K.: Systematische Netzanalysen—Grundlagen für die Festlegung von Referenzmodellen und Standards: Erkenntnisse aus Versuchsprojekten der RWE AG mit der Universität Karlsruhe. *Tagungsband der Fachkonferenz "Powerline für EVU,"* Institute for International Research (IIR), Cologne, Germany, April 21–23, 1999.

[H24] Langfeld, P., et al.: Powerline communication system design strategies for local loop access. *Proceedings of the Workshop "Kommunikationstechnik,"* Ulm, Germany, July 1999, *Technical Report ITUU-TR-1999/02*, pp. 21–26.

[H25] Dostert, K.: Möglichkeiten und Grenzen neuer, alternativer TK-Festnetzzugänge im Spannungsfeld bestehender Dienste und Randbedingungen. *Tagungsband des Funkschau-Intensivseminars "Powerline Communication und EMV,"* Munich, Germany, July 1999.

[H26] Dostert, K., and Zimmermann, M.: PLC-Meßtechnik: Neue Methoden und Meßsystemkonzepte. *Tagungsband des Funkschau-Intensivseminars "Powerline Communication und EMV,"* Munich, Germany, July 1999.

[H27] Dostert, K., and Gebhardt, M.: Grundlagen und Problematik der PLC-Funkfeldmeßtechnik. *Tagungsband des Funkschau-Intensivseminars "Powerline Communication und EMV,"* Munich, Germany, July 1999.

[H28] Dostert, K., and Zimmermann, M.: Ein parametrisches Modell für die Übertragungsfunktion von Energieverteilnetzen im Frequenzbereich von 0.5-20MHz. *Kleinheubacher Berichte*, Vol. 42, 1999, pp. 342–354.

[H29] Giebel, T., and Rohling, H.: OFDM and regularity aspects of power line applications. *Proceedings of the 1st International OFDM-Workshop*, Hamburg, Germany, Sept. 1999, pp. 4.1–4.4.

[H30] Galda, D., and Rohling, H.: Coded OFDM for power line applications. *Proceedings of the 1st International OFDM-Workshop*, Hamburg, Germany, Sept. 1999, pp. 5.1–5.4.

[H31] Dostert, K.: Power Line Carrier (PLC)—Neue Möglichkeiten im Kabelnetz. *Elektrizitätswirtschaft*, Vol. 98, No. 23, 1999, pp. 78–81. VWEW-Verlag, ISSN 0013-5496.

[H32] Dostert, K., et al.: EMV und Telekommunikation auf Stromnetzen. *EMC-Kompendium*, 2000, pp. 261–263, ISBN 3-9804947-7-2.

[H33] Studie Powerline (Studienergebnisse zu EMV-Problemen) im Auftrag der RegTP: http://www.regtp.de/tech_reg_tele/start/fs_06.html.

[H34] Special Issue on Powerline Communications, *Int. Journal of Electronics and Communications (AEÜ)*, 54, No. 1, 2000.

[H35] Lampe, L., and Huber, J.: Bandwidth efficient power line communications based on OFDM. *Int. Journal of Electronics and Communications (AEÜ)*, 54, No. 1, 2000, pp. 2–12.

[H36] Dostert, K., and Gebhardt, M.: Characterization of the radiation from PLC by means of the coupling factor. *Int. Journal of Electronics and Communications (AEÜ)*, 54, No. 1, 2000, pp. 41–44.

[H37] Dostert, K., and Zimmermann, M.: The low voltage power distribution network as last mile access-network—signal propagation and noise scenario in the HF-range. *Int. Journal of Electronics and Communications (AEÜ)*, 54, No. 1, 2000, pp. 13–22.

[H38] Hensen, C., and Schulz, W.: Time dependency of the channel characteristics of low voltage power-lines and its effects on hardware implementation. *Int. Journal of Electronics and Communications (AEÜ)*, 54, No. 1, 2000, pp. 23–32.

[H39] Dostert, K., and Zimmermann, M.: An analysis of the broadband noise scenario in powerline networks. *Proceedings of the 4th International Symposium on Powerline Communications and Its Applications*, Limerick, Ireland, April 2000, pp. 131–138.

[H40] Dostert, K., and Langfeld, P.: OFDM system synchronization for powerline communications. *Proceedings of the 4th International Symposium on Powerline Communications and Its Applications*, Limerick, Ireland, April 2000, pp. 15–22.

[H41] Dostert, K., and Zimmermann, M.: Die Kanalkapazität von Power Line Kanälen unter Berücksichtigung von Beschränkungen der Sendeleistung und der nutzbaren Frequenzbereiche. *Kleinheubacher Berichte*, Vol. 43, 2000, pp. 58–66.

[H42] Dostert, K.: EMC Aspects of high speed powerline communications. *Proceedings of the 15th International Wroclaw Symposium and Exhibition on Electromagnetic Compatibility*, Wroclaw, Poland, June 27–30, 2000, pp. 98–102, ISBN 83-901999-0-4.

8.8 Conference Proceedings

[CP1] *Proceedings of the 4th IEEE International Symposium on Spread Spectrum Techniques and Applications (ISSSTA '96)*, Mainz, Germany, Sept. 1996.

[CP2] *Tagungsband der Fachkonferenz "Powerline,"* Institute for International Research (IIR), Düsseldorf, Germany, June 2–3, 1997.

[CP3] *Proceedings of the 1997 International Symposium on Power Line Communications and Its Applications*, Essen, Germany, April 1997.

[CP4] *Proceedings des Telekom-Anwender-Kongresses '97*, Cologne, Germany, Dec. 2–3, 1997.

[CP5] *Tagungsband der Konferenz: "Power Line Telecommunications—Forschung trifft Markt,"* EUTELIS PTF e.V., Düsseldorf, Germany, Nov. 25, 1997.

[CP6] *Tagungsband des Funkschau-Intensivseminars "Powerline,"* Munich, Germany, June 29, 1998.

[CP7] *Proceedings of the 1998 International Symposium on Power Line Communications and Its Applications*, Tokyo, Japan, March 1998, ISBN 90-74294-18-3.

[CP8] Neue Dienste auf Energieverteilnetzen: Technische Machbarkeit und wirtschaftlicher Nutzen. *Kongreßband der Konferenz "Powerline für EVU,"* Ueberreuter Manager-Akademie, Berlin, Germany, April 27–28, 1998.

[CP9] *Proceedings of the 5th IEEE International Symposium on Spread Spectrum Techniques and Applications (ISSSTA '98)*, Vols. 1–3, Sun City, South Africa, Sept. 1998, ISBN 0-7803-4281-X.

[CP10] *Proceedings of the 1998 European Utilities Summit*, Amsterdam, The Netherlands, Dec. 1998.

[CP11] *Proceedings of the 3rd International Symposium on Power-Line Communications and Its Applications (ISPLC '99)*, Lancaster, UK, March/April 1999, ISBN 90-74249-22-1.

[CP12] *Tagungsband der Fachkonferenz "Powerline für EVU—Innovative Energiedienstleistungen über Stromnetze,"* Institute for International Research (IIR), Cologne, Germany, April 21–23, 1999.

[CP13] *Congreßband I der ONLINE '99*, Düsseldorf, Germany, April 1–4, 1999, ISBN 3-89077-192-0.

[CP14] *Tagungsband des Funkschau-Intensivseminars "Powerline Communication und EMV,"* Munich, Germany, July 1999.

[CP15] *Proceedings of the 4th International Symposium on Power-Line Communications*, Limerick, Ireland, April 5–7, 2000.

[CP16] Session, L: Power-line communication-EMC problems. *Proceedings of the 15th International Wroclaw Symposium and Exhibition on Electromagnetic Compatibility*, Wroclaw, Poland, June 27–30, 2000, pp. 98–120, ISBN 83-901999-0-4.

8.9 Books on Powerline Communication

[B1] Dostert, K.: *Power Line Kommunikation*. Franzis-Verlag, Poing, Germany, 2000, ISBN 3-7723-4423-2.

8.10 WWW Links Related to PLC

[W1] http://www.polytrax.com

[W2] http://www.siemens.de/plc

[W3] http://www.regtp.de

[W4] http://www.iad-de.com

[W5] http://www.homeplug.org

[W6] http://www.ul.ie/~isplc2000/index.html

[W7] http://www.itrancomm.com/Tech1.html

[W8] http://www.ascom.ch/plc/

[W9] http://www.keyintelecom.com/

[W10] http://www.cogency.com/products/index.html

[W11] http://www.intellon.com/index.asp

[W12] http://www.enikia.com/

[W13] http://www.oneline.de/

[W14] http://www.ptf.de/

[W15] http://www-iiit.etec.uni-karlsruhe.de/~plc/

[W16] http://www-iiit.etec.uni-karlsruhe.de/~dostert/

[W17] http://www.etsi.org/plt/

[W18] http://www.regtp.de/tech_reg_tele/start/fs_06.html

Index

A

A/D converter, 163
A-band, 74
Access
    channels, 271
    channel capacity, 272
    domain, 253, 273, 284, 289
    domain (last mile), 273
    impedance, 81, 92, 93, 221, 245, 268–69
    statistics, 269
Accumulated sum, 165
Accumulation, 159
Accumulator, 147, 177, 151, 157, 211
    word width, 158
Acquisition, 109, 113
Active bandpass filter, 85
Active
    correlator, 206
    filter, 208, 225
    receiving coupler, 245
Adaption mechanisms, 286
Adaptive
    frequency allocation, 285
    threshold, 161
Adder, 159
Additive white Gaussian noise (AWGN), 263
    environment, 198
    scenario, 190
    channel, 108
Admissible transmission power, 78
Admissible transmission voltage, 89
ADSL, 232–33, 237, 244
Aerial cables, 18
AGC, 226
AGC amplifier, 178
Air-core coil, 13, 82
All-digital solutions, 137
Allowable transmit power, 53
AM broadcasting, 297
Amateur radio, 2, 280, 313
Amplitude
    distribution, 78, 298
    modulation, 46
AM, 94
    shift keying (ASK), 62, 94
    spectrum, 33
    weighting, 144
Analog
    front end, 246
    switch, 124, 134
Analog/digital conversion, 132
    converter, 147
Antenna efficiency, 282
Antenna function, 288
Aperiodic autocorrelation, 179, 181
Apparent power, 55
Application microcontroller, 225, 227
Application specific integrated circuit (ASIC), 67, 152, 161
ARQ, 219–20
ASIC, 163, 169, 174–75, 177–80, 183–84, 205–8, 221, 226–27
    development, 171
Asymmetric digital subscriber line (ADSL), 299
Asynchronous
    impulsive noise, 264
    motor, 56
Attenuation, 242, 247, 252–53
    constant, 10
    fluctuations, 75
    values, 238
Audiofrequencies, 59
Audio-frequency
    carrier, 67
    converter, 68
    generator, 67–68
    signals, 55
Autocorrelation, 108, 146
    function, 101, 148
    maximum, 105
Automatic
    gain control (AGC), 163, 179, 210
    meter reading, 136
    retransmission request (ARQ), 198
Autonomous correlative synchronization, 215

B

B-, C-, and D-bands, 74
Background
    clock, 67
    noise, 239, 264, 267–68, 272, 274, 295
Band filter, 70
Bandpass
    filter, 163
    signals, 94
Band-spreading, 201
    modulation, 94, 99, 294
Bandwidth
    efficiency, 310
    efficient, 81
Base station, 232, 234
Bibliography, 315
BiCMOS technology, 174
Bidirectional
    communication link, 235
    transmission, 81
Binary
    codes, 63
    phase shift keying (BPSK), 96, 283
    pseudonoise sequence, 100, 294
Bit
    decision, 147
    error measurements, 136
    error probability, 126, 190, 200
    error rate (BER), 99, 109, 136, 195, 219–20
    loading, 301
    mapping, 306
    rate, 78
Block codes, 188, 200
Block length, 188–89
Bluetooth, 230
Books on powerline communication, 328
BPSK, 303

Index

Broadband, 247
  character, 268
  coupling filters, 45
  equalization, 310
  modulation, 94, 99, 186
  signal, 76
  single-carrier modulation, 293
  techniques, 294
Broadcast
  bands, 268
  radio station, 294
  signals, 288
  stations, 264
Broadcasting, 313
  corporations, 285
Brown out, 114
Building
  — automation, 107, 156, 162–63, 174, 176, 183, 186, 226
  — automation system, 99, 119, 174
  installations, 282, 291
  — installation networks, 32
Burst errors, 265

C

Cable
  television, 233
  TV, 237
Capacitance per length, 8
Capacitive load, 60
Carrier
  modulation and demodulation, 302
  sense multiple access (CSMA), 74
  suppression, 78
  transmission over powerlines (CTP), 8
Carrier-frequency
  barriers, 20
  modulation, 315
  signaling, 313
  technology, 44
  transmission, 43
Carrier-suppressing modulation, 81, 96
CARRY-SAVE
  adder array, 167
  adders, 167
CDMA, 295–96
CENELEC
  A-band, 265, 268
  bands, 239, 243
  standard EN 50065, 3, 73
Channel
  adapted design, 201
  analysis, 230
  analysis and modeling at low frequencies, 318
  analysis, modeling, for high-speed PLC, 323
  capacity, 2, 52, 107, 271, 273–75, 312
  characteristics, 237, 239
  coded, 306
  coding, 99, 162, 188–90, 200
  coding methods, 151
  emulator, 263
  model, 239, 251, 260
  number, 122
Characteristic impedance, 10, 22, 26, 237, 261, 268–69, 281, 288
Charge injection, 136
Check word, 195, 198
Checksums, 160
Chimney approach, 277
Chimneys, 276, 285
Chip
  decision, 123
  duration, 101, 105, 113, 154, 213
  interval, 132
Chip set, 97
Chirp signal, 58
Clearable integrator, 128–29, 132
Clearable resonance circuit, 123, 131
CMOS
  gate array, 171
  technology, 113, 172, 174, 206–7
Coarse synchronization, 216
Coaxial cable, 24
Code
  concatenation, 195
  words, 198
  transmission, 198
Code-division multiple access (CDMA), 100, 294
Coding methods, 51
Coexistence, 276, 281, 284–85
Coherence, 150
Coherent, 104
  reception, 150
Collision, 121
Collision-free, 106
Colored background noise, 33, 264
Common mode, 131
  propagation, 272, 291
  rejection, 280, 284
Communication over powerlines, 43
Commutator sparking, 38
Comparator, 113
Compensation measures, 12
Complete binary codes, 64
Complex
  filters, 278
  frequency response, 247
  propagation factor, 259
  transfer function, 246, 248
  transmission techniques, 285
Concatenation scheme, 198
Conditioning
  inductor, 291
  method, 256
Conference proceedings, 327
Consumer products, 310
Continuous
  Fourier transform, 155
  phase, 205
  phase FSK, 184
  value, 142
Controllable rectifier, 69
Convolution, 108, 143, 249
Core magnetization, 240
Coreless coils, 288
Corona losses, 7
Correlation, 108, 146
  principle, 151
  process, 147
  reference, 173
  signal, 148
  value, 150, 218
Correlative
  processing, 177–78
  reception, 205
  signal processing, 177, 205
  synchronization, 217
Correlator, 146–47, 151, 172
  output, 157
  structure, 144
Coupler, 163, 178, 211
Coupling, 237
  capacitors, 241
  equipment, 84
  filters, 45
  points, 43
  transformer, 83, 87
  unit, 20, 67
CRC, 220–21
CREST factor, 300–2
Crossbar system, 84, 92, 245, 269, 286, 288
Cross-correlation, 146
Crosstalk, 185, 288, 291
Current harmonics, 57
Cyclic
  counter, 177
  prefix, 309–10
  read-out, 111
  redundancy check (CRC), 220

D

D/A converter, 138, 141
Data
  bit vector, 192

Data (*Continued*)
  detection, 303
  packet, 114
  rate, 94
  remote network, 44
  vector, 95
Database, 92
Data-transmission capacity, 275
Direct digital signal synthesis (DDS), 243–244
Deattenuated resonance circuit, 110, 124–27
Decision
  criterion, 203
  distance, 150
  hardware, 158
  logic, 145, 153, 158, 182
Decoding, 162, 194
Decoupling factor, 272, 274, 281–82, 284, 290, 292, 297, 312
  measurement, 283
DECT, 230
Degree of branching, 61
Degree of coupling, 288
Deregulated electricity market, 136
Deregulation, 1, 55, 62, 127, 229
Detection
  and ranging, 99
  losses, 178
  quality, 180
Deutsche Telekom, 2
DFT, 302–3
  algorithm, 157
Dielectric losses, 16
Differential
  mode, 292
  mode signaling, 257, 289
  to common-mode conversion, 292
Digital
  audio broadcasting (DAB), 299
  correlator, 97, 147–48
  correlator structures, 151
  data communications, 44
  FH receiver, 172
  frequency synthesis, 139
  frequency synthesizer, 97, 177
  modulation, 52
  phase-locked loop (PLL), 208, 211, 216
  samples, 141
  signal processing, 147, 245
  signal processor (DSP), 147, 156, 163, 219, 303
  signal synthesis, 137
  signal synthesizer, 83
  switching signals, 134
  video broadcasting (DVB), 299
Dimmer interference, 189

Dirac impulse, 141, 144, 156
  sequence, 141
Direct sequencing, 295
Directional couplers, 246
Discrete
  convolution, 250
  Fourier transform (DFT), 155
  impulse response, 250
  time, 142
Distribution network, 75
Dominant echoes, 260
Double-sideband AM (DSB-AM), 47–48
Downlinks, 232
DSP, 310
  system, 246
Dump impulse, 124
Dump operation, 124
Duplex operation, 115
Dynamic range, 131, 150

E

Echo model, 252, 258, 260–63
Echo path, 259, 261
Echo-attenuation, 258
Echo-based model, 256
8-bit multiplier, 169
Effective magnetic cross section, 88
Electric power market, 220
Electrical energy market, 55
Electrical field
  level limits, 279
  strength, 271, 274, 282–83
Electricity market, 127
Electromagnetic
  compatibility (EMC), 2, 61, 98, 230, 239, 270, 276, 313
  coupling, 289
  decoupling factor, 272
  fields, 276, 280
Electronic
  meters, 136
  power meter, 220
  ripple control receiver, 64
EMC, 255, 257, 271, 278, 289, 291, 294
  performance, 239
  problems, 245, 251
  reasons, 301
  standards, 281
EMI
  suppression, 186, 195
  suppression equipment, 221
EN 50065, 201, 206, 208, 216, 226
Energy
  density, 37
  density spectrum, 38

distribution networks, 18
information systems, 313
market deregulation, 234
per bit, 80
signal, 38
Energy-related value added services, 84, 86, 97, 107, 119, 156, 162–63, 174–75, 200–1, 219–20, 226–27
Energy-specific value added services, 32
Entertainment, 230
Envelope, 110
Equalization, 296, 299, 307
Equalizer, 100
Equalizing, 308
Equivalent gate functions, 164
Error
  bursts, 190
  control, 63
  correction, 63, 194
  correction and detection, 221
  correction coding, 188
  detection, 195
  detection capabilities, 219
  probability, 241
  words, 191, 193, 196–97
Etching processes, 172
Ethernet technology, 74
ETSI, 231
Euclidian distance, 96
European standard EN 50065, 176, 200
European standards, 73
Experimental modems, 320
Externally commuted converter, 68
Externally commuted rotating converter, 69

F

Factor of quality, 124, 126
False alarm, 161, 178, 180
  probability, 179
Far-field conditions, 281
Fast
  Fourier transform (FFT), 266, 303
  frequency hopping, 172, 219
  hopping, 105
  multiplication, 166
  parallel multiplication, 166, 168
  parallel multiplier, 166
  transients, 83
Faulty switching, 63
FEC, 219–21
Feedback, 51
Feedback channel, 51
Ferrite cores, 23, 240
Ferrite ring cores, 86

Index

FFT, 304
FH
    modem prototype, 120
    waveform, 101, 124, 127, 130, 134, 138
Field
    strength, 280
    tests, 161
    trials, 86, 169, 204, 224
Filtering property, 143
Final adders, 169
FIR finite impulse response filter, 251, 253
Fissured spectrum, 271, 296
Flat rate, 229
Flat transmission characteristic, 296
Fourier transform, 37
Forward error correction (FEC), 188, 265
Four sector structure, 29
Four-frequency shift keying, 53
Fourier
    coefficient, 266, 303, 306–8
    series, 129
    transform, 143–44, 154–56, 246
Four-sector cable, 237
Four-wire trunk, 8
FPGA, 312
Frequency
    allocation, 236, 275, 296, 313
    dependence, 92, 269
    division, 111
    division multiplexing (FDM), 299
    domain, 143
    hop rate, 204
    hopping (FH), 53, 101, 176, 186, 201, 299
    modulation (FM), 50
    multiplexing, 46
    resolution, 156
    shift keying (FSK), 62, 97
    stability, 69
    synthesizer, 119, 123, 132, 137–38, 144, 243
Frequency-agile
    methods, 107, 201
    modulation, 235
    signals, 127
    system, 137
    transmission system, 137
    waveforms, 107–8, 137, 155, 163, 201
Frequency-dependent variations, 91
Frequency-selective
    attenuation, 92, 190
    fading, 296
FSK frequencies, 98
Full adders, 167

Full-duplex, 134
Full-duplex capability, 161
Functional samples, 107
Fundamentals of communications, 317

G

Galvanic separation, 82
Gaussian distribution, 78
General telephone network, 44
Generating matrix, 192
Generator matrix, 192, 198
Geometric addition, 132, 147, 149, 213
Global
    interest, 314
    reference, 109, 183
    synchronization, 103, 106, 110, 112, 114–16, 118–19, 121
    time pattern, 205
Globally available reference, 104
Glow discharges, 17
Graceful degradation, 296
Ground cable, 18, 22
Ground loops, 245
Grounding conditions, 75
Group delay, 299, 307
Guard interval, 309

H

Half adder, 167
Half-duplex modem, 162, 174
Half-duplex operation, 161, 176
Hamming
    code, 191–92, 195, 198, 219–21
    distance, 191, 195
    limit, 191
Harmonics, 34
HF
    balun, 23
    barriers, 284
    measuring system, 243
    signal coupling, 244
H-field loop antenna, 281
High-energy transients, 83
High-frequency
    attenuation, 21
    coupling, 240
    interference, 7
Highpass filter, 82
High-quality resonance circuit, 123
High-speed
    communication, 257, 265
    indoor digital networks, 274
    indoor PLC, 289
    networking, 278

PLC, 270, 278, 299, 308, 311
PLC systems, 312–13
powerline networking, 279
High-voltage lines, 315
High-voltage overhead lines, 17
Home automation, 1, 230
Home networks, 278
HomePlug, 231, 310
    alliance, 2, 278
    powerline alliance, 230
Hop rate, 106, 134, 138
House
    connection, 234
    connection point, 269, 289–90
    service connection, 82, 92

I

IDFT, 303
IFFT, 304, 307
IHDN, 312
Impedance measurement, 92, 247
Impulse
    amplitude, 40
    noise, 265, 267
    response, 241, 245–49, 262, 307, 309, 312
    width, 40
Impulse-noise
    modeling, 265
    parameters, 265
    synthesis, 265–66
Impulsive
    interference, 17, 151, 189
    interference model, 33
    noise, 33, 39, 188, 190, 265, 275, 296, 307, 312
Incoherence, 151
Incoherent, 132
    reception, 145–46, 154, 190, 206, 208, 211
Indoor, 257
    applications, 255
    channel, 255
    channel characteristics, 254
    channel capacity, 273
    installations, 255
    link, 239, 256, 274
    mains conditioning, 290
    PLC, 289
    PLC channels, 272
    PLC link, 281
    powerlines, 176
    systems, 284
Inductance per length, 8
Industrial modem, 224
Information
    Society Technologies (IST), 2
    transmission, 32

In-home digital network (IHDN), 275–276, 311
Inhouse data transmission, 95
Initial synchronization, 109
In-phase, 149, 177
    branch, 149
    component, 21
    part, 151
    position, 145
Input latches, 169
Instantaneous frequency, 101
Instruction set, 207
Integrity classes, 220
Intelligent evaluation, 180
Interarrival time, 40, 265
Interference
    measurements, 252
    resistant, 53
    scenario, 16, 32, 238, 263, 265, 312
Internet access, 2, 229, 310, 314
    over the wall plug, 229, 274
    via the standard wall plug, 311
    from the wall plug, 312
Intersymbol interference (ISI), 309
Inverse discrete Fourier transform (IDFT), 302
Inverse fast Fourier transform (IFFT), 156, 266
iPLATO, 41, 247, 263, 265–66, 269
Isolated gate bipolar transistor (IGBT), 69
Isolated waveguide, 21

J

Jitter, 115–16, 183, 185, 208, 211
Joule's heat, 7

K

K-factor, 282

L

Last meter solution, 230
Last mile, 1–2, 32, 222, 273, 314
    solution, 232–33
Leakage
    conductance per length, 8
    losses, 7
Level
    estimation unit, 210
    limitation, 296
    limit measurements, 297
Limits of radiation, 285
Line
    commuted converter, 68
    constants, 8
    properties, 31

Linear AM, 47
Line-of-sight radios, 233–34
L-N coupling, 256
Load distribution, 54
Local loop, 1, 275
Local reference, 217
Lock-in amplifier, 110, 127, 131–35, 172
Lock-in principle, 136, 146
Lock-in receiver, 138
Logical bus, 232
Long-, medium-, and short-wave broadcasting, 285
Long-distance links, 254
Long-term stationary, 243
Long-wave, 2
    broadcast band, 73–74
Loop antenna, 297
Loose coupling, 84
Loss factor tan δ, 26
Lossless line, 13
Low spectral power density, 294
Low-data-rate transmission, 107
Lowpass
    character, 40, 238
    characteristic, 237, 251, 253, 257, 284
    effects, 254
Low-speed powerline communication, 127
Low-voltage, 2
    transformer, 32

M

MAC operations, 148, 151, 163
Macro cell, 219
Magnetic
    field strength, 297
    flux, 82
    materials, 23
    trunk flux, 88
Magnitude comparison, 161
Mains
    control engineering, 227
    decoupling factor, 281–82
    impedance, 85, 89
    period, 110
Majority decision, 102
Manipulation safety, 208
Master-slave
    principle, 107
    protocol, 84, 161
Matched filter, 104, 108, 110, 117, 123–26, 129, 131, 145, 207, 216
    function, 126, 146, 151
    reception, 101, 144
    receiver, 114

Matched filtering, 101, 107, 176, 206
Matrix elements, 167
Mean value, 128
Measurement
    database, 273
    filter, 76
    regulations, 75
    results, 241
    standard, 74
    systems, 241
    techniques, 237
Medium-voltage, 2
    overhead lines, 18
Medium-wave, 2
Memory
    mapping, 219
    reading, 139
    reading process, 134
    read-out, 138
Memoryless modulation schemes, 77
MFH scheme, 235
MFH series device, 226
Microcontroller, 113, 132, 138, 161–63, 172–73, 177–79, 183–85, 188–92, 194, 200, 206, 210–11, 219, 225, 244–45, 312
Microelectronic systems, 99, 151
Microprocessor, 64, 69, 164, 173
Military communications, 99
Mixed-Signal ASIC, 206, 211, 219, 225
Mixing procedure, 122
Mobile radio, 99
Modem, 210
    operation, 81
Modified
    chimney approach, 281
    frequency hopping (MFH), 103
    frequency-hopping modulation, 163
    Hamming codes, 188
Modulation, 92
    schemes, 293
Modulator, 70
Monofrequency, 62
Motor generator, 67–68
M-out-of-n codes, 64
Multicarrier
    modulation, 293, 296
    scheme, 207
Multifrequency method, 55, 62–63
Multilayer metallization, 172
Multipath
    delay spread, 309
    propagation, 259
Multiple
    access, 114, 134, 161, 299

access system, 113, 118–19
reflections, 261
Multiplication
matrix, 168
scheme, 166
Multiplier, 164
Multiply and accumulate (MAC), 147
Multisync devices, 36
Multi-user protocol, 232

N

Narrowband
conventional modulation, 46
coupling unit, 69
demodulation, 100
filtering, 63
FM, 50
interference, 33, 99–100, 185–86, 294
interferer, 106, 201, 204, 294
modulation, 94
noise, 264, 275
signal, 76
National grid, 44
Natural oscillation, 126
Natural power, 12
Navigation, 99
NB 30, 237
Near-field conditions, 281
Near-field problem, 282
Network
analyzer, 245–46
analyzing equipment, 247
conditioning, 280, 284–86, 289–90, 313
modeling, 204
Neutral
conductor, 74
point shift, 116, 183
Night-storage heating, 61
Noise
immunity, 185, 188, 202–4, 208, 226
level, 267
power density, 89, 190, 199, 272
power spectral density, 270, 274
Nominal mains voltage, 59
Nominal power, 53
Non-Gaussian interferes, 105
Nonharmonics, 34
Notches, 260, 262
Numeric field calculation, 24

O

Object and group addressing, 64
OFDM, 235, 277–78, 293–94. 296, 299–300, 302, 305–9, 312
multicarrier modulation scheme, 313
receiver, 307
scheme, 246
signal spectrum, 304
signals, 156
subchannel, 304, 310
symbol, 310
transmitter, 306
ON resistance, 124
On/off switching, 62
1-tap equalizer, 308–9
One's-complement, 160
Operational amplifier, 127
Operations management, 43
Optical fibers, 289
Optimal reception techniques, 144
Optimizing modulation schemes, 271
Optimum
energy distribution, 43
receiver, 107
receiver technology, 144
reception, 123
symbol processing, 203
Optocoupler, 113
Orthogonal, 146–47, 294
block, 105
FH waveforms, 104, 112
frequency division multiplexing (OFDM), 98
spreading code, 294
waveforms, 101, 110–11, 119, 140, 208, 299
Orthogonality, 97–98, 129–30, 148, 185, 299
OTP (one-time programmable) EPROM, 162
Out-of-band
interference, 98
noise, 205, 216
Output spectrum, 75
Overflow errors, 158
Overhead line, 10
Overhead wiring, 251
Oversampling, 139
Overtones, 34, 56

P

Parallel
feed, 82
multiplication, 151
resonance, 57
resonant circuit, 124
Parameter variations, 208
Parasitic capacitance, 245
Parity matrix, 191–92, 198
Parity-check matrix, 193
Partial
interference, 58
products, 169
sum, 154
Path attenuation, 23
Pay per view, 235
Pay-TV, 235
Peak
power, 53
value detector, 75, 76
to-mean-power ratio, 77
to-rms, 300
PE-N coupling, 257, 274
Perfect orthogonality, 103, 118
Periodic
dimmer interference, 188
impulses, 34
impulsive interference, 195, 198–99
impulsive noise, 264
spread-spectrum signals, 247
Permissible PLC transmission level, 280
PH receiver, 106
Phase, 247
constant, 10
continuity, 216
continuous, 176
continuous FSK, 98
continuous modulation, 98
details, 262
detection, 97
hops, 98, 140, 185, 205, 294
independent reception, 132
locked loop (PLL), 97
modulation, 96
response, 100, 242
Photoresist development, 172
Pipeline, 154
registers, 167
Pipelining, 168
algorithm, 165
principle, 156
registers, 167
PLC
modulation schemes, 297
service providers, 277
signal characteristics, 297
system on silicon, 312
forum, 284, 313
related electronic circuit design, 318
Point-to-multipoint network, 232
Polysilicon paths, 172
Polyvinyl chloride (PVC), 19, 25
Postal lines, 44
Power
density, 108

Power (*Continued*)
  plant, 43
  plant peak power, 61
  spectral density (PSD), 17, 33, 38, 98, 264, 283, 298
  supply utility (PSU), 1
  tariffs, 235
Powerline
  analyzing tool iPLATO, 245–46
  ASIC, 164
  cell structure, 233
  channel capacity, 269, 273
  channel emulation, 265
  channel model, 257
  channel modeling, 263
  communication devices, 127
  modem, 119, 126, 132, 161, 163–64, 169
  noise scenario, 264
Powernet-EIB, 118–19, 163, 175–76, 184, 186, 188
Powernet-EIB modem, 187
Preamble, 114, 160, 176, 178–79
  chip, 179–80, 182–83
  chip decision, 160
  code, 179
  detection, 157, 162, 177, 179, 181
  length, 181–82
  pattern, 160
  reception, 179
Premagnetization, 83, 87–89
Probability density function, 78
Processing gain, 100, 295
Programmable address counter, 138
Propagation constant, 10
Propagation speed, 60
Protection circuitry, 240
Protection earth (PE), 255–56
Protective earth, 290–91
  and neutral (PEN), 24
Protective ground, 74
Pseudonoise direct sequencing, 100
Pseudonoise sequence, 249
Pseudorandom sequences, 136
PTF, 237

Q

QPSK, 304–5
QPSK carrier modulation, 306
Quadrature
  amplitude modulation (QAM), 52, 271, 303
  branch, 149
  component, 212
  part, 151
  position, 145
  receiver, 145–47, 149–50, 176
  receiver principle, 154, 213
  receiver structure, 130
  signal, 149, 177
Quantization errors, 139
Quantization noise, 139
Quartz
  crystal, 69
  oscillator, 111, 216
  stable clock, 138
Quasi-contemporaneity, 110, 112
Quasi-peak
  detection, 298
  value, 78, 297
Quasi-stationary conditions, 60

R

Radiated field, 276, 278, 281, 284, 290
Radiation, 242, 274–75, 281, 285, 288, 291–92, 301
Radio
  bands, 267
  reception, 280
  relay system, 94
  services, 267, 276
  transmitters, 280
Random
  bit stream, 160
  variables, 265
  access memory, 137
Rapid prototyping, 171
Reactive impedance, 87
Reactive power, 12
Read-only memory, 137
Realizable data rate, 270
Receiver branch, 210
Reconstruction lowpass, 210, 216
Redundancy, 64, 103, 188, 192, 203, 219
Reference
  signal, 145
  signal samples, 215
  signal synthesis, 172
  waveforms, 177
Reflecting taps, 260
Reflection, 258, 262, 269
  factor, 22
  scenario, 261
Regulation, 276, 281, 285, 297, 300, 303
  issues, 316
  bodies, 230
Remote meter reading, 115, 119, 136, 221
  system, 127
Remote
  monitoring, 44
  power-meter reading, 134
  control technology, 220
Repetition periods, 265
Repetition rates, 264
Reproducibility, 127
Residual error probability, 220
Resistance per length, 8
Resistance to interference, 103
Resonant circuit, 124
RF
  coupling problems, 287
  coupling unit, 241
  signal coupling, 239
  technology, 317
  shorts, 289
Right-circular matrix, 250
Rigid coupling, 84
Ring architecture, 219
Ring core, 85, 89, 178
Ring counter, 161
Ring structure, 212
Ripple control
  receiver, 221
  system (RCS), 20, 34, 54
  telegram, 62
  transmitter, 70
  waveform, 57
Ripple-carrier signaling, 313, 315
Ripple-carry adders, 168
Robust data transmission, 105
Robustness, 203, 294, 296, 299, 308
Rounding, 158

S

Sample memory, 221, 227
Sampling theorem, 139
Satellite communications, 99
Saturation, 288
  effects, 86
  induction, 82, 87–89
Selective
  attenuation, 99, 106, 185, 201, 204, 294
  DFT, 207, 211
  discrete Fourier transform, 154–55
  interference, 106
Serial feeding, 82
Series
  production, 137, 230, 310
  resonance, 57
  resonant circuit, 65, 70
SFSK, 186, 220–21, 226
Sharp notches, 255
Shielded twisted pair, 244
Shielding procedures, 280
Short-wave, 2

Index

broadcast bands, 252
broadcast stations, 252
radios, 264
Sidelobes, 35, 179
Signal
   amplitude loss, 83
   attenuation, 81
   coupling, 81, 227, 241, 244
   filtering, 67
   level limitation, 77
   overlapping, 103
   power density spectrum, 270
   space, 304
   spectral power density, 274
   synthesis, 173, 246, 302
   synthesizer, 161–62, 207
   to-noise ratio, 17, 54, 105–6, 109, 116–18, 139, 239, 244, 252, 272–74
   to-noise power ratio, 241
Simulation, 262–63
Single-carrier
   broadband schemes, 296
   modulation, 294
   schemes, 296
Single-frequency method, 55, 62
Single-sideband AM (SSB-AM), 49
Sinusoidal interferer, 106
16-QAM, 304
Skin effect, 7
Slow hopping, 105
Smart appliances, 235
Smart home, 184
SNR, 273, 303
   values, 257
Soil impact, 16
Spectral
   bandwidth, 75
   coefficients, 303
   compression, 106
   efficiency, 94, 98, 104, 273, 275, 277, 296
   forming function, 266
   redundancy, 94, 185
   resources, 236, 285
Spectrum
   analyzer, 37, 243, 255, 267, 272
Speed of light, 60, 115
Spread
   antennas, 276
   FSK (SFSK), 119, 163, 185
Spreading code, 294
Spread-spectrum
   method, 101
   modulation, 293
   techniques (SST), 53, 99, 294, 299, 316
Square root, 159
Squarer, 153, 158–59, 164, 169
   algorithm, 170
Standard
   cell, 183
   cell arrays, 174
   DSP processors, 164, 312
   EIB, 186
   interfaces, 227
   ripple control receiver, 227
Standardization, 230, 276, 278, 312
Standards, 316
Standardization, 255
Star shape, 75
Start impulse, 65, 66
Start/stop method, 65
Static compensation, 136
Static power converter, 67, 69
Stimuli, 247
Stray field, 85
Strip line model, 28
Strong coupling, 83, 290
Subcarriers, 310
Subchannel, 306–8, 271, 277, 293, 299–300
Supply limit length, 60
Supply radius, 19
Suppressed carrier, 96
Suppressor diode, 83, 240
Switching power supply, 57, 290
Switching equipment, 43
Switching signal, 129
Symbol, 202–3, 216, 218, 304, 306
   and bit detection, 307
   decision, 213
   generator, 306
   processing, 202–3
   registers, 215
Symbol-processing
   multicarrier scheme, 202, 204, 206, 208, 219
   multicarrier system, 211, 225
   multicarrier technology, 207–8
   system, 210–11
Symbol-to-bit correspondence, 304
Symmetric
   binary channel, 189
   coupling, 239
   series coupling, 70
Symmetrical coupling, 286
Symmetrical signal coupling, 284
Synchronization, 65, 100, 104, 145, 151, 162, 205–6
   equipment, 123, 144
   error, 115–16, 118, 185, 216, 218
   loss, 114
   method, 110
   problem, 109
   reference, 114
   signal, 211
Synchronous
   demodulator, 127, 129
   motor, 35, 66, 68
   rectifier, 127, 129
   voltage-to-frequency converter (SVFC), 132, 134
Syndrome, 192, 194–97
Synthesizer design, 139
System concepts, 320
System reliability, 185
Systems theory, 317

T

Tariff switching, 61
Telecommunication, 1
   access, 232
   markets, 229
   monopoly, 229
   systems, 164
   and energy markets, 311
Telecontrolling, 51, 61
Telegram, 62
Telegram structure, 64
Telemetering, 43–44
Test results, 320
Theoretical channel capacity, 270
Three-phase current, 6
Three-phase system, 6
Threshold decider, 182
Thue-Morse Code, 181
Thue-Morse sequence, 183
Thyristor, 69
Time
   controlled energy switching, 67
   dependence, 92
   dispersive channel, 310
   division multiplexing, 55, 115, 119
   domain, 143
   inverted, 108
   variance, 265, 269
   variant, 100, 264
Tracking, 109
Training sequence, 306–8
Transfer function, 91, 241–42, 245, 247, 249, 255–57, 262, 270, 281, 307, 312
Transformer station, 20
Transient interferer, 63
Transients, 82
Transition frequencies, 208
Transmission
   attenuation, 91
   band, 58
   channel properties, 109
   characteristics, 251
   errors, 64

Transmission (*Continued*)
   matrix, 250
   power density spectrum, 270
   properties, 18
   protocol, 225, 227
   vector, 303
Transmitter output stage, 92
Transmitter power stage, 301
Transmitting-power density, 276
Troubleshooting, 43
Tuning fork generator, 69
Two's-complement, 140, 165–67, 169
Two-terminal network, 57
Two-way rectification, 127
Two-wire
   model, 21
   network, 292
   system, 21

U

Unbiased estimation, 247
Uncorrectable error, 198
Underground cable, 237, 251
Undetected errors, 198
Unintended radiation, 280, 286
Universal communication, 225
Unwanted radiation, 313
Uplink, 232

V

Value-added services, 75
Very high speed PLC, 284
VHDL, 174
Voltage harmonics, 57
Voltage-to-frequency converter (VFC), 132
Vulcanized polyethylene (VPE), 19

W

Wave propagation, 115
Wave propagation effects, 7
Wave table, 111, 137
Waveform, 154
   duration, 131, 145, 148, 154, 157, 187, 211, 303
   memory, 211, 215
   reception, 152
Wavelength, 60, 262
Weak coupling, 83, 272
Weighted comparison, 182
Weighting function, 144
White Gaussian noise, 17, 105, 109
White noise, 34
Wideband signals, 298
Wire-bound data transmission, 237
Wireless services, 276, 277–78, 281, 312
Worldwide standards, 290, 314
WWW links related to PLC, 329

Z

0/180-degree phase hops, 100
Zero locations, 98
Zero-crossings, 34, 109, 112, 114, 162, 185, 208
   detection, 211
   detector, 178
   instants, 110
Zero-phase angle, 60